AF643473

Mathematical Ecology

Mathematical Ecology

edited by
Peter Allen (Brussels), Werner Ebeling (Berlin), Manfred Peschel (Berlin), Peter Schuster (Vienna), Yuri M. Svirezhev (Moscow)

Mathematical Ecology deals with mathematical models of evolution processes, in the biosphere as a unity of growth and structure-building, with spring-up of new species, their interaction and possible extinction, with the impacts of human activities on the environment and the corresponding consequences for the biosphere.

The whole field seems to be a certain amalgam of suitable system philosophy, in which real world phenomena are considered through an ecological pair of spectacles with the help of system methodology and mathematics.

All publications published herein are of interdisciplinary interest for ecology, biology, economy and technical engineering.

Dynamical Systems and Environmental Models

Proceedings of an International Workshop cosponsored
by IIASA and the Academy of Sciences of the GDR
held on the Wartburg, Eisenach (GDR), March 17–21, 1986

Edited by
Hans Günter Bothe
Werner Ebeling
Alexander B. Kurzhanski
Manfred Peschel

THEORIA CUM PRAXI

Akademie-Verlag Berlin 1987

Editors:

Dr. habil. Hans Günter Bothe
Akademie der Wissenschaften der DDR
Karl-Weierstraß-Institut für Mathematik

Prof. Dr. Werner Ebeling
Humboldt-Universität zu Berlin
Sektion Physik

Prof. Dr. Alexander B. Kurzhanski
International Institute for Applied
Systems Analysis (IIASA)

Prof. Dr. Manfred Peschel
Akademie der Wissenschaften der DDR
Zentrum für wissenschaftlichen Gerätebau
Reproduction of the original authors' manuscript.

ISBN 3-05-500334-9

Erschienen im Akademie-Verlag Berlin, DDR-1086 Berlin, Leipziger Straße 3—4
© Akademie-Verlag Berlin 1987
Lizenznummer: 202 · 100/521/87
Printed in the German Democratic Republic
Gesamtherstellung: VEB Druckerei „Thomas Müntzer“, 5820 Bad Langensalza
Einbandgestaltung: Ralf Michaelis
LSV 1095
Bestellnummer: 763 719 9 (9044)
03500

Introductory Remarks

Advanced attempts on modelling the development of complex systems require to an increasing degree concepts and results from theoretical dynamics. This general fact applies especially to environmental models, and it is widely accepted that each of the both sides - the theoretical research and the work with realistic models - can gain a lot of ideas, fruitful motivations and improvements concerning the methods by knowing about the spirit, the problems and the methods of the other side. Nevertheless, in general there is a serious gap between the world of purely theoretical research and that of applications. This situation is rather stable, and it would not be realistic to expect that it will change drastically in the near future. But it seems to be an important task to look for already existing bridges and for developments which promise to lead to such bridges. This was the general aim of the workshop.

So the scientific activities of the participants belong to different disciplines, and accordingly broad is the spectrum of their papers which are presented here.

On the one end of this spectrum there are contributions to the mathematical theory of dynamical systems (Section I). They concern concepts and ideas which have proved to be (or promise to become) fruitful in connection with applications. These topics of pure mathematics are followed by the presentation of general mathematical models (in Section II) which are motivated by questions arising in attempts to get an insight into certain evolutions.

Section III is devoted to mathematical models for more or less concrete evolution processes in ecology, while in Section IV mathematical methods are applied to describe facts which are known from experience or experiments. Here, in some cases, mathematics proves to be useful not only for a phenomenological description of processes in society and nature, but it helps to discover mechanisms which cause these phenomena and their laws.

List of Participants

K.F. Albrecht, GDR
T. Aldenberg, The Netherlands
E. Amann, Austria
A. Arneodo, France
Th. Bley, GDR
H.G. Bothe, GDR
M. Bröhl-Kerner, FRG
B. Bruhn, GDR
P. Brunovsky, Czechoslovakia
G. Czajkowski, Poland
W. Ebeling, GDR
P. Erdi, Hungary
H.I. Freedman, Canada
R. Funke, GDR
K.P. Hadeler, FRG
Z. Harnos, Hungary
M. Heilig, Austria
H.P. Herzel, GDR
L. Kairiukstis, IIASA
P. Kindlmann, Czechoslovakia
B.P. Koch, GDR
M. Kubiček, Czechoslovakia
Z. Kubičková, Czechoslovakia
E. Labos, Hungary
A. Lasota, Poland
R.W. Leven, GDR
L. Maistrenko, USSR
H. Malchow, GDR
S. Markov, Bulgaria
V. Mazenko, USSR
W. Mende, GDR
A. Molcanov, USSR
J.S. Nicolis, Greece
L.F. Olsen, Denmark
M. Peschel, GDR
B. Pompe, GDR
K.R. Schneider, GDR
E.E. Shnol, USSR
K. Sigmund, IIASA
U. Svedin, Sweden
W.M. Terhorst, FRG
E. van de Vrie, The Netherlands

Organizing Committee

F. Auert, Karl-Weierstraß-Institute for Mathematics
E. Herbst, IIASA
H. Pierau, Karl-Weierstraß-Institute for Mathematics
D. Ruchhöft, Karl-Weierstraß-Institute for Mathematics

Contents

Part I: Mathematical Theory of Dynamical Systems

H.G. BOTHE
Hausdorff dimension of attractors: An example 9

P. KINDLMANN
K-stability as an ecologically more relevant stability concept 14

A. LASOTA
Stability, periodicity and chaos from the statistical point of view 24

K.R. SCHNEIDER
Singularly perturbed autonomous differential systems 32

A.N. SHARKOVSKY, YU.K. MAISTRENKO
Difference equations with continuous time as mathematical models of the structure emergencies 40

E.E. SHNOL
Stability of equilibria in critical cases 50

Part II: General Mathematical Models

E. AMANN
Permanence in population dynamics 58

W. EBELING, A. ENGEL, R. FEISTEL, V.G. MAZENKO
Models of evolution processes including age structure 67

F. FELBER, A. HUNDING, H. MALCHOW
Pattern formation in an ionic reaction-diffusion system: A contribution to the morphogene prepattern theory of mitosis 76

H. HERZEL, TH. SCHULMEISTER
Chaotic dynamics and fluctuations in a biochemical system 85

R.W. LEVEN, B.P. KOCH, G.S. MARKMAN
Periodic, quasiperiodic and chaotic motion in a forced predator-prey ecosystem 95

J.S. NICOLIS
Chaotic dynamics of logical paradoxes 105

M. PESCHEL, M. VOIGT, W. MENDE
Control of growth processes 114

Part III. Mathematical Models for Concrete Processes

A. ARNEODO, F. ARGOUL, P. RICHETTI, J.C. ROUX
The Belousov-Zhabotinskii reaction: A paradigm for theoretical studies of dynamical systems 122

P. BRUNOVSKY, T. KMET
The nitrogen transformation cycle in water 132

G. CZAJKOVSKI
Modelling of polymer growth 139

H.I. FREEDMAN, G.S.K. WOLKOWICZ
A mathematical model of group defence in predator-prey systems 149

K.P. HADELER
Vector-transmitted diseases in structured populations 154

Z. KUBÍČKOVÁ, M. KUBÍČEK, M. MAREK
Nonlinear behaviour in mathematical models of anaerobic digesters 162

E. LABOS, E. NOGRADI
Examples of computer-aided exploration and design of dynamical systems in neurosciences: Design of optimal nets 172

S. MARKOV, T. KOSTOVA-VASILEVSKA
A dynamical model of synaptic transmission 182

K. SIGMUND
Gradients for replicator systems 186

Part IV: Mathematical Description of Environmental Processes

T. ALDENBERG
The limiting factor concept in relation to stability, sensitivity and bifurcation in plankton models 196

G. BARNA, P. ERDI
Effects of structured environment to dynamic behaviour: Some illustrations 209

TH. BLEY, B. HEINRITZ
Modelling and control of yeast growth in biotechnical systems 213

H. BOSSEL
A dynamic simulation model of tree development under pollution stress 220

H. BRÖHL-KERNER
A biota model to assess the influence of human impacts on the global carbon cycle 230

L. KAIRIUKSTIS
An integrative approach to the solution of ecological problems with particular attention to the regional case 240

W. MENDE, K.-F. ALBRECHT
Application of the evolon model on evolution and energy growth processes 253

L.F. OLSEN
Non-linear dynamics of the epidemics of some childhood deseases in Copenhagen, Denmark 265

E.M. VAN DE VRIE
Modelling an estimating transport and fate of heavy metals in water column and sediment layer in some enclosed branches of the sea in the S.W. Netherlands 273

POSTER SESSION

T. CZIERZYNSKI
Mathematical software developed by the Karl-Weierstraß-Institute of Mathematics 282

HAUSDORFF DIMENSION OF ATTRACTORS : AN EXAMPLE

H. G. Bothe *)

The aim of this paper is to show by a simple example that the connection between the Ljapunov exponents and the Hausdorff dimension of an attractor is not so close as sometimes assumed. The mathematical background for the facts presented here (see [1]) covers a more general class of attractors (1-dimensional hyperbolic or higher dimensional expanding attractors of a diffeomorphism).

Let S^1 be the unit circle which is regarded as the real line $\mathbb{R}$ modulo 1, i.e. S^1 is obtained from $\mathbb{R}$ by identifying points whose difference is an integer. So for points $t_1, t_2 \in S^1$ a sum $t_1+t_2 \in S^1$ is defined, and points $t \in S^1$ can be multiplied by real numbers. By $\mathbb{D}^2$ we denote the unit disk in the plane $\mathbb{R}^2$. The cartesian product $V = S^1 \times \mathbb{D}^2$ is a solid torus. For $t \in S^1$ the disk $\{t\} \times D^2$ will be denoted by D_t.

We consider a (not necessarily continuous) embedding

$$f: V \rightarrow \text{Int } V$$

which has the following properties: (1) The restriction f_t of f to to a disk D_t is a linear embedding of D_t into the interior of the disk D_{2t} which contracts all distances by a factor $\lambda(t) > 0$ (i.e. f_t is a similarity mapping). (2) There is a positive δ such that for each $t \in S^1$ the distance between the two disks $f(D_{t/2})$, $f(D_{t/2+1/2})$ in D_t is at least δ.

By $t \mapsto \lambda(t)$ the embedding f defines a (not necessarily continuous) positive function $\lambda: S^1 \rightarrow \mathbb{R}$. If f is continuous, then to apply the mapping $f: V \rightarrow \text{Int } V$ means to stretch V to a longer solid torus of twice the original length, where during the stretching each disk D_t is contracted to a smaller circular disk of radius $\lambda(t)$ (see Fig. 1).

The set

$$\Lambda = \bigcap_{n=0}^{\infty} f^n(V)$$

attracts all orbits $(x, f(x), f^2(x), \ldots)$ $(x \in V)$: it is the attractor of f. As easily seen the intersection $\Lambda \cap D$ of Λ with any disk D_t is a Cantor set C_t. We are interested in the

*) Karl-Weierstrass-Institut für Mathematik
DDR 1086 Berlin, Mohrenstr. 39

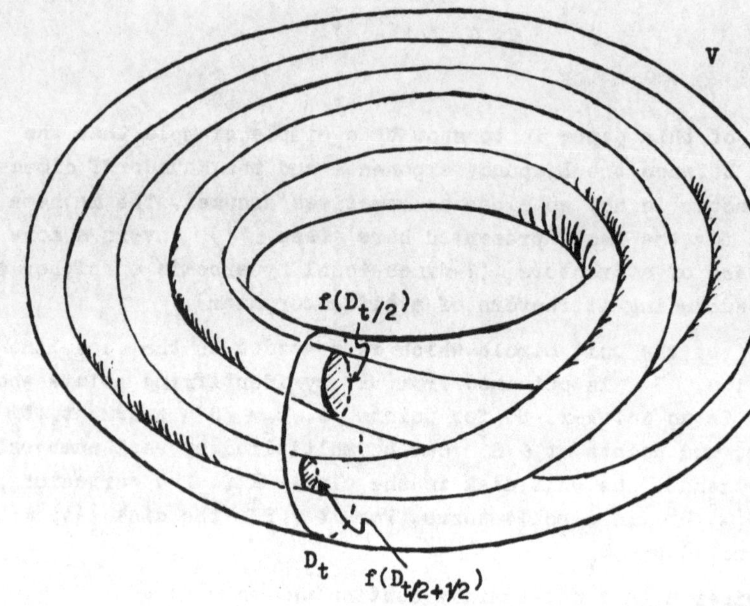

Fig. 1

Hausdorff dimension $\dim_H C_t$ of these sets. If these dimensions are independent of t, then the Hausdorff dimension $\dim_H \Lambda$ is $\dim_H C_t + 1$.

Under our assumptions (1), (2) for each $t \in S^1$ the dimension $\dim_H C_t$ is already determined by the function λ. Some relations between λ and $\dim_H C_t$ are stated in the following theorem.

Theorem 1. (A) If for a positive real number p_λ the functional equation

$$\lambda^p(t)\xi(t) + \lambda^p(t+1/2)\xi(t+1/2) = \xi(2t) \qquad (*)$$

with $p = p_\lambda$ has a (not necessarily continuous) solution $\xi: S^1 \to \mathbb{R}$ whose infimum $\inf \xi$ is positive, then for all $t \in S^1$

$$\dim_H C_t = p_\lambda .$$

(B) If λ is a Lipschitz function, then there is a positive real number p_λ for which (*) has a solution ξ with $\inf \xi > 0$.

(C) Let $\mathcal{L}$ be the space of all real functions λ on S^1 for which there is a positive real number p_λ such that (*) with $p = p_\lambda$

has a solution ξ satisfying $\inf \xi > 0$. The topology in $\mathcal{L}$ is defined by the metric $\mathrm{dist}(\lambda_1,\lambda_2) = \sup_{t\in S^1} |\lambda_1(t) - \lambda_2(t)|$, i.e. we consider the topology of uniform convergence. Then p_λ (which for $\lambda \in \mathcal{L}$ is uniquely determined by $\mathcal{L}$) depends continuously on λ, i.e. $\lambda \mapsto p_\lambda$ is a continuous function on $\mathcal{L}$. (By (B) the space $\mathcal{L}$ contains all Lipschitz continuous functions on S^1.)

<u>Corollary.</u> If λ belongs to $\mathcal{L}$, then $\dim_H C_t$ does not depend on t.

For proofs of these facts see [1].

Now, for a continuous function λ in $\mathcal{L}$, we try to reduce the functional equation (*) to an ordinary equation for p_λ, i.e. we look for a real function $P(p)$ for which p_λ is the maximal solution of $P(p) = 0$. This can be done approximatively as follows: We first approximate λ (with respect to the topology of uniform convergence) by a step function λ^* which, for some $n \geq 1$, is constant on each subarc $[(i-1)2^{-n}, i2^{-n})$ of S^1 $(i = 1,2,\dots,2^n)$. If λ^*_i denotes the value of λ^* on $[(i-1)2^{-n}, i2^{-n})$, then each solution $\xi_1,\dots,\xi_{2^n}$ of the following 2^n linear equations (**) yields a solution of (*).

$$\begin{aligned} \lambda_i^{*p}\xi_i + \lambda_{i+2^{n-1}}^{*p}\xi_{i+2^{n-1}} &= \xi_{2i} \\ \xi_{2i} &= \xi_{2i-1} \end{aligned} \qquad (i = 1,2,\dots,2^{n-1}) \qquad (**)$$

Indeed the solution of (*) corresponding to $\xi_1,\dots,\xi_{2^n}$ is the step function on S^1 whose values on the arcs $[(i-1)2^{-n}, i2^{-n})$ are ξ_i. The matrix A on the left hand side of (**) is non negative, and, by the theorem of Perron-Frobenius (see [2], XIII,2), we get a positive solution of (**) if and only if 1 is an eigenvalue of A and no eigenvalue is of modulus greater than 1. Let P_{λ^*} be $\det(A - E)$ regarded as a function of p. Since $0 < \lambda_i^* < 1$ the powers λ_i^{*p} are decreasing functions of p, and, applying the Perron-Frobenius theorem once more, we find that (**) has a positive solution if and only if $p = p_{\lambda^*}$ is the maximal solution of $P_{\lambda^*}(p) = 0$. Since $P_{\lambda^*}(0) = -1$, $\lim_{p\to\infty} P_{\lambda^*}(p) = 1$, this solution p_{λ^*} exists and is positive. Hence λ^* belongs to $\mathcal{L}$, and by (C) $|p_{\lambda^*} - p_\lambda|$ is small provided $\max_{t\in S^1} |\lambda^*(t) - \lambda(t)|$ is small.

A similar approximative determination of $\dim_H C_t$ from a different point of view (not using (*)) is done in [3].

For $n = 2$ we have

$$P_{\lambda^*}(p) = (\lambda_1^{*p} - 1)(\lambda_4^{*p} - 1) - \lambda_2^{*p}\lambda_3^{*p}.$$

In this case we consider the step function λ'^* which is defined by $\lambda'^*(t) = \lambda(t + \frac{1}{4})$. Then

$$P_{\lambda'^*}(p) = (\lambda_2^{*p} - 1)(\lambda_1^{*p} - 1) - \lambda_3^{*p}\lambda_4^{*p},$$

and for almost all values of $\lambda_1^*, \ldots, \lambda_4^*$ we get $p_{\lambda'^*} \neq p_{\lambda^*}$.

It can be proved that for a function $\lambda : S^1 \to \mathbb{R}$ which is smooth and differs from λ^* only near the points $t = i/4$ $(i = 0,1,2,3)$ as indicated in Fig. 2 the difference $|p_\lambda - p_{\lambda^*}|$ is small (see [1]). Therefore, if $\lambda' : S^1 \to \mathbb{R}$ is defined by $\lambda'(t) = \lambda(t + \frac{1}{4})$, the inequality $p_{\lambda'^*} \neq p_{\lambda^*}$ implies $p_{\lambda'} \neq p_\lambda$, and we have proved the following theorem.

<u>Theorem 2.</u> There is a smooth function λ in $\mathcal{L}$ and a point t_o in S^1 such that for the function $\lambda' : S^1 \to \mathbb{R}$ which is defined by $\lambda'(t) = \lambda(t + t_o)$ we have $p_{\lambda'} \neq p_\lambda$.

<u>Remark.</u> By (C) this phenomenon is stable in the sense that the set of all functions λ in $\mathcal{L}$ for which the assertion of the theorem holds true is C^0 open in $\mathcal{L}$.

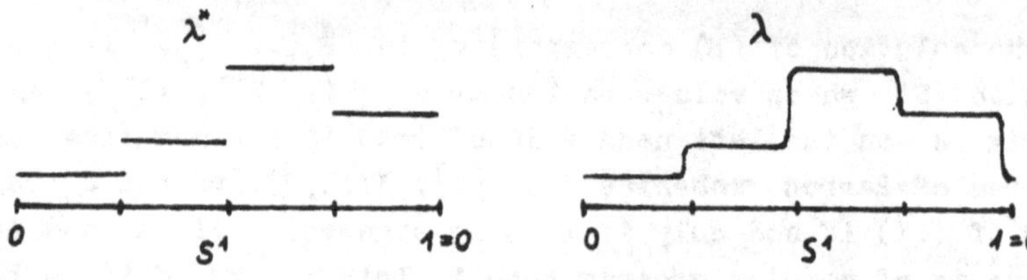

Fig. 2

For $t_o \in S^1$ let $\rho_{t_o} : V \to V$ be the rotation of V given by

$$\rho_{t_o}(t,x) = (t + t_o, x) \qquad ((t,x) \in S^1 \times \mathbb{D}^2 = V).$$

If $f : V \to V$ is an embedding satisfying (1),(2) and if $\lambda : S^1 \to \mathbb{R}$ is the corresponding function, then the mapping

$$f' = \rho_{-2t_o} f \rho_{t_o} : V \to \operatorname{Int} V$$

is an embedding which also satisfies (1),(2). The function $\lambda' : S^1 \to \mathbb{R}$ belonging to f' is given by $\lambda'(t) = \lambda(t + t_o)$.

This leads to the following corollary in which $\mathcal{F}$ denotes the space of all C^1 embeddings $f: V \to \mathrm{Int}\, V$ satisfying (1),(2) with the C^0 topology.

<u>Corollary.</u> $\mathcal{F}$ contains a non empty open subset $\mathcal{F}^*$ such that for each $f \in \mathcal{F}^*$ there is an embedding $f' = \varrho_{-2t_0} f \varrho_{t_0} \in \mathcal{F}$ for which the Hausdorff dimension of the corresponding attractor is different from the Hausdorff dimension of the attractor belonging to f.

<u>Remark.</u> Probably $\mathcal{F}^*$ is dense in $\mathcal{F}$.

The Ljapunov exponents of an attractor Λ are defined with respect to an f-invariant ergodic probability measure on Λ. The measure of this kind which is most closely connected with the dynamics on Λ, i.e. with the restriction of f to Λ, is the so called Bowen-Ruelle measure μ_Λ which in our case can be characterized by the condition that its projection to S^1 is the Lebesgue measure on S^1. The ordinary Ljapunov exponents, i.e. those with respect to μ_Λ, for an attractor Λ of an embedding $f: V \to \mathrm{Int}\, V$ satisfying (1),(2) are $\log 2$ (with multiplicity 1) and $\int_0^1 \log \lambda(t)dt$ (with multiplicity 2). Hence, the Ljapunov exponents of the attractor belonging to an embedding $f \in \mathcal{F}$ do not change if we apply (as in the corollary) a rotation to f. This shows that the attractors of the mappings f, f' in the corollary have the same Ljapunov exponents but different Hausdorff dimensions.

Besides the geometric Hausdorff dimension considered here another dimension of attractors Λ which dependes on the measure μ_Λ on Λ is defined as follows

$$\dim_{\mu_\Lambda} \Lambda = \inf \left\{ \dim_H \Gamma \, ; \, \Gamma \text{ a Borel set in } \Lambda, \mu_\Lambda(\Gamma) = 1 \right\} .$$

A formula for $\dim_{\mu_\Lambda} \Lambda$ proved in [4], Corollary 5.1 implies

$$\dim_{\mu_\Lambda} \Lambda = 1 - \log 2 \Big/ \int_0^1 \log \lambda(t)dt .$$

Therefore our corollary to Theorem 2 shows that in many simple cases $\dim_{\mu_\Lambda} \Lambda$ differs from the (geometric) Hausdorff dimension of Λ.

References

[1] H.G.Bothe: On the Hausdorff dimension of expanding attractors. (Preprint 1986)

[2] F.R.Gantmacher: Matrizenrechnung. Berlin 1959

[3] M.Schulz: Hausdorff-dimension von Cantormengen mit Anwendungen auf Attraktoren. (Dissertation Berlin 1986)

[4] L.-S. Young: Dimension, entropy and Lyapunov exponents. Ergodic Theory and Dynam. Syst. 2 (1982), 109-124

K-STABILITY AS AN ECOLOGICALLY MORE RELEVANT STABILITY CONCEPT

Pavel Kindlmann [1)]

1. Introduction

The term stability is frequently used both in ecology and in mathematics. Whereas in mathematics there exist a lot of rigid definitions, the understanding of this term in ecology is often vague. The stability concept commonly used in mathematical models of ecological systems is the Liapunov stability. However, it does not completely fit the needs of ecologists. In this paper, a new mathematical definition of stability, which fits the demands of ecologists better is proposed and applied to several mathematical models of population dynamics of one and two species.

A thorough discussion of different stability concepts in ecology was given by Kindlmann and Lepš [4]. Here we shall stress only some major points of this discussion, in order to elucidate the motivation of the new stability concept definition:

In most mathematical models, differential equations are used. It means that the variables are supposed to be changing continuously in time. In the field ecology the continuous measuring of variables (with the exception of a few ones) is almost impracticable. Moreover, the ecological systems are subjected to periodic changes (circadian, diurnal). This phenomenon is particularly important in organisms with non-overlapping generations. Hence, it is often meaningful to measure the variables in particular intervals only, in terrestrial plants and vertebrates usually once a year [1]. Consequently, in many cases the discrete models and the corresponding stability concepts are more appropriate for ecological application.

In mathematical models of ecosystem population dynamics it is usually supposed that under normal conditions the system remains in "equilibrium". However, the environmental conditions are never constant in nature and even the endogeneous dynamics of each community causes fluctuations in values of particular variables. The quantitative determination of equilibrium usually causes difficulties in the field studies. One possibility to approximate this equilibrium is to take the average over some period which is considered to be "normal". The more realistic approach is to consider an interval (range of values) instead of a point. Some ecosystems are able to exist only far from the equili-

1) Pavel Kindlmann, Department of Biomathematics, Biological Research Centre, Czechoslovak Academy of Sciences, Branišovská 31, 370 05 České Budějovice, Czechoslovakia

brium. As a nice example may serve snow avalanche paths [5], where the great species diversity is caused by avalanches which disturb the system from time to time. If these avalanches were not there, many species would die out.

The type of Liapunov stability commonly used is a local property of the system. It assures only the existence of some neighbourhood of the equilibrium, in which the system is stable, without saying anything about the size of this neighbourhood. In ecology, however, we are interested in the consequences of fairly great perturbations.

In biological systems we meet fairly great perturbations which act permanently (climatic changes, biological variability etc.) and this fact is not incorporated in the Liapunov stability concept either.

This does not mean that the Liapunov stability concept is always unrealistic. Under certain circumstances (especially in simple systems and under constant environmental conditions - e.g., in the laboratory prey-predator systems) this concept is quite well founded. But even in this case one must be careful with the interpretation.

2. The definition of K-stability

Bearing in mind the criticism above, we shall define an ecologically more relevant stability concept [4]. In the definition we shall use difference equations, as they are more appropriate in many cases. Obviously, a similar definition could be stated for their differential counterparts. We shall assume that the system remains in some invariant set G under normal conditions rather than in equilibrium. The behaviour within this set will not be of any interest to us - it may be a periodical, an almost periodical or even a chaotic motion. The stability of this invariant set will then be studied. Local stability will be replaced by the global one, i.e., we define a set H, within which the system is stable. This means that the system, after having been displaced by some external force to some point $x \in H-G$, never leaves the set H and finally reaches the set G. The existence of permanently acting perturbations leads either to stochastic models, or to some perturbed differential or difference equations. None of the various stability concepts of perturbed equations in mathematics (see, e.g., [2,3]) is satisfactory for the use in ecological applications. All of them assume that the "perturbation function" is small in a certain sense: that its norm approaches zero as the state of the system approaches equilibrium, or that there exists some constant (maybe very small), so that the norm of the perturbation is smaller than this constant. But in ecology the norm of the "normal" perturbations may be fairly high. Therefore, there should be some predetermined constant K, serving as an upper bound for the norm of "normal" perturbations. All the subsets of the set H may be divided into three categories:

DEFINITION 1. Let N be the set of natural numbers, R_m the m-dimensional Euclidean vector space, function f: $N \times R_m \to R_m$, K a positive number, $G \subset H \subset R_m$. We shall denote the set $P = \{g;\ g: N \times R_m \to R_m,\ \sup_{n \in N,\ y \in H} ||g(n,y)|| \leq K\}$, where $||.||$ is the Euclidean norm. We shall consider the system

$$y_{n+1} = f(n,y_n) + g(n,y_n) \qquad (1)$$

Then we define: the set $G \subset H$ is

a) <u>K-attractive</u> in H for the system (1), if for all the solutions of the system (1) it holds: 1) for all $m \in N$, $y_m \in H$ and for all $g \in P$ there exists an $n \in N$, $n > m$ so that $y_n \in G$; 2) if $m \in N$, $y_m \in G$, then for all $g \in P$ and $n \in N$, $n > m$, it holds $y_n \in G$.

b) <u>K-transient</u> for the system (1), if for all $m \in N$, $y_m \in G$ and for all $g \in P$ there exists such $n \in N$, $n > m$ so that it is not $y_n \in G$.

c) <u>K-semiattractive in H</u> for the system (1), if it is neither K-attractive, nor K-transient in H.

DEFINITION 2. The set $M \subset H$ is called the <u>minimal K-attractive set in H</u> for the system (1), if and only if it is K-attractive and if for all the K-attractive sets $G \subset H$ of the system (1) it holds $M \subset G$.

The difference between this stability concept and the common ones is rooted in the fact that the constant K is a priori defined (and therefore it may be fairly great) and does not depend on x. This stability concept distinguishes two types of perturbations: "normal", which are bounded and permanently acting (performed by the function g) and "extreme" - catastrophes like fire, human activities etc., which may displace the system outside the regions of its "normal" motion - outside the K-attractive set G. There are several ecologically interesting questions, which can be answered only by means of this stability concept [4].

We shall state some simple lemmas on K-stability, which will be used in the applications. Their simple proofs will be left to the reader.

LEMMA 1. Decomposition of exp(H) into K-attractive, K-semiattractive and K-transient sets is a decomposition of exp(H) into pairwise disjoint sets.

LEMMA 2. Let f: $N \times H \to G$. If $G' = \{x \in R_m;\ \exists\ y \in G:\ ||x - y|| \leq K\} \subset H$, then G' is K-attractive in H for the system (1).

LEMMA 3. Let there exist an $x' \in G$, so that for some $n' \in N$ there holds at least one of the conditions

a) $||f(n',x')|| + K > \sup_{y \in G} ||y||$,

b) $0 < ||f(n',x')|| - K < \inf_{y \in G} ||y||$.

Then G is not K-attractive in H for the system (1).

LEMMA 4. If there exists an $x' \in G$, so that for any $n \in N$ it holds $||f(n,x') - x'|| \leq K$, then G is not K-transient in H for the system (1).

LEMMA 5. Let there hold one and only one of the conditions

a) $\sup_{x \in G,\ n \in N} (||f(n,x)|| - ||x||) = S < -K < 0$,

b) $\inf_{x \in G,\ n \in N} (||f(n,x)|| - ||x||) = I > K > 0$.

Then G is K-transient in H for the system (1).

For K-stability there also can be proved theorems of Liapunov type. We shall bring one for illustration:

THEOREM. Let $K > 0$, $H \subset R_m$, let for each $n \in N$, for each $g \in P$ and let for each $y \in H$ it holds $f(n,y) + g(n,y) \in H$. Furthemore, let for some function $V: N \times R_m \to R$ and for all $n \in N$, for all $x \in H$ it holds $V(n,x) \geq 0$ and $V(n+1, f(n,x)) - V(n,x) < 0$. We shall denote

$$M = \sup_{x,y \in H,\ ||x-y|| < K,\ n \in N} ||V(n,y) - V(n,x)||. \tag{2}$$

For some $\varepsilon > 0$ let

$$G' = \{ x \in H;\ \forall n \in N:\ V(n+1, f(n,x)) - V(n,x) > -M - \varepsilon \}, \tag{3}$$

$$L = \sup_{x \in G',\ n \in N} V(n,x), \tag{4}$$

$$G = \{ x \in H;\ \forall n \geq 0:\ V(n,x) \leq M + L \}. \tag{5}$$

Then G is K-attractive in H for the system (1). (The situation for the case of autonomous systems is illustrated in Fig. 1.)

Proof: We shall verify the conditions of the definition 1 a).
1) Let $y_m \in H - G$. Then for each $g \in G$ and for all $n \in N$, $n \geq m$, if $y_n \in H - G$, then it is satisfied

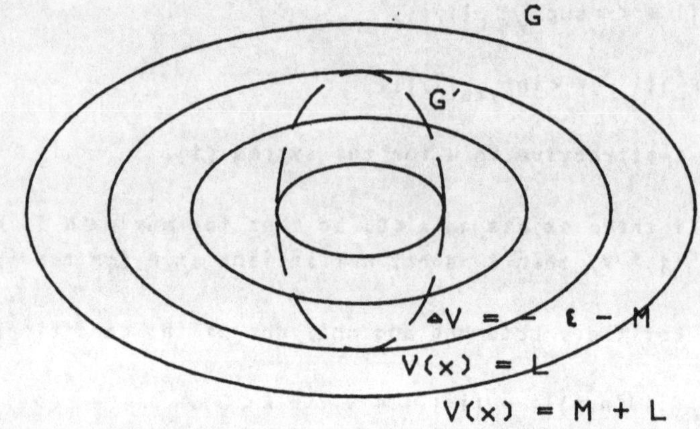

Fig. 1. Construction of the set G in the Theorem in the case of autonomous systems. Full lines: $V(x) = \text{const.}$, broken line: $V(x) = V(f(x)) - V(x) = \text{const.}$

$$V(n+1,\ f(n,y_n) + g(n,y_n)) - V(n,y_n) \leq$$
$$\leq V(n+1,\ f(n,y_n)) + M - V(n,y_n) \leq -\varepsilon \tag{6}$$

(as $G' \subset G$). Therefore there exists an $n \in N$, $n \leq (V(m,y_m) - M - L)/\varepsilon$, so that $y_{m+n} \in G$, as $V(m + n,\ y_{m+n}) \leq M + L$.
2) Let $y \in G$ and $M + L \geq \sup_{n\in N} V(n,y) > L$. Then from (4) it follows $y \in G - G'$ and from (2) and (3) we have

$$V(n+1,\ f(n,y) + g(n,y)) < V(n,y) \leq M + L \tag{7}$$

for all $n \geq 0$, i.e., $f(n,y) + g(n,y) \in G$. If $y \in G$ and $\sup_{n\in N} V(n,y) \leq L$, then due to (2) and as $V(n+1,\ f(n,y)) - V(n,y) \leq 0$, it holds $V(n+1,\ f(n,y) + g(n,y)) \leq M + L$. I.e., again $f(n,y) + g(n,y) \in G$.

3. Application of K-stability to one- and two-species models

We shall analyze the behaviour of a wide class of "one-species models", i.e., of the case where $f,g\colon R_1^+ \to R_1^+$ in the system (1). Many published functions f satisfy the following set of conditions (see, e.g. [3], [6], [7], [8], [9]):

1. The function f is continuous and $f(0) = 0$.
2. $f'(0) > 1$.
3. There exists an $x^1 > 0$ such that $f(x)$ is increasing for $x < x^1$ and decreasing for $x > x^1$.
4. There exists an $x^2 > x^1$, such that $f(x)$ is concave for $x < x^2$ and convex for $x > x^2$.

5. $\lim_{x\to\infty} f(x) = 0$.
We shall analyze closed connected sets $G \subset H$ from the point of view of K-stability for the system (1) under the conditions above. We shall denote by $x^3 = K + \max_{x>0} f(x)$ and by x^4 the unique number, for which $f(x^4) + K = x^4$.

It is wise to distinguish three cases for given f and K (Fig.2):
a) For each $x > 0$ it holds $f(x) - K < x$.
In this case any set G is either K-transient or not K-transient in the dependence of whether it does not contain a number from the interval $\langle 0;x^4\rangle$ or whether it does (see Lemmas 4. and 5.). If G contains a number from this interval and moreover, it does not contain 0, then - according to the Lemmas 1. and 3. - G is K-semiattractive. If $0 \in G$ and $\sup_{y\in G}|y| < x^3$, then G is again K-semiattractive (Lemmas 1. and 3.). If finally $0 \in G$ and $\sup_{y\in G}|y| \geq x^3$, then G is K-attractive (Lemma 2.). Evidently, in this case the minimal connected K-attractive set in H (if $\langle 0;x^3\rangle \subset H$) for the system (1) is the interval $\langle 0;x^3\rangle$. If the condition $\langle 0;x^3\rangle \subset H$ is not satisfied, then there does not exist any K-attractive set $G \subset H$ for (1).
b) There exists a unique number x^5, such that $f(x^5) - K = x^5$.
The same holds in this case as in the previous one. Moreover, if $\inf_{y\in H}|y| = \inf_{y\in G}|y| = x^5$, $\sup_{y\in G}|y| \geq x^3$ and if $f(\sup_{y\in H}|y|) = f(x^5)$, then G is K-attractive in H (Lemma 2.). This makes a substantial difference, as for H satisfying $\inf_{y\in H}|y| = x^5$, $f(\sup_{y\in H}|y| = f(x^5)$, $x^3 \leq \sup_{y\in H}|y|$, the minimal K-attractive set in H is the interval $\langle x^5; x^3\rangle$. Again the interval $\langle 0;x^3\rangle$ is the minimal K-attractive set in H, if it is contained in H. For any other set H there does not exist any K-attractive set in H for (1).
c) There exist two different numbers x^5 and x^6, satisfying $f(x^5) - K = x^5$, $f(x^6) - K = x^6$.
The analysis in this case is much more complicated. We shall show only one example, useful for biological applications. Let x^7 be the unique number, for which $f(x^7) = f(x^5)$. If $x^3 < x^7$, then from Lemma 2. it follows that $G = \langle x^5;x^3\rangle$ is K-attractive in $H = \langle x^5;x^7\rangle$. One could easily show that this does not hold for any greater set H.

There may be derived biologically interesting conclusions from this analysis: Let the population dynamics of some species obey the equation $x_{n+1} = f(x_n)$. Let the norm of perturbations may reach maximally the value K. Then the population cannot survive, if its initial number does not exceed some critical value x^5, or if the case a) is true. If the initial number of individuals exceeds x^5 and the case a) is not true, then the population survives. An important feature is the fact that this conclusion holds also for equations with a very complicated (e.g., chaotic) behaviour. A numerical example is shown in Fig.3.

Fig. 2. Three possible cases for given function f satisfying 1.-5. and constant K.

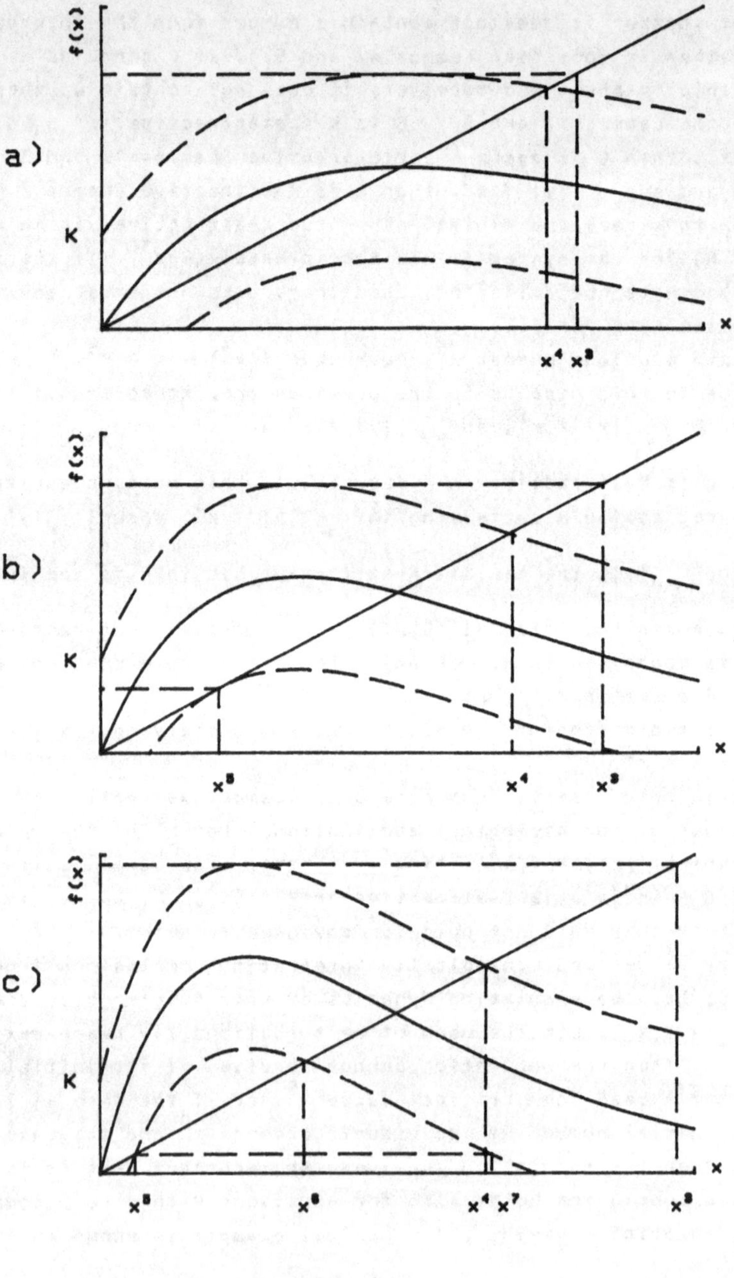

Finally, we shall show one example of two species system analysis. We shall consider a predator-prey system

$$x_{n+1} = f(x_n) - q(x_n, y_n)y_n + g_1(n, x_n, y_n),$$
$$y_{n+1} = dy_n + h(x_n, y_n)y_n + g_2(n, x_n, y_n), \qquad (8)$$

where x_n and y_n are numbers of prey and predator individuals respectively and where the functions f,q and h satisfy:

a) The function f satisfies conditions 1.-5. above.

b) $q(0,y) = 0$ for each $y \geq 0$ and there exists an $M > 0$ so that for each $x > 0$ and for each $y \geq 0$ it holds $M > q(x,y) > 0$.

c) The function h is continuous and $h(0,y) = 0$ for each $y \geq 0$, $H(x,y) > 0$ for each $x > 0$ and $y \geq 0$.

d) $0 < d < 1$.

e) There exists an $y_{max} > 0$, so that for each $y \leq y_{max}$ and for each $x > 0$ it holds $y(d+h(x,y)) < y_{max} - K$.

f) There exists an $x > 0$, so that $f(x) - x > K + My_{max}$.

g) There exists an $y_{min} > 0$, so that for each $y \geq y_{min}$ and for each $x \geq x_a$ it holds $y(d+h(x,y)) > y_{min} + K$. Here, x_a is the smallest number, for which $f(x_a) - K - My_{max} = x_a$.

h) $\sup_{n \in N,\ (x,y) \in R \times R} \|g(n,xy)\| \leq K$.

Let x_c be the unique number, for which $f(x_c) = f(x_a)$, $x_c \neq x_a$, $x' = K + My_{max} + \max_{x>0} f(x)$. Then $G = \langle x_a; x'\rangle \times \langle y_{min}; y_{max}\rangle$ is K-attractive in $H = \langle x_a; x_c\rangle \times \langle y_{min}; y_{max}\rangle$ for the system (8).

Proof: For each $(x,y) \in H$ it holds $y_{min} \leq y \leq y_{max}$; moreover, $x \geq x_a$. Therefore, from e),g) and h) it follows that for each $n \in N$ and for each

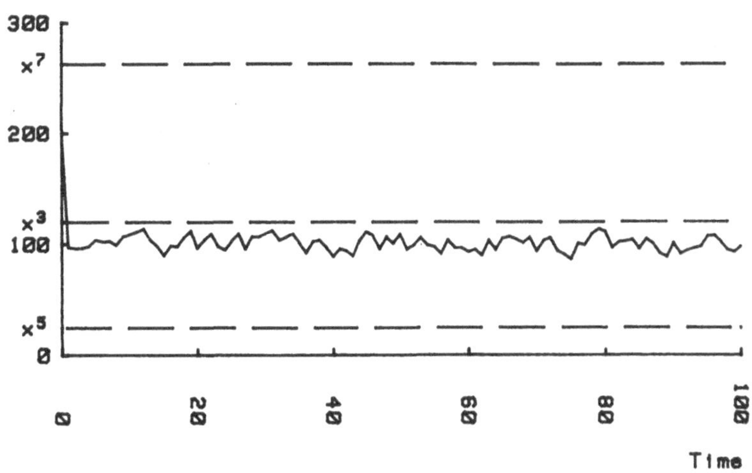

Fig. 3. A numerical example of the behaviour of a function satisfying 1.-5. Here, $f(x) = \max(x(1 + .53(1-x/100), 0)$, $x_0 = 200$, K=10.

$(x_n, y_n) \in H$ it is $y_{min} \le y_{n+1} \le y_{max}$. Furthermore, the conditions a) and f) actually mean that the function f with perturbation smaller in norm than $K + My_{max}$ falls into the case c) of the analysis of the function f. As for each $(x,y) \in H$ it holds $0 < g(x,y) < M$ and $y \le y_{max}$, necessarily $- K - My_{max} < - y\alpha(x,y) + g_1(x,y) < K + My_{max}$. From the analysis of the function f it follows that for each $n \in N$ and for each (x_n, y_n) it holds $x_a \le x_{n+1} \le x'$, which completes the proof. A numerical example is shown in Fig. 4.

References

[1] Connel, J.H.; Sousa, W.P.: On the evidence needed to judge ecologi- stability or persistence. Amer. Nat. 121(1982), 6, 789-824

[2] Hahn, W.: Stability of Motion. Springer, Berlin 1967

[3] Kindlmann, P.: Soustavy diferenčních rovnic a jejich využití v modelování populačních systémů. [Systems of difference equations and their use in modelling of population systems.] PhD Thesis, Charles University, Praha, 1985

[4] Kindlmann, P. and Lepš, J.: What is stability? A mathematician's and ecologist's point of view. Syst. Anal., Modelling and Simul., in press.

[5] Rejmánek, M.: Perturbations-dependent coexistence and species diversity in ecosystems. In Schuster, P. (ed.): Stochastic Phenomena and Chaotic Behaviour in Complex Systems. Springer, Berlin 1984, pp. 220-230.

[6] Scudo, F.C.: Discrete time models in ecology; history and perspectives. Automazione e utilizzazione delle risorse. XIV Convegno Internazionale di Automazione e Strumentazione - Milano, 23-24 Novembre 1976. FAST, Milano 1976.

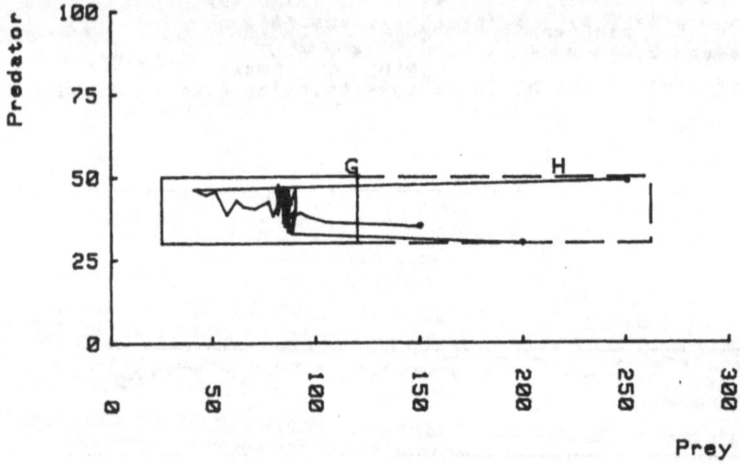

Fig. 4. A numerical example of the behaviour of a system, which satisfies conditions a)-h). Here, $x_{n+1} = x_n + .53x_n(1 - x_n/100) - y_n(1-\exp(-x_n))/10$,
$y_{n+1} = y_n/2 + 20(1 - \exp(-x_n))$, K = 5, initial conditions marked by •

[7] Svireżev, Ju. M. and Logofet, D.O.: Ustojčivosť biologičeskich soobščestv. Nauka, Moskva 1978.
[8] Turelli, M. and Petry, D.: Density-dependent selection in a random environment - an evolutionary process that can maintain stable population dynamics. Proc. Nat. Acad. Sci. US - Biol. 77(1980) 12, 7501-7505
[9] Ullyett, G.C. : Competition for food and allied phenomena in sheep and blowfly populations. Phil. Trans. Roy. Soc. B 260 (1950), 77-174

STABILITY, PERIODICITY AND CHAOS FROM THE STATISTICAL POINT OF VIEW

Andrzej Lasota [1)]

0. Introduction

In many interesting dynamical systems the behaviour of trajectories is extremely irregular and chaotic. In order to understand and to predict such behaviour we have a few criteria in our disposal. Some of them, however, imply the existence of chaos only on a small part of the phase space. Consider, for example, the mapping $S: [0,1] \to [0,1]$ given by the formula

$$S(x) = 4x^2(1 - x^2) \tag{0,1}$$

It is easy to verify that S has a periodic point of period 3 and thus according to the known results of Sharkovsky [7] and Li - Yorke [5] it admits periodic orbits of any given period and behaves chaotically on an uncountable set. On the other hand almost all of the trajectoties converge to zero (conjecture !) and consequently this chaotic behaviour is restricted to a set of measure zero.

The purpose of this lecture is to present criteria which allow to determine the behaviour of dynamical systems globally. They will be formulated in terms of the convergence of densities and thus may be considered as a statistical description of dynamical systems. It is significant that the same technique may be used to study dynamical systems generated by purely deterministic locally expanding mappings and by arbitrary mappings in the presence of noise.

1. Frobenius-Perron operator

Let $(X, \mathcal{A}, m)$ be a σ-finite measure space. As usually by $L^1 = L^1(X)$ we denote the space of all real valued integrable functions on X and by $D = D(X)$ the set of all densities, i.e.

$$D = \left\{ f \in L^1:\ f \geq 0,\ \|f\| = 1 \right\}, \qquad \|f\| = \int_X |f(x)|\, m(dx) .$$

A mapping $S: X \to X$ is called nonsingular if S is measurable and if the condition $m(A) = 0$ with $A \in \mathcal{A}$ implies $m(S^{-1}(A)) = 0$. Given a nonsingular mapping S we define the corresponding Frobenius-Perron operator $P_S: L^1 \to L^1$ by setting

1) Institute of Mathematics, Silesian University, ul. Bankowa 14, 40-007 Katowice, Poland .

$$\int_A P_S f(x)m(dx) = \int_{S^{-1}(A)} f(x)m(dx) \quad \text{for} \quad f \in L^1, \; A \in \mathcal{A}. \tag{1.1}$$

Due to the nonsingularity of S the integral on the right hand side of (1.1) vanishes whenever $m(A) = 0$. Therefore, according to the Radon-Nikodym theorem, condition (1.1) defines for every $f \in L^1$ a unique element $P_S f \in L^1$.

The operator P_S has an important statistical interpretation. Namely assume that a large number of particles move independently (without interaction) in the space X and that the position of a particle starting at time $t = 0$ from a point x is given at time $t = n$ by $S^n(x)$. Assume moreover that at time $t = 0$ the positions of particles are distributed according to a density f. Then this density evolves in time and its evolution is described by $\{P_S^n f\}$.

In the case when X is an interval on the real line formula (1.1) may be substantially simplified. Setting $A = (a,x)$ and differentiating with respect to x we obtain

$$P_S f(x) = \frac{d}{dt} \int_{S^{-1}(a,x)} f(y)dy . \tag{1.2}$$

<u>Example 1.1.</u> Consider a piecewise monotonic transformation $S: [0,1] \to [0,1]$ which satisfies the following conditions: There exists a partition

$$0 = a_0 < \dots < a_q = 1$$

of $[0,1]$ such that S restricted to (a_{i-1}, a_i) is a C^2 function. Moreover

$$|S'(x)| \geqslant \lambda > 1, \qquad |S''(x)| \quad [S'(x)]^2 \leq M < \infty \tag{1.3}$$

with some constants λ, M and

$$S(a_{i-1}, a_i) = (0,1) \quad \text{for} \quad i = 1,\dots,q, \tag{1.4}$$

Under these conditions it is easy to calculate P_S. In fact demoting by S_i the restriction of S to the interval (a_{i-1}, a_i) we have

$$S^{-1}(0,x) = \bigcup_j (a_{j-1}, \psi_j(x)) \quad \cup \quad \bigcup_k (\psi_k(x), a_k)$$

where $\psi_i = S_i^{-1}$ and the first union is over all intervals in which S_i is increasing whereas the second is over intervals in which S_i is decreasing. Thus according to (1.2)

$$P_S f(x) = \sum_{i=1}^{q} |\psi_i'(x)| f(\psi_i(x)) . \tag{1.5}$$

Using (1.3) it is easy to verify (see [3]) that for every C^1 function $f \in D$ the logarithmic derivative of $P_S f$ satisfies

$$\frac{|(P_S^n f)'|}{P_S^n f} \leqslant K \quad \text{for} \quad n \geqslant n_o(f) \tag{1.6}$$

with some constant K independent on f.

2. Properties of the Frobenius-Perron operator

An operator $P: L^1 \to L^1$ is called a Markov operator if it satisfies the following two conditions

$$Pf \geqslant 0 \quad \text{for} \quad f \geqslant 0, \quad f \in L^1, \tag{2.1}$$

$$\|Pf\| = \|f\| \quad \text{for} \quad f \geqslant 0, \quad f \in L^1. \tag{2.2}$$

Conditions (2.1) and (2.2) imply that

$$\|Pf\| \leqslant \|f\| \quad \text{for} \quad f \in L^1. \tag{2.3}$$

Using formula (1.1) it is easy to show that the Frobenius-Perron operator P_S is a Markov operator.

The behaviour of the iterates $\{P_S^n\}$ allows to determine many properties of the transformation S such as preservation of measure, ergodicity, mixing and exactness. Here we make only a few remarks concerning these problems. Let a density f_* be given. Define the corresponding measure m_* by

$$m_*(A) = \int_A f_*(x)m(dx) \qquad \text{for} \quad A \in \mathcal{A}. \tag{2.4}$$

We say that m_* is invariant if

$$m_*(S^{-1}(A)) = m_*(A) \qquad \text{for} \quad A \in \mathcal{A}. \tag{2.5}$$

Comparing (1.2),(2.4) and (2.5) we obtain immediately the following

Proposition 2.1. The measure m_* defined by (2.4) is invariant if and only if $P_S f_* = f_*$.

The measure m_* is called ergodic if for every set $A \in \mathcal{A}$ the condition $A = S^{-1}(A)$ implies that either $m_*(A) = 0$ or $m_*(A) = 1$. In other words m_* is ergodic if X does not contain measurable invariant subsets except the trivial ones. It may be proved that ergodicity is related with the uniqueness of solutions of $P_s f = f$. We have namely the following (see[3])

Proposition 2.2. The measure m_* given by (2.4) is invariant and ergodic if and only if the density $f = f_*$ is the unique (in the set D) solution of $P_S f = f$.

If a measure m_* is invariant and ergodic, then by the Birkhoff individual ergodic theorem

$$\lim_{n\to\infty} \frac{1}{n}\sum_{k=0}^{n-1} g(S^k(x)) = \int_X g(x)m_*(dx)$$

for every m_*- integrable function g and the convergence in (2.6) holds m_*- almost everywhere. Thus from the statistical properties of S we may derive corollaries concerning the behaviour of individual trajectories.

3. Asymptotical stability

Let P be a Markov operator. We say that P is asymptotically stable if there exists a unique density f_* such that

$$\lim_{n\to\infty} \| P^n f - f_* \| = 0 \qquad \text{for every} \quad f \in D. \tag{3.1}$$

From (3.1) it follows that $Pf_* = f_*$ and that f_* is the unique fixed point of P. Thus, in particular, when the Frobenius-Perron operator is asymptotically stable, then m_* given by (2.4) is invariant and ergodic .

A convenient criterion for the asymptotical stability may be formulated by the use of lower bound functions. A function $h \in L^1$ is called a lower bound function if

$$P^n f \geqslant h + \varepsilon_n f \qquad \text{for} \quad f \in D \tag{3.2}$$

where the remainder $\varepsilon_n f$ satisfies $\|\varepsilon_n f\| \to 0$ as $n \to \infty$. A lower bound function h is nontrivial if $h \geqslant 0$ and $\|h\| > 0$. We have the following [3]

Theorem 3.1. A Markov operator P is asymptotically stable if and only if it has a nontrivial lower bound function.

Observe that in applications it is not necessary to verify condition (3.2) for all possible $f \in D$. Due to the fact that Markov operators are contractive (inequality (2.3)) it is sufficient to verify (3.2) for all f belonging to a dense subset of D.

Example 3.1. Consider the mapping $S: [0,1] \to [0,1]$ described in the Example 1.1. From inequality (1.6) it follows that

$$P_S^n f(y) \leqslant P_S^n f(x) e^{K|x - y|} \qquad \text{for} \quad x,y \in [0,1] \ .$$

Since $P_S^n f$ is a density there is a point $y \in [0,1]$ such that $P_S^n f(y) \geqslant 1$ and concequently

$$P_S^n f(x) \geqslant e^{-K} \qquad \text{for} \quad x \in [0,1] \ , \quad n \geqslant n_0(f) \ .$$

The last inequality shows that the constant $h = e^{-K}$ is a nontrivial lower bound function for P_S and that P_S is asymptotically stable.

The fact that in Examples 1.1 and 3.1 the transformation S maps the unit interval [0,1] into itself is not essential. Analogous results may be obtained for expanding transformations on the real line and on manifolds (see [3]).

4. Asymptotical decomposition

Let $P: L^1(X) \to L^1(X)$ be a given Markov operator. We say that P has an asymptotical decomposition if its iterates can be written in the form

$$P^n f(x) = \sum_{i=1}^{r} \lambda_i(f) g_{\alpha^n(i)}(x) + \varepsilon_n f(x) \quad \text{for } f \in L^1 \tag{4.1}$$

where: r is a fixed integer,

α^n is the n-th iterate of a permution α of the sequence of numbers $(1,\ldots,r)$,

$g_1,\ldots,g_r$ are densities with mutually disjoint supports ($g_i g_j = 0$ for $i \neq j$) and such that

$$Pg_i = g_{\alpha(i)} \quad \text{for } i = 1,\ldots,r , \tag{4.2}$$

$\lambda_1,\ldots,\lambda_r$ are linear functionals on L^1,

the remainder $\varepsilon_n f$ satisfies $\|\varepsilon_n f\| \to 0$ as $n \to \infty$.

Observe that the summation part in (4.1) is periodic in n with period which does not exceeds r!. Thus if P has an asymptotical decomposition, then for every $f \in L^1$ the sequence $\{P^n f\}$ is asymptotically periodic, i.e. $\{P^n f\}$ is a sum of a periodic sequence and a sequence which converges to zero.

A necessary and sufficient condition for the existence of an asymptotical decomposition may be formulated by the use of upper bound functions. A function $h \in L^1$ is called an upper bound function if

$$P^n f \leq h + \varepsilon_n f \quad \text{for } f \in D \tag{4.3}$$

where the remainder $\varepsilon_n f$ satisfies $\|\varepsilon_n f\| \to 0$ as $n \to \infty$. Again it is sufficient to verify (4.3) on a dense subset of D. From the results of J. Komornik [2] it is easy to derive the following

<u>Theorem 4.1.</u> A Markov operator P has an asymptotical decomposition if and only if P has an upper bound function.

Thus we may say shortly that for Markov operators boundness implies periodicity. This fact has many interesting applications. In this section we will discuss two of them.

<u>Example 4.1.</u> Consider a mapping $S: [0,1] \to [0,1]$ which satisfies all the assumptions described in Example 1.1 except (1.4). In this case the operator P_S has the form

$$P_S f(x) = \sum_{i=1}^{q} |\psi_i'(x)| f(\psi_i(x)) 1_{\Delta_i}(x) \tag{4.4}$$

where $1_{\Delta_i}(x)$ denotes the characteristic function of the interval $\Delta_i = S_i(a_{i-1}, a_i)$. It can be verified [3] that there exists a constant K, independent on on f, such that for every $f \in D$ of bounded variation

$$\mathrm{Var}\, P_S^n f \leqslant K \qquad \text{for} \quad n \geqslant n_o(f).$$

Since $P_S^n f(y) \leqslant 1$ for some $y \in [0,1]$, this implies

$$P_S^n f(x) \leqslant K + 1 \qquad \text{for} \quad x \in [0,1], \quad n \geqslant n_o(f)$$

and shows that the constant $h = K + 1$ is an upper function. Thus, according to Theorem 4.1, the operator P_S has an asymptotical decomposition. This fact was first observed by G.Keller [1].

Example 4.2. Consider the family of transformations $S_\lambda : [0,1] \to [0,1]$ given by the formula

$$S_\lambda(x) = \lambda x(1 - x)$$

with $\lambda \in [0,4]$. From the results of M. Misiurewicz [6] it follows that there exists an uncountable set $\Lambda \subset [0,4]$ such that for every $\lambda \in \Lambda$ the operator P_{S_λ} has an upper bound function and admits an asymptotical decomposition.

5. The influence of noise

In this section $X = R^d$ (d-dimensional Euclidean real space) or $X = T^d$ (d-dimensional torus). Consider a dynamical system with random perturbations given by the formula

$$x_{n+1} = S(x_n) + \xi_n, \qquad n = 0,1,\ldots \tag{5.1}$$

where $S: X \to X$ is a Borel measurable transformation and $\{\xi_n\}$ is a sequence of independent random variables with the same density distribution function g. We also assume that the initial value x_o is a random vector independent on $\{\xi_n\}$.

Denote by f_n the density of x_n. It is well known that in the recurrence formula

$$f_{n+1} = Pf_n, \qquad n = 0,1,\ldots \tag{5.2}$$

the operator P may be written in the form

$$Pf(x) = \int_X f(y) g(x - S(y)) dy. \tag{5.3}$$

It is difficult to verify directly that P has an upper bound function.

Thus in order to understand the behaviour of $\{P^n f\}$ we will use another version of Theorem 4.1. Assume $X = T^d$ and denote by N the set of nonnegative integers. Consider all possible sequences $P^n f$ for $f \in D$. This family will be shortly denoted by $\{P^n f\}_{N,D}$.

We say that the family $\{P^n f\}_{N,D}$ is asymptotically uniformly integrable if there exist a positive valued function $\delta: (0,\infty) \to (0,\infty)$ and integer valued function $n_o: D \to N$ such that for every $\varepsilon > 0$

$$\int_A P^n f(x)dx \leq \varepsilon \quad \text{whenever} \quad m(A) \leq \delta(\varepsilon) \text{ and } n \geq n_o(f). \quad (5.4)$$

From the results of J. Komornik [2] it follows immediately

<u>Theorem 5.1.</u> If the family $\{P^n f\}_{N,D}$ is asymptotically uniformly integrable, then P has an asymptotical decomposition.

In order to apply Theorem 5.1 to the operator P given by (5.3) we will use the fact that g is an integrable function. This implies the existence of a function $\delta_g: (0,\infty) \to (0,\infty)$ such that

$$\int_A g(x)dx \leq \varepsilon \quad \text{if } m(A) \leq \delta_g(\varepsilon). \quad (5.5)$$

Now consider $P^n f$ for given $f \in D$ and $n \geq 1$. Setting $f_{n-1} = P^{n-1}f$ we have

$$\int_A P^n f(x)dx = \int_X f_{n-1}(y)dy \int_A g(x - S(y))dx$$

$$= \int_X f_{n-1}(y)dy \int_{A-S(y)} g(x)dx .$$

Since $m(A) = m(A-S(y))$ inequality (5.5) implies

$$\int_A P^n f(x)dx \leq \varepsilon \int_X f_{n-1}(y)dy = \varepsilon . \quad (5.6)$$

The last inequality shows that the family $\{P^n f\}_{N,D}$ with P given by (5.3) is asymptotically uniformly integrable. In this case $n_o(f) \equiv 1$. Consequently the operator P describing the evolution of densities in the presence of noise has an asymptotical decomposition. We may say shortly that noise implies statistical periodicity.

The fact that $X = T^d$ simplifies the arguments but is not essential. Analogous results can be proved for any measurable $X \subset R^d$ and quite general type of noise [4]. It can also be shown that the large amplitude of noise corresponds to the small r in the decomposition formula (4.1). In particular if $g(x) > 0$ for $x \in X$, then P is asymptotically stable.

References

[1] Keller, G., Stochastic stability in some chaotic dynamical systems, Monatsh. Math. 94(1982), 313 - 333.

[2] Komornik, J., Asymptotic periodicity of the iterates of weakly constrictive Markov operators, Tôhoku Math. J. (In press).

[3] Lasota, A. and Mackey, M. C., Probabilistic Properties of Deterministic systems, Cambridge Univ. Press, Cambridge 1985.

[4] Lasota, A. and Mackey, M. C., Statistical periodicity and noise, Physica D (to be published).

[5] Li, T. Y. and Yorke, J. A., Period three implie chaos, Amer. Math. Monthly 82(1975), 985-992.

[6] Misiurewicz, M., Absolutely continuous measures for certain maps of an interval, Publ. Math. IHES 53(1981), 17 - 51.

[7] Sharkovsky, A.N., Coexistence of cycles of a continuous map of a line into itself, Ukr. Mat. Zh. 16(1964), 61 - 71.

SINGULARLY PERTURBED AUTONOMOUS DIFFERENTIAL SYSTEMS

Klaus R. Schneider [1)]

1. Introduction

To motivate the following considerations we recall the famous Belousov-Zhabotinskii reaction [13, 15]: When four or five chemical compounds (for example sodium bromate, malonic acid, sulfuric acid and ferroin) are mixed in the appropriate concentration ranges and at the appropriate temperature, the Belousov - Zhabotinskii system organizes itself into temporal dissipative structures of macroscopic dimension: synchronous oscillations, i.e. spatially homogeneous time periodic oscillations with the same phase can be observed in a well-stirred reactor. The Belousov - Zhabotinskii reaction has played a stimulating role for the developpment of the theory of dissipative structures and has caught the attention both of chemical engineers and of mathematicians. Field and Noyes invented a kinetic model which they called the "Oregonator". It bases on five important steps of the underlying FKN-mechanism and reads [3]

$$\begin{aligned} \varepsilon \frac{d\xi}{dt} &= \xi + \eta - q\xi^2 - \xi\eta , \\ \frac{d\eta}{dt} &= -\eta + 2h\varrho - \xi\eta , \qquad (1.1) \\ p\frac{d\varrho}{dt} &= \xi - \varrho . \end{aligned}$$

Here we have $\xi = [H\,Br\,O_2]$, $\eta = [Br^-]$, $z = [Ce^{+4}]$, $0 < \varepsilon, q, p^{-1} \ll 1$. The system (1.1) is a special case of the following class of differential systems

$$\begin{aligned} \frac{dx}{dt} &= f(x,y,\varepsilon,\alpha) , \\ \varepsilon \frac{dy}{dt} &= g(x,y,\varepsilon,\alpha) , \end{aligned} \qquad (1.2)$$

$x \in R^n$, $y \in R^m$, $\alpha \in R^k$, $\varepsilon \in R$ small. A system of the type (1.2) is called a singularly perturbed autonomous differential system since the solution of the Cauchy problem in general can't be constructed as a power series in ε which converges as $\varepsilon \to 0$ uniformly in x.
An other kinetic model of the B-Z reaction derived by R.J. Wolfe [16] represents a system (1.2) where x is a 3-vector and y an 11-vector. The occurence of slow and fast variables is a general feature of dynamical systems describing biochemical processes. A detailed model of even relatively simple biochemical processes contains a large number of state variables. On the other hand, there is a lot of biochemical processes which can be described qualitatively by low-dimensional systems. Thus,

1) Karl-Weierstraß-Institut für Mathematik der AdW der DDR
DDR - 1086 Berlin, Mohrenstr. 39

a main problem in studying singularly perturbed systems is the problem of reducing the order of the original system (1.2). A usual way for simplifying the system (1.2) bases on the quasi-stationary assumption or the slaving principle: The fast variables tend to their stationary values in a short time. Thus, if we are interested in the long time behaviour of (1.2), we may restrict ourselves to the dynamics of the slow variables. Assuming

$$g(x,y,0,\alpha) = 0 \tag{1.3}$$

has a unique isolated solution $y = \varphi(x,\alpha)$, this principle says that the degenerated system

$$\frac{dx}{dt} = f(x,\varphi(x,\alpha),0,\alpha) \tag{1.4}$$

characterizes the qualitative behaviour of the system (1.2) for small ε. In some cases this principle works, but it is also possible to give examples where it fails. The aim of our paper is to derive conditions under which the degenerated (reduced) system (1.4) yields information about the behaviour of the system (1.2). Especially we are interested in the possibility to use the degenerated system in order to find out chaotic regions for the system (1.2).

2. On the reduction of singularly perturbed systems

As mentioned above, a fundamental problem in studying singularly perturbed systems is the problem of reducing the order of those systems without loss of information about the qualitative behaviour of their trajectories. In this connection integral manifolds play a crucial role.

A set $G_{\alpha,\varepsilon} \subset R^n \times R^m$ is said to be an invariant set of the system (1.2) if a solution of (1.2) passing for $t = t_0$ through a point of $G_{\alpha,\varepsilon}$ belongs for all t to $G_{\alpha,\varepsilon}$. We call an invariant set $G_{\alpha,\varepsilon}$ of (1.2) an integral manifold if $G_{\alpha,\varepsilon}$ is a topological manifold.

Using the fast time τ, $t = \varepsilon\tau$, the system (1.2) can be rewritten in the form

$$\begin{aligned} \frac{dx}{d\tau} &= \varepsilon f(x,y,\varepsilon,\alpha) , \\ \frac{dy}{d\tau} &= g(x,y,\varepsilon,\alpha) . \end{aligned} \tag{2.1}$$

For $\varepsilon = 0$, the system (2.1) has the same topological structure of the trajectories as the system (1.2).

Let G_x, G_y and G_α resp. be simply connected sets of R^n, R^m and R^k resp., let $\varepsilon_0 > 0$. Further, let f and g be defined and sufficiently smooth on $G_x \times G_y \times (-\varepsilon_0,\varepsilon_0) \times G_\alpha$. We are interested in the qualitative behaviour of the trajectories of (2.1) in a compact set $K \subset G_x \times G_y$ where int K is simply connected. To be able to determine the qualitative behaviour of the (n+m)-dimensional system (2.1) by means of an n-dimensional system we introduce the following crucial hypothesis:

(A). The equation (1.3) has a solution of the form $y = \varphi(x,\alpha)$ such that the following properties hold true:

(i) φ is defined on $G_x \times G_\alpha$ and sufficiently smooth.

(ii) There is a $\vartheta > 0$ such that the spectrum σ of the matrix $A(x,\alpha) := g_y(x,\varphi(x,\alpha),0,\alpha)$ satisfies

$$|\operatorname{Re} \sigma\{A(x,\alpha)\}| \geq \vartheta \quad \text{for } (x,\alpha) \in G_x \times G_\alpha .$$

(iii) For fixed $\alpha \in G_\alpha$, the graph E_α of the function φ has a non-empty intersection with int K.

Let E_α^K be a simply connected subset of $E_\alpha \wedge$ int K, let $N_\alpha^{K,E}$ be a neighbourhood of E_α^K in int K. Without loss of generality we may assume $\varphi \equiv 0$. Modifying f and g outside $N_\alpha^{K,E}$, under the assumptions above we can restrict our investigations to a system of the following type

$$\frac{dx}{d\tau} = \varepsilon f(x,y,\varepsilon,\alpha), \qquad\qquad (2.2)$$
$$\frac{dy}{d\tau} = A(x,\alpha)y + h(x,y,\varepsilon,\alpha)$$

where the following hypotheses hold

(A_1). $f : R^n \times R^m x(-\varepsilon_0,\varepsilon_0) \times G_\alpha \rightarrow R^n$,
$h : R^n \times R^m x(-\varepsilon_0,\varepsilon_0) \times G_\alpha \rightarrow R^m$, $A : R^n \times G_\alpha \rightarrow L(R^n,R^m)$
are sufficiently smooth.

(A_2). f, h, A and all derivatives are bounded.

(A_3). $h(x,0,0,\alpha) \equiv 0$, $h_y(x,0,0,\alpha) \equiv 0 \quad \forall (x,\alpha) \in R^n \times G_\alpha$.

(A_4). $|\operatorname{Re} \sigma(A(x,\alpha))| \geq \vartheta > 0 \quad \forall (x,\alpha) \in R^n \times G_\alpha$.

__Theorem 2.1.__ Let the hypotheses (A_1) - (A_4) be valid. Then there are open subsets $G_{\alpha_1} \subset G_\alpha$, $(-\varepsilon_1,\varepsilon_1) \subset (-\varepsilon_0,\varepsilon_0)$ such that there exists a unique integral manifold $M_{\alpha,\varepsilon}$ of (2.2) for $(\varepsilon,\alpha) \in (-\varepsilon_1,\varepsilon_1) \times G_{\alpha_1}$ with the properties:

(i) $M_{\alpha,\varepsilon}$ admits the representation $y = \Psi(x,\varepsilon,\alpha)$ where $\Psi : R^n x(-\varepsilon_1,\varepsilon_1) \times G_{\alpha_1} \rightarrow R^m$ is continuously differentiable. Ψ satisfies uniformly in $(x,\alpha) \in R^n \times G_{\alpha_1}$

$$\lim_{\varepsilon \to 0} \Psi(x,\varepsilon,\alpha) = \varphi(x,\alpha), \quad \lim_{\varepsilon \to 0} \Psi_x(x,\varepsilon,\alpha) = \varphi_x(x,\alpha).$$

(ii) There is a $k > 0$ such that if any solution $(\bar{x}(t,\varepsilon,\alpha), \bar{y}(t,\varepsilon,\alpha))$ of (2.2) with $(\varepsilon,\alpha) \in (-\varepsilon_1,\varepsilon_1) \times G_{\alpha_1}$ satisfies

$$\sup_{t \in R} |\bar{y}(t,\varepsilon,\alpha) - \Psi(\bar{x}(t,\varepsilon,\alpha),\varepsilon,\alpha)| \leq k,$$

then we have $\bar{y}(t,\varepsilon,\alpha) = \Psi(\bar{x}(t,\varepsilon,\alpha),\varepsilon,\alpha)$.

__Remark 2.1.1.__ A proof of Theorem 2.1 basing on the corresponding Poincaré map was given by R. Purfürst [10]. Another proof using center manifold theory can be found in a forthcoming paper of the author where the differentiability assumptions of R. Purfürst are relaxed. Relating results are due to N. Fenichel [2], D. Henry [6], H.W. Knobloch/B.

Aulbach [7].

Remark 2.1.2. Assuming $\mathrm{Re}\,\sigma(A(x,\alpha)) < -\vartheta \;\; \forall (x,\alpha) \in R^n \times G_{\alpha_1}$ and $\varepsilon > 0$ then there is a neighbourhood $U_{\alpha,\varepsilon}$ of $M_{\alpha,\varepsilon}$ such that all trajectories of (2.2) passing through a point of $U_{\alpha,\varepsilon}$ tends to $M_{\alpha,\varepsilon}$ as $t \to +\infty$.

Remark 2.1.3. The integral manifold $M_{\alpha,\varepsilon}$ is a local integral manifold of the system (1.2). Thus, under the assumptions of Remark 2.1.2 the long time behaviour of (1.2) near $M_{\alpha,\varepsilon}$ is determined by the dynamics on the (slow) manifold $M_{\alpha,\varepsilon}$, i.e. by the differential system

$$\frac{dx}{dt} = f(x, \psi(x,\varepsilon,\alpha), \varepsilon, \alpha) . \tag{2.3}$$

3. On the relationship between the degenerated system (1.4) and the system (1.2)

Essentially, we pay attention to the problem under what conditions some properties of the degenerated system (1.4) remain true for the system (1.2). Using "regular" perturbation theory we obtain the following result.

Theorem 3.1. Assume the hypotheses (A_1) - (A_4) hold. Then we have:
(i). If the degenerated system (1.4) has a hyperbolic equilibrium point x_0, then the system (1.2) has also a hyperbolic equilibrium point near $(x_0, \varphi(x_0,\alpha))$ for sufficiently small ε. If we additionally assume $\mathrm{Re}\,\sigma(A(x_0,\alpha)) < -\vartheta$, then the corresponding equilibrium point of the system (1.2) is asymptotically stable for $\varepsilon > 0$. (This assertion coincides with a theorem of A.N. Tichonov [14] with respect to autonomous systems.)
(ii). If the degenerated system (1.4) has a non-degenerated closed orbit $x = p_0(t)$ with the primitive period T_0, then the system (1.2) has also a non-degenerated closed orbit near $(p_0(t), \varphi(p_0(t),\alpha))$ with the primitive period T_ε near T_0 for sufficiently small ε. (This proposition coincides with a theorem of A.D. Anosov [1].)
(iii). If the degenerated system (1.4) has a hyperbolic closed orbit whose stable and unstable manifolds intersect transversally (for definitions see [5], [9]), then the same holds true for the system (1.2) for sufficiently small ε.

Let J be an invariant set of a dynamical system, let M_J^+ (M_J^-) be the stable (unstable) manifold of J. If $M_J^+ \cap M_J^- \neq \emptyset$ then a trajectory in $M_J^+ \cap M_J^-$ is said to be a homoclinic orbit of J. If M_J^+ and M_J^- intersect transversally then a trajectory in $M_J^+ \cap M_J^-$ is called a transversal homoclinic orbit. Using this concept the last part of Theorem 3.1 can be formulated as follows.

Theorem 3.1'. Assume the hypotheses of Theorem 3.1 hold, then the existence of a transversal homoclinic orbit of the degenerated system (1.4) implies the existence of a transversal homoclinic orbit of the system (1.2) for sufficiently small ε.

It is well-known (Smale-Birkhoff Homoclinic Theorem see [5]) that the existence of a transversal homoclinic orbit γ implies the existence of a hyperbolic invariant Cantor set J_C near γ such that the dynamics on J_C is topologically conjugate to the shift automorphism on a closed subset of the set of bi-infinite sequences on a finite set of symbols (for definitions see [5], [8]). Therefore, the set J_C contains a countable set of closed orbits of arbitrarily long primitive periods, an uncountable set of bounded nonclosed orbits, a dense orbit.
In case of existence of hyperbolic invariant sets one expects long chaotic transients before the orbits behave like their ω-limit sets.

Although, in general, we cant say that J_C is an attracting set, there are examples with the property that the existence of J_C is an indicator for the existence of a strange attractor for systems which are close to the considered one. In what follows we formulate conditions ensureing the existence of an invariant Cantor set J_C with the properties described above for the system (1.2).

Let $\tilde{d}(x,\varepsilon,\alpha) := f(x,\psi(x,\varepsilon,\alpha),\varepsilon,\alpha)$. Then the system (2.3) reads

$$\frac{dx}{dt} = \tilde{d}(x,\varepsilon,\alpha) \tag{3.1}$$

where for $\varepsilon = 0$ the system (3.1) represents the degenerated system (1.4). We assume:

(H_0). For $\varepsilon = 0$ and $\alpha = \alpha_0 \in G_\alpha$ the system (3.1) has a homoclinic orbit γ_0 whose α- and ω-limit set is an equlibrium point x_0.

Without loss of generality we may assume $x_0 = 0$. Our main purpose is to give additional conditions guaranteeing the existence of a hyperbolic invariant Cantor set J_C for a set $P := \{(\varepsilon,\alpha) \in (-\varepsilon_1,\varepsilon_1) \times G_{\alpha_1} : \alpha = w(\varepsilon), |\varepsilon| \leqq \varepsilon_2 < \varepsilon_1\}$.

Since lack of space we describe only one way to get such a result.

We suppose further:

(H_1). $x = 0$ is a hyperbolic equilibrium point of (3.1) for $\varepsilon = 0$, $\alpha = \alpha_0$.

Under this assumption we may rewrite the system (3.1) in the form

$$\begin{aligned} \frac{dx_1}{dt} &= A^-(\varepsilon,\alpha)x_1 + d_1(x,\varepsilon,\alpha) , \\ \frac{dx_2}{dt} &= A^+(\varepsilon,\alpha)x_2 + d_2(x,\varepsilon,\alpha) \end{aligned} \tag{3.2}$$

for $|\varepsilon| + |\alpha - \alpha_0| < \delta_0$ where δ_0 is a positive sufficiently small number and

$$|d| = o(|x|), \ \mathrm{Re}\,\sigma(A^-(\varepsilon,\alpha)) < 0, \ \mathrm{Re}\,\sigma(A^+(\varepsilon,\alpha)) > 0 . \tag{3.3}$$

Let $M^s_{\alpha,\varepsilon}$ and $M^u_{\alpha,\varepsilon}$ resp. be the stable and unstable manifold resp. of the equilibrium point $x_0 = 0$ of the system (3.2). By the hypothesis (H_0) we have $M^s_{\alpha_0,0} \cap M^u_{\alpha_0,0} = \gamma_0$, but under the assumption (H_1) one can easily verify that $M^s_{\alpha_0,0}$ and $M^u_{\alpha_0,0}$ do not intersect transversally. In the sequel we consider the simplest case of non-transversal intersection.

(H_2). $\dim (TM^{s,x}_{\alpha_0,0} \cap TM^{u,x}_{\alpha_0,0}) = 1 \quad \forall x \in \gamma_0$.

Here TM^x denotes the tangent space of the differentiable manifold M at the point x.

Since γ_0 is not a transversal homoclinic orbit we need an additional assumption: suppose that the system (3.2) has a hyperbolic invariant Cantor set J_C near γ_0 for $(\alpha,\varepsilon) \neq (\alpha_0,0)$. To simplify matters we suppose that (3.2) is a 3-dimensional system.

(H_3). $\sigma(A^+(0,\alpha_0)) = \{\lambda\}$, $\sigma(A^-(0,\alpha_0)) = \{ -\varkappa \pm i\beta, \varkappa > 0, \beta \neq 0\}$.

(H_4). $\lambda - \varkappa > 0$.

By a known result of L.P. Šilnikov [11] we have

Theorem 3.2. Let d be sufficiently smooth. Further assume (H_0), (H_3), (H_4) to be valid. Then there is a hyperbolic invariant Cantor set J_C near γ_0 of the system (3.2) for $\varepsilon = 0$, $\alpha = \alpha_0$.

Remark 3.2.1. In case dim x > 3 an analogous result holds [12].

From assumption (H_2) it follows that we can not expect $M^s_{\alpha,\varepsilon} \cap M^u_{\alpha,\varepsilon} \neq \emptyset$ $\forall |\varepsilon| + |\alpha - \alpha_0| < \delta_0$. Our next aim is to give conditions ensureing the existence of a function $w : (-\varepsilon_2,\varepsilon_2) \to G_\alpha$ such that $M^u_{w(\varepsilon),\varepsilon} \cap M^s_{w(\varepsilon),\varepsilon} = \gamma_\varepsilon \neq \{0\}$ and $\gamma_\varepsilon \to \gamma_0$ as $\varepsilon \to 0$. For sake of simplicity we again assume dim x = 3.

Let $\gamma_0 := \{ x \in R^3 : x = p_0(t), -\infty < t < \infty \}$, where $p_0(t)$ is a solution of (3.2), let

$$\frac{dx}{dt} = D(t)x \tag{3.4}$$

be the linearization of the system (3.2) at $p_0(t)$ for $\varepsilon = 0$, $\alpha = \alpha_0$. From $\lim_{t\to\pm\infty} p_0(t) = 0$ it follows

$$\lim_{t \to \pm\infty} \sigma(D(t)) = \{ \lambda , -\varkappa \pm i\beta\} \quad . \tag{3.5}$$

It is clear that $p_0' := \frac{dp_0}{dt}$ is a solution of (3.4). From (3.5) it follows that there are a solution u(t) of (3.4) and bounded nonzero functions $v_+ : R \to R$ and $v_- : R \to R$ such that

$$\lim_{t \to +\infty} [u(t)e^{\varkappa t} - v_+(t)] = 0 \quad ,$$

$$\lim_{t \to -\infty} [u(t)e^{\varkappa t} - v_-(t)] = 0 \quad .$$

Let

$$k(t) := \exp\left[-\int_0^t \operatorname{div} d(p_0(s),0,\alpha_0)ds\right],$$

$$l_j(t) := \det\left[p_0'(t),u(t),\frac{\partial d}{\partial \alpha_j}(p_0(t),0,\alpha_0)\right].$$

Using a result of J. Gruendler [4] we can prove

Theorem 3.3. Let d be sufficiently smooth. Further assume (H_0) - (H_2) hold. Additionally we suppose

(H_5). There is a j, $1 \le j \le k$, such that

$$\int_{-\infty}^{\infty} l_j(t)dt \neq 0 .$$

Then there exists a function $w_j : (-\varepsilon_3,\varepsilon_3) \to R$ such that for $w : (-\varepsilon_3,\varepsilon_3) \to R^k$; $w_i(\varepsilon) := \alpha_{i,0}$ for $1 \le i \le k$, $i \neq j$, $M^s_{w(\varepsilon),\varepsilon} \cap M^u_{w(\varepsilon),\varepsilon} = \gamma_\varepsilon$ and $\gamma_\varepsilon \to \gamma$ as $\varepsilon \to 0$.

From Theorem 3.1 - Theorem 3.3 we conclude:

Theorem 3.4. Let d be sufficiently smooth. Further suppose the degenerated system (1.4) satisfies for $\alpha = \alpha_0$ (H_0) - (H_5). Then there exists a function $w : (-\varepsilon_3,\varepsilon_3) \to R^k$ such that for $\alpha = w(\varepsilon)$, $\varepsilon \in (-\varepsilon_3,\varepsilon_3)$ the system (3.2) has a homoclinic orbit γ_ε with the property that in each neighbourhood of γ_ε there is a hyperbolic invariant Cantor set J_C.

References

[1] Anosov, D.V.: On limit cycles of systems of differential equations with a small parameter at the highest derivative (Russian). Mat. Sbornik 50 (1960), 299-334.

[2] Fenichel, N.: Geometric singular perturbation theory for ordinary differential equations. J. Diff. Equs. 31 (1978), 53-98.

[3] Field, R.J. and R.M. Noyes: Oscillations in chemical systems. IV. Limit cycle behavior in a model of a real chemical reaction. J. Chem. Phys. 60 (1974), 1877-1884.

[4] Gruendler, J.: Homoclinic solutions for autonomous systems with a hyperbolic equilibrium. (Preprint)

[5] Guckenheimer, J. and P. Holmes: Nonlinear oscillations, dynamical systems, and bifurcation of vector fields. Springer-Verlag, New York 1983.

[6] Henry, D.: Geometric theory of semilinear parabolic equations. Lect. Not. Math. 840, Springer-Verlag, Berlin 1978.

[7] Knobloch, H.W. and B. Aulbach: Singular perturbations and integral manifolds. J. Math. Phys. Sci. 18 (1984), 415-424.

[8] Moser, J.: Stable and random motions in dynamical systems. Annals of Math. Stud. 77 (1973).

[9] Palis, J. and W. de Melo: Geometric theory of dynamical systems. Springer-Verlag, New York 1982.

[10] Purfürst, R.: Invariante Mannigfaltigkeiten und Hopf-Bifurkation bei singulär gestörten Systemen. Institut für Mathematik der AdW der DDR, Report R-Math-05/82.

[11] Šilnikov, L.P.: On a case of the existence of a countable set of periodic motions (Russian). Dokl. Akad. Nauk SSSR 160 (1965), 558-561.

[12] ---: About the problem of the structure of an extended neighbourhood of a hyperbolic equilibrium point of saddle-focus typ (Russian). Mat. Sbornik 81 (1970), 92-103.

[13] Smoes, M.L.: Chemical waves in the oscillatory Zhabotinskii system. A transition from temporal to spatio-temporal organization. In "Dynamics of Synergetic Systems" (Ed.) H. Haken, Springer Series in Synergetics, Vol. 6, Springer-Verlag, Berlin 1980.

[14] Tichonov, A.N.: Systems of differential equations with a small parameter at the derivatives (Russian). Mat. Sbornik 31 (1952), 575-586.

[15] Tyson, J.J.: The Belousov-Zhabotinskii-reaction. Lect. Not. Biomathematics, Vol. 10, Springer-Verlag, Berlin 1976.

[16] Wolfe, R.J.: Temporal oscillations in a kinetic model of the Belousov-Zhabotinskii reaction. Arch. Rat. Mech. Anal. 67 (1978), 225-250.

DIFFERENCE EQUATIONS WITH CONTINUOUS TIME AS MATHEMATICAL MODELS OF THE STRUCTURE EMERGENCES

A.N.SHARKOVSKY
Yu.L.MAISTRENKO

Institute of Mathematics
Ukrainian Academy of Sciences
Kiev-4
Repin Str.3
USSR

Asymptotic behavior of difference equations with continuous argument (time) of the following form $x(t+1)=f(x(t))$; $x(t):\mathbb{R}^+\to\mathbb{R}$, is considered. The state space Φ of the dynamical system $\mathcal{F}$: $\varphi\mapsto f\circ\varphi, \varphi\in\Phi$, generated by the equation is infinite dimensional. Due to this important fact it is possible to use the solutions of the equation for simulation of the processes of formation both the simple and the complex structures. For example, the solutions may be considered as a model of cascade process of the emergence of decreasing size structures. Such properties of the structures,arising in the model, are determined in terms of spectral decomposition for the non-wandering set of interval mapping $f: x\mapsto f(x)$. We define classes of relaxational and turbulent structures indicating that they have property of automodelity. Primary attention is paid to the so called fractal structures, whose graphs have fractional Hausdorff dimention.

Introduction. The problems of the structure formation take one of the central places in the investigation of a number of important scientific problems, among which we can distinguish the description of the atmosphere circulation and weather forecast,global climate changes and the relation "a man - an environment", the problem of origin of life on the Earth and organization of human society, etc. [4]. The determination of the general laws of the structure emergences has the primary significance for perception of the turbulence, whose structural properties have become more evident recently [5,7] .

There are no many mathematical means allowing to describe the process of structure emergence with sufficient extent of completeness. Certain hopes on positive shifts in this question are connected with the recently formed new field of science - synergetics , whose subject includes the processes of self-organization in physics, chemistry, biology and mathematic methods of their investigation as well. In spite of importance of synergetics, its role is mainly reduced tillmore to the pointing out the question about self-organization as one of the most essential properties of all substance levels. At present stage of development of synergetics a great importance has the elaboration of its own methods and with this purpose tracing of new models of the structure emergences. What's more, borth those which are directed on special classes of problems and those which have universal character and allowing to describe the mechanisms of the structure emergences at a qualitative level.

The proposing models - difference equations with continuous argument (time) - belong to the second group of the models. In the simpliest case such equations may be presented in the following form

$$x(t+1) = f(x(t)), \ t \in \mathbb{R}^+, \tag{1}$$

where $x(t): \mathbb{R}^+ \to \mathbb{R}$ - is unknown variable and $f: \mathbb{R} \to \mathbb{R}$ - is given non-linear function. In this equation two important properties are joined: non-linearity and infinite dimension (the state space of dynamical system $\mathcal{F}: \varphi \longmapsto f \circ \varphi$, generated by equation (1), as it is easy to see, is infinite dimensional). At the same time the model (1) is relatively simple for analysis: equation (1) solutions properties to a considerable extend are determined by the properties of the interval mapping $f: x \longmapsto f(x)$. In fact, putting the initial problem

$$x(t) = \varphi(t), \ t \in [0,1], \tag{2}$$

the solution of (1), (2) has the representation

$$x(t) = f^n \circ \varphi(t-n), t \in [n, n+1), \ n = 0, 1, \dots \tag{3}$$

where f^n - n-th iterate of f.

Without dwelling on real problems, leading to difference equations with continuous argument, we denote just, that such equations naturally arise when the processes in bounded mediums with non-linear boundary affect are considered. A number of boundary value problems for hyperbolic systems of partial differential equations are reduced in particular to the equations of the mentioned type (cf., for example, [3,8]).

1. Equations with quadratic non-linearity. The analysis of dif-

ference equations with continuous argument can be started from the computing experiment. We can examplify it with equation

$$x(t+1) = \lambda x(t)(1-x(t)), \quad t \in \mathbb{R}^{+}, \tag{4}$$

where λ - parameter from the interval $[0,4]$. The quadratic mapping of the equation (4) $x \longmapsto \lambda x(1-x)$, $x \in [0,1]$, is one of the more investigated, A number of papers are devoted to its dynamics, among which we note [1,6]. Using the computer graphic it is easy to observe the behavior of the equation (4) solutions with different λ. The graphs of the solutions are shown at Fig.1 a)- d) (initial function is chosen as $\varphi(t) = (\frac{\lambda}{4} - (1-\frac{1}{\lambda}))\sin \pi t + 1 - \frac{1}{\lambda}$, $t \in [0,1]$) . It is interesting also to see in what way the graph of solution changes under continuous change of the parameter (Fig.2).

We can note from Fig.1 a)-d), that graph of every solution $x(t)$ converges at $t \to \infty$ to some limit function $\rho(t)$. The fragments of these limit functions are shown at Fig.3 a)-d).

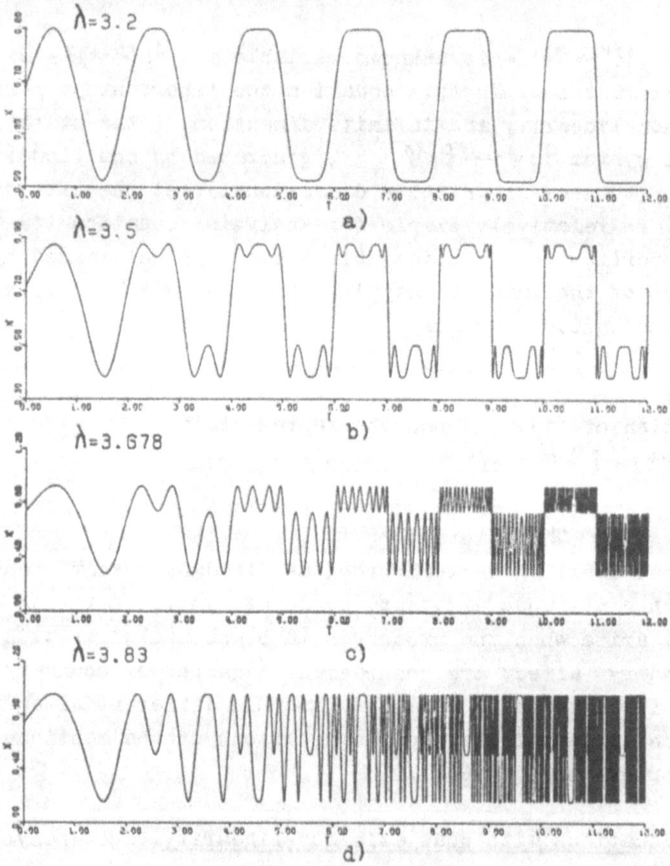

Fig.1

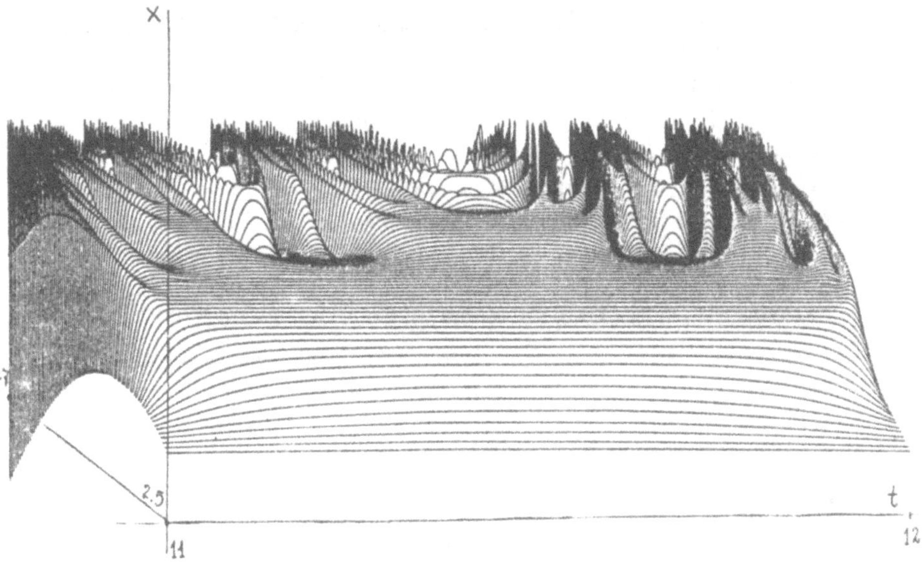

Fig.2

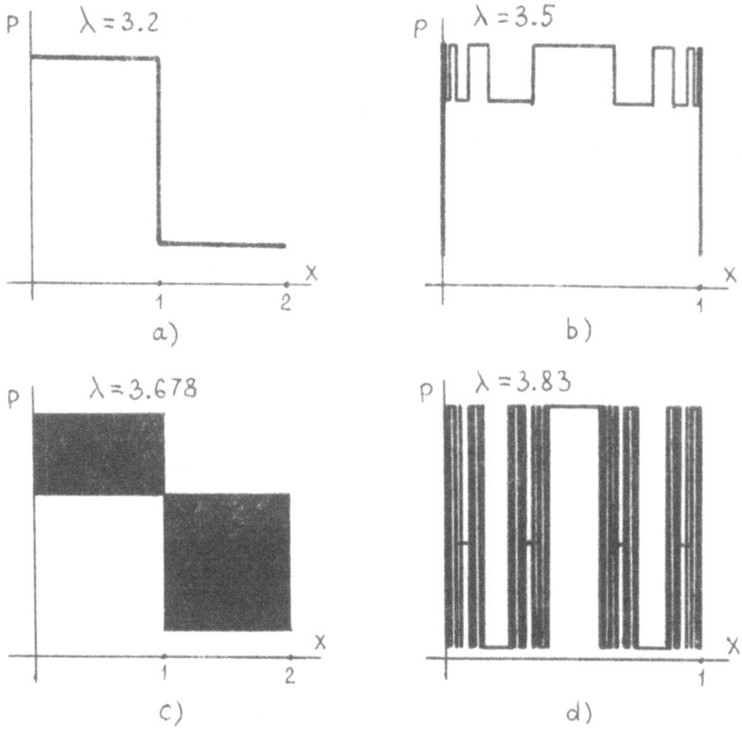

Fig.3

2. <u>Asymptotic behavior of solutions. Limit semi-group and attractor</u>. Let consider equation (1) with initial condition (2) and suppose that $f \in C^{\circ}(I,I)$, $\varphi \in C^{\circ}([0,1],I)$ and concordance condition $\varphi(1)=f(\varphi(0))$ is satisfied, where I - is some closed bounded interval. Then by virtue of representation (3) the solution $x(t)$ exists, is unique and belongs to the class $C^{\circ}(\mathbb{R}^{+},I)$.

As it is evident after the representation (3), the analysis of the equation (1) solutions is equivalent to the analysis of the dynamic system ($\mathcal{F}^{n}$, C° , $\mathbb{Z}^{+}$) trajectories (here $\mathcal{F}$: $\varphi \mapsto f\circ\varphi$, $C^{\circ}=C^{\circ}([0,1],I)$, $\mathbb{Z}^{+}=\{0,1,\dots\}$). Really, if $\mathcal{F}^{n}[\varphi]$, $\varphi \in C^{\circ}$, $n \in \mathbb{Z}^{+}$,- is a trajectory of the system, then the solution $x(t)$ of equation (1) with the initial function $\varphi(t)$ comes out by "gluing" at every n of right ends $\mathcal{F}^{n}[\varphi](t)$ with left ends $\mathcal{F}^{n+1}[\varphi](t)$: $x(t+n)=\mathcal{F}^{n}[\varphi](t)$, $t \in [0,1]$, $n \in \mathbb{Z}^{+}$. In fact the situation is the same as in case of difference equations with discrete time $x_{n+1}=f(x_n)$, $n \in \mathbb{Z}^{+}$, the analysis of which can be substituted by the analysis of corresponding one - dimensional dynamic systems (f^{n} , I , $\mathbb{Z}^{+}$). In case of equations with continuous argument the arising dynamic system turns out to be infinite dimensional (states space is functional).

Phase space C° is non-compact, so the trajectory $\mathcal{F}^{n}[\varphi]$ can have no any C° - limits. In fact, it takes place in most cases . The disconnection of periodic points f set $Perf$ is necessary and sufficient condition for this.

Thus, we have to consider more general phase space, in which the trajectories, responding to equation (1) solutions already had the limits. Such a space, to a great deal reflecting the asymptotic properties of solutions, appears to be the space of upper semicontinuous functions $\Psi:[0,1]\to 2^{I}$ (2^{I} - the space of closed subsets from I) with topology set by the Housdorff distance between the graphs of functions in $\mathbb{R}^2$:

$$\Delta(\Psi_1,\Psi_2)=\max\{\sup_{z\in gr\Psi_2}\rho(z,gr\Psi_1);\ \sup_{z\in gr\Psi_1}\rho(z,gr\Psi_2)\} \quad (5)$$

where $gr\,\Psi_i$ - is the graph of the function $\Psi_i(t)$ at $t \in [0,1]$, $\rho(z,gr\Psi_i)$ - is the distance in $\mathbb{R}^2$ from the point z up to the set $gr\,\Psi_i$. Let denote this space $C^{\Delta}=C^{\Delta}([0,1],I)$. It, as it is easy to see, is compact already. The geometric sence of convergence in metric (5) of the trajectory $\mathcal{F}^{n}[\varphi]$ at $n\to\infty$ to some element $\Psi \in C^{\Delta}$ is as following : with increasing n the function $\mathcal{F}^{n}[\varphi](t)$, $t \in [0,1]$, graph still more exactly merges with the graph $\Psi(t)$, $t \in [0,1]$, moreover with the entire graph, but not its part.

In [3] for mapping $f: I \to I$ the notion of separator $D(f)$ as the set, every point of which generates unstable according to Lyapunov trajectory of the semi-group $\{f^n\}$, have been introduced. With this notion the notion of Julia set $\mathcal{J}(f)$ is nearly connected: in fact $\mathcal{J}(f)$ is closure of $D(f)$, so in those situations, when separator is closed, this two sets are equal.

Let introduce to the consideration the following set $\mathcal{T} = \{t \in [0,1] : \varphi(t) \in D(f)\}$.

Definition 1. We say, that function $\varphi \in C^0$ satisfies the condition of intersection, if for any $t^* \in \mathcal{T}$ and any $\delta > 0$ set $\varphi(\mathcal{U}_\delta(t^*))$ ($\mathcal{U}_\delta(t^*)$ δ-neighbourhood of t^*) includes a neighbourhood of the point $\varphi(t^*)$.

In future we'll consider solutions with initial functions only, satisfying the intersection condition. The initial functions set we denote by C^0_{tv}.

We define the limit semi-group. Let

$$f^\Delta = \lim_{n \to \infty} f^{n!}, \tag{6}$$

where the symbol $\lim$ is mentioned as the limit in the space $C^\Delta(I,I)$. When the conditions are general enough, the limit (6), i.e. mapping f^Δ exists and has the properties

$$f^\Delta \circ f^\Delta = f^\Delta, \quad f^\Delta \circ f^n = f^n \circ f^\Delta, \quad n = 0, 1, \ldots \tag{7}$$

We'll assume, that conditions guaranteeing (6) and (7) are fulfilled*). Then the set $\{f^n \circ f^\Delta\}$ is a semi-group of the mappings, which we name limit. It occurs, that for limit semi-group there are just two possibilities: semi-group is periodic (and then it is finite group) or almost periodic.

Definition 2. We'll say that the solution $x(t)$ of the equation (1) converges in Housdorff metric to the function $p \in C^\Delta(\mathbb{R}^+, I)$ as $t \to \infty$, if

$$\Delta_{[T,\infty]}(x(\cdot), p(\cdot)) \to 0, \quad T \to \infty,$$

*) The condition, guaranteeing the existance of f^Δ, may be proposed as following: the mapping f doesn't have ω-intervals, ω-limit sets of which are different from periodic or closure of almost periodic trajectories (interval $I' \subset I$ is called ω - interval of f, if the trajectories, crossing the interval, have the same ω- limit set). For carrying out (7) it is enough to require, that function f satisfies intersection condition ($\forall x^* \in D(f), \delta > 0$ $f(\mathcal{U}_\delta(x^*))$ includes some neighbourhood of $f(x^*)$).

where $\Delta_{[T,\infty]}(\cdot,\cdot)$ - is a distance in space $C^{\Delta}([T,\infty),I)$. The solution $x(t)$ we'll define as asymptotically periodic (asymptotically almost periodic), if its limit function $p(t)$ is periodic (almost periodic).

<u>Theorem</u>. Let limit semi-group $\{f^{n}\circ f^{\Delta}\}$ be periodic of the period m (almost periodic). Then the solution $X(t)$ of the equation (1) with the initial function $\varphi\in C^{\circ}_{tv}$ is asymptotically m - periodic (asymptotically almost periodic) and converges at $t\to\infty$ to the function

$$p(t)=f^{n}\circ f^{\Delta}\circ\varphi(t-n),\ t\in[n-1,n),\ n\in\mathbb{Z}^{+} \quad . \qquad (8)$$

Let us explain the result of the theorem in terms of dynamic systems theory. With the dynamic system ($\mathcal{F}^{n}, C^{\Delta}, \mathbb{Z}^{+}$), we'll consider limit dynamic system ($\mathcal{F}^{n}_{\Delta}, C^{\Delta}, \mathbb{Z}^{+}$), $\mathcal{F}_{\Delta}:\varphi\to f\circ f^{\Delta}\circ\varphi$, and two responding to each other trajectories $\mathcal{F}^{n}[\varphi]$ and $\mathcal{F}^{n}_{\Delta}[\varphi]$, $\varphi\in C^{\circ}_{tv}$. By virtue of theorem $\mathcal{F}^{n}[\varphi]$ tends asymptotically to $\mathcal{F}^{n}_{\Delta}[\varphi]$ at $n\to\infty$. As it is valid for arbitrary $\varphi\in C^{\circ}_{tv}$, then in phase space C^{Δ} the subset

$$\mathfrak{M} = \bigcup_{\varphi\in C^{\circ}_{tv}} f^{\Delta}\circ\varphi$$

is an attractor, attracting all the trajectories crossing the elements of C°_{tv} . According to the theorem the motions on the attractor are simple enough - they are periodic or almost periodic, though the attractor elements as functions acting from $[0,1]$ into 2^{I} can be arranged in a rather complicated way. Which from the motions are realised on $\mathfrak{M}$ depends upon limit semi-group $\{f^{n}\circ f^{\Delta}\}$: if semi-group is finite, then the motions on $\mathfrak{M}$ are periodic, if infinite - almost periodic.

Thus, a considerable information about solution $x(t)$ behavior is contained in limit function $p(t)$ of the (8) form. In typical situations the range of values of $p(t)$, as function from $\mathbb{R}^{+}$ into 2^{I} is finite; it can be described in term of spectral decomposition of the non-wandering set of the interval mapping f . The set of that $t\in\mathbb{R}^{+}$, in which $p(t)$ is an interval, nor the point, is an important characteristic also. We can show that at $[0,1]$ it is exactly the set $\mathcal{T}$, and on the $\mathbb{R}^{+}-\bigcup_{n>0}\{t\in[n,n+1):t-n\in\mathcal{T}\}$.

3. <u>Ordered structures emergences</u>. The view at the solutions converges to the limit functions, as the process of structure formation is justified due to the following causes. At first, every solution $x(t)$ with the increasing t in great deal "forgets" the details of the initial data generating the solution. In fact,

just the information about how the set $\mathcal{T}$ is organised and what values $\varphi(t)$ has in arbitrarily small neighborhoard of $\mathcal{T}$ is preserved in limit at $t \to \infty$. In this connection, it is important that in many typical situations $\mathrm{mes}\,\mathcal{T} = 0$. Secondly, from the topological point of view the limit functions can be organized in a very complicated way. For example, the situations often occur, when they have Cantor structure (their graphs have the fractional Housdorff-Besicovich dimention). Thirdly, the set of all limit functions is massive enough, it is, at least, countably parametric (if, of course, $D(f) \neq \emptyset$). It is important also that under sufficiently general conditions the solutions and the corresponding limit functions are stable with respect to pertubations both the initial data and the right parts of the equation (1).

Let $p(t)$ is a function belonging to $C^{\Delta}([0,T], I)$.

Definition 3. We'll say that function $\psi(t)$ of the class $C^{0}(\mathbb{R}^{+}, I)$ forms the structure p at $t \to \infty$, if sequence $t_1, t_2, \ldots \to +\infty$ exists, for which

$$\Delta_{[0,T]}(\psi(t_t + \cdot), p(\cdot)) \to 0, \quad t \to \infty .$$

We'll name the structure p periodic (almost periodic), if t_t can be choosen in such a way, when the sequence $\{t_{t+1} - t_t\}$ was periodic (almost periodic).

Consider solution $x(t)$ of the difference equation (1) with the initial function φ. Then according to the theorem $x(t)$ forms at $t \to \infty$ periodic (if the limit semi-group is infinite) or almost periodic (if the limit semi-group is infinite) structure

$$p = f^{\Delta} \circ \varphi. \tag{9}$$

($x(t)$ forms at $t \to \infty$ other structures also, for example, of the $f^{t} \circ f^{\Delta} \circ \varphi$, $t = 1, 2, \ldots$, form. However, all of them may be obtained from p by means of iterates of f).

One of the characteristics of the structure p of the (9) form is the cardinality of the set $\mathcal{T}$ which is namely the set of multivaluence points of p. If $\mathcal{T}$ is finite and not empty, we call the structure p relaxational, if infinite - turbulent.

Essential information about the complexity of turbulent structure is carried out by Housdorff - Besicovich dimention of its graph $\dim_H \mathrm{gr}\,p$ (for relaxational structures always $\dim_H \mathrm{gr}\,p$). The structures, for which $1 < \dim_H \mathrm{gr}\,p \leq 2$, are of the most interest. In this case we name the structure fractal.

If the mapping f - is unimodal of the C^3 - class with the negative Shvartz derivative, then it is possible to propose the simple tests for the existance of structures of that or another type. Let the spectral decomposition of non-wandering set $\Omega(f)$ has the form $\Omega_1 \cup \ldots \cup \Omega_{m^*}$, $m^* \leq \infty$. The structures ρ of the (9) form are fractal, when Ω_{m^*} - is a cycle of intervals (and then $dim_H gr\rho = 2$) or when one of the elements of the spectral dimention Ω_m, $m < m^*$, - is homeomorphic to Cantor set.

The structures, generated by the solutions of difference equations, are intrinsically ordered. The degree of the order can be characterized by means of the notion of automodelity (self-similarity).

Definition 4. We'll call the structure $\rho(t)$, $t \in [0,T]$, locally automodel, if for every $t^* \in (0,T)$ is possible to find the sequence of inserted to each other neighbourhoods $\mathcal{U}_0 \supset \mathcal{U}_1 \supset \ldots$, retracted into the point t^* , and two sequences preserving orientation homeomorphisms $\sigma_t : \mathcal{U}_t \to \mathcal{U}_{t+1}$, $h_t : V_t \to V_{t+1}$ $(V_t = \rho\,\mathcal{U}_t)$, $t = 0,1,\ldots$, of such form, then for every t the diagram

$$\begin{array}{ccc} \mathcal{U}_t & \xrightarrow{\rho} & V_t \\ \sigma_t \downarrow & & \downarrow h_t \\ \mathcal{U}_{t+1} & \xrightarrow{\rho} & V_{t+1} \end{array}$$

is commutative. We'll name the structure automodel if for some t^* the neighbourhood $\mathcal{U}_0$ can be choosen equal to the whole interval $(0,T)$.

It can be shown, that at the additional condition of C^1 - smoothness for the functions f and φ , the structures, generated by the equation (1) solutions (i.e. of the (9) form), are locally automodel.

REFERENCES

1. Collet,P.,and Eckman Y.-P.: Iterated Maps on the Interval as Dynamical Systems. Birkhäusen, Boston 1980.
2. Julia,G.: Memoire sur l'iteration des functions rationelles. J.Math. Pures Appl. 1918, p.47-245.
3. Maistrenko Yu.L., and Sharkovsky A.N.: Turbulence and Simple Hyperbolic Systems. In: Mathematical Analysis. Teubner-texte zur Mathematik. Leipzig. 79 (1985), 220-235.
4. Moiseyev N.N., Alexandrov V.V., and Tarko M.: A Man and the Biosphere. Nauka, Moscow 1985 (Russian).
5. Roshko A.: Structure of Turbulent Shear Flows : A New Look. AIAA Journal. 14 (1976) 1o, 1349-1357.

6. Sharkovsky A.N.: Difference Equations and Population Dynamics. Preprint Inst. of Math. Acad. of Sci. Ukr.SSR,No18, Kiev 1982 (Russian).
7. Structural Turbulence : Coll. of Sci. Papers. Novosobirsk 1982 (Russian).
8. Vereikina M.B., Maistrenko V.L., Romanenko E.Yu., and Sharkovsky A.N.: Boundary Value Problems for Hyperbolic Partial Systems. Formation of the Structures and Transition to Chaos. Preprint Inst. of Math. Acad. of Sci.Ukr.SSR. No29. Kiev 1985.

STABILITY OF EQUILIBRIA IN CRITICAL CASES

E.E. Shnol [1)]

1. THE SUBJECT

Assume that a real system of differential equations has an equilibrium (stationary solution) $u \equiv s$ so that

$$\frac{du}{dt} = f(u); \quad u = (u_1, \ldots, u_n); \quad f = (f_1, \ldots, f_n); \quad f(s) = 0 \qquad (1)$$

By linearizing system (1) in the neighbourhood of s, we have

$$\dot{v} = \Gamma v, \qquad \Gamma = f'(s) \quad \left(\Gamma_{jk} = \frac{df_j}{du_k}(s) \right), \qquad \dot{} = \frac{d}{dt} \qquad (2)$$

One speaks of a critical case of the problem of stability,when the question whether the equilibrium s is stable cannot definitively be answered by the linear approximation (2). It follows that in a critical case:

1. For all the eigenvalues of the matrix Γ $\quad \mathrm{Re}\, \lambda_j(\Gamma) \leq 0$ (otherwise the linear approximation guarantees instability).

2. There are $\lambda_j(\Gamma)$ lying exactly on the imaginary axis such that $\mathrm{Re}\, \lambda_j(\Gamma) = 0$, $j = 1, \ldots, n$ (otherwise the condition 1 will guarantee asymptotic stability). More precisely, there are n_0 zero λ and n_2 pairs of pure imaginary λ $(= \pm i\omega)$.

2. FORMULATION OF THE PROBLEM

In critical cases the stability of equilibria is influenced by nonlinear terms (quadratic and subsequent terms of Taylor expansions of $f_j(u)$). Therefore we find it necessary (A) to indicate what nonlinear term coefficients are essential, and (B) to find conditions (inequalities) guaranteeing asymptotic stability which would be as close to necessary and sufficient ones as possible.

Three remarks. 1.(Obvious). If the order of a system is not limited, then there are infinitely many various critical cases (all n_0 and n_2 are possible).

2. (Less obvious). Critical cases correspond to a degeneracy in the original problem of stability, namely to a situation when some of the ine-

1) Research Computing Centre of the USSR Academy of Sciences, 142292 Pushchino, Moscow Region, USSR

qualities (Re $\lambda_j < 0$) ensuring asymptotic stability turn into equalities. If some λ_j are pure imaginary, the situation is repeated, i.e. by transforming some of the inequalities providing for asymptotic stability into equalities, one obtains a new more degenerate problem. Thus, when n_0 and n_2 are fixed, there is an infinite set of critical cases corresponding to different degeneracy levels ν [1)] .

3. Arbitrarily small changes of $f_k(u)$ may lead to violating the conditions Re $\lambda_j = 0$ or other conditions of the equality type defining a fixed critical case. Therefore the results of a mathematical study can be used in applications only when it is pointed out how things stand for the systems close to "critical" ones.

3. A BRIEF SURVEY

Two simplest critical cases (one zero λ or a pair of $\lambda = \pm i\omega$) were studied by A.M.Ljapunov in his famous dissertation "The general problem of stability of motion" (1892) (see [10]). Later many researchers (mostly from the USSR) investigated various more complicate cases (see,e.g. [12], [9]). To date all 20 critical cases with the degeneracy levels up to three and particular cases of higher degeneracy have been examined , as well as some infinite series (containing problems with any ν). The most interesting series consists of cases without zero λ ("critical" $\lambda = \pm i\omega$). Here for all n_2 the stability conditions have been found, which are valid in the absence of the simplest resonances between the frequencies ω_i: $\omega_j \neq \omega_k$, $\omega_j \neq \omega_k + \omega_l$, $\omega_j \neq \omega_k + \omega_l + \omega_m$ (k, l and m may coincide). If these resonances are present, the situation sharply changes (see Section 9). In most cases studied there are estimates of "stability degree" for the situations close to critical ones.

Remark. The larger the degeneracy level ν , the rearer the corresponding critical case in "general" systems of differential equations. From this point of view a critical case is more important, the smaller ν is.

Critical cases are usually investigated by reducing equations (through linear and nonlinear changes of variables) to a so called Poincaré-Dulac normal form (see [2]). Then the stability conditions are naturally written in terms of the normal form coefficients. According to the modern views this reduction is an independent procedure , ideologically and algorithmically separated from the main problem [2)].

4. NORMALIZATION AND SEPARATION OF "CRITICAL" VARIABLES

Assume that system (1) corresponds to a critical case. Reduce it to

1) ν is the number of conditions of the equality type giving this case.

2) But there are various normal forms and in complicate cases the most suitable should be chosen.

the normal form up to an order m and omit the residual terms ($= o(|u|^m)$). Then we shall have the polynomial system

$$\dot{z}_j = P_j(z); \quad j = 1, \ldots, n'; \quad n' = n_0 + n_2 \tag{3a}$$

$$\dot{y}_k = Q_k(y,z); \quad k = 1, \ldots, n'' \tag{3b}$$

Here the variables z_j correspond to the "critical" eigenvalues $\lambda_j(\Gamma)$ ($\lambda_j = i\omega$ or $\lambda_j = 0$) and y_k to the remaining values λ_k with Re $\lambda_k < 0$ [1]. In the absence of multiple

$$P_j(z) = \lambda_j z_j + O(|z|^2); \quad Q_k(y,z) = \lambda_k y_k + O(|y|^2 + |z|^2)$$

For some suitable m the stability of the equilibrium in system (1) is completely defined by system (3a), which is model for this critical case. Its real dimension is equal to $n_0 + 2n_2$.

Remarks. 1. It is natural to choose the smallest m such that an answer would be independent of higher terms. It can turn out that the stability will not be influenced by all the terms of order $\leq$ m. In these cases it is worth preserving only essential terms in the model system.

2. Whether a model system is chosen correctly becomes clear only after the stability and instability theorems for the original system are proved.

5. MAIN CRITICAL CASES (THE DEGENERACY LEVEL $\nu = 1$)

Assume that there is only one "critical" λ : $\lambda_1 = 0$. Then the model system reduces to a single (real) equation $\dot{x} = ax^2$. It is obvious that the equilibrium in this equation is unstable if $a \neq 0$. The second critical case corresponding to $\nu = 1$ is $\lambda_{1,2} = \pm i\omega$ (for the other λ_j, Re $\lambda_j < 0$). The model system is a complex equation $\dot{z} = i\omega z + Az|z|^2$. Just so clear is the answer: the equilibrium z = 0 is asymptotically stable for Re A < 0 and is unstable for Re A > 0 [2].

We see that the two main critical cases are drastically different. **In the former the equilibrium is unstable "as a rule" and in the latter with "half probability".**

More profound understanding of this fact is given by the bifurcation theory. We only want to recall that if the system $\dot{u} = f(u)$ has the zero eigenvalues, the small perturbations of f(u) may lead to disappearing the equilibrium we are studying (this also can split up into several equilibria).

---

1) One complex variable z_j (or y_j) corresponds to a pair of the complex conjugate λ.

2) It is sufficient to write the equation for $r = |z|^2$: $\dot{r} = ar^2$, a = 2ReA.

Remarks. 1. To calculate the coefficient a in case 1 ($n_0 = 1$) or the coefficient A in case 2 ($n_2 = 1$), we do not need to normalize fully the original system up to second or third order, respectively. But nevertheless when $n \geqslant 3$, the calculation becomes too complicate, needing the use of computer, particularly for case 2 (see [3]).

2. The lesson given by case 1 is general. Always when there are zero eigenvalues, an equilibrium is unstable "as a rule", i.e. under the condition of the type $\varphi \neq 0$. Here φ is a function of the linear and quadratic coefficients of the original system. The stability is only possible when $\varphi = 0$, i.e. in the presence of additional nonlinear degeneracy (and higher ν).

6. CRITICAL CASES OF THE DEGENERACY LEVEL $\nu = 2$

They are five in all. The first three are obvious, being determined only by the number of "critical" eigenvalues:

a) $n_0 = 2$, b) $n_0 = n_2 = 1$, c) $n_2 = 2$.

The remaining cases correspond to the additional (nonlinear) degeneracy:

d) $n_0 = 1$ ($\lambda = 0$) at $a = 0$ and e) $n_2 = 1$ ($\lambda = \pm i\omega$) at $\mathrm{Re}\, A = 0$.

These two cases are as simple [1)] as the main cases, and many authors (and Ljapunov himself!) do not think of them as of separate problems. Note that for $\lambda = 0$ and $a = 0$ account should be taken of the third order terms and for $\lambda = \pm i\omega$ and $\mathrm{Re}\, A = 0$ of the fourth and fifth order terms. In cases a) and b) instability is usual (see Remarks in Section 5). Let us precise this. For $n_0 = n_2 = 1$ only $\lambda = 0$ "works" so that the model system is $\dot{x} = ax^2$, $\dot{z} = i\omega z$, and the equilibrium is unstable at $a \neq 0$. For $n_0 = 2$ the matrix Γ "normally" has a Jordan cell,so that in linear approximation the system is unstable . The question is whether the nonlinear terms can suppress this (weak) instability or not. The answer is: as a rule they cannot (but sometime they do suppress, see Section 9). When we have reduced the system to the normal form up to third order, we write down the equalities for the "critical" variables x and y.

$$\begin{cases} \dot{x} = y + d_0x^2 + dxy + gy^2 + \dots \\ \dot{y} = a_0x^2 + bxy + cy^2 + \dots \end{cases} \tag{4}$$

Assertion [11]. The equilibrium of system (4) is unstable for $a_0 \neq 0$ (independently of the other quadratic coefficients).

---

1) Recall that this simplicity is the result of using "drastic" means including the normalization of the equations.

The most interesting problem with $\nu = 2$ corresponds to $n_0 = 2$, we shall consider it in more detail.

7. KAMENKOV PROBLEM: TWO PAIRS OF PURE IMAGINARY EIGENVALUES

Assume that system $\dot{u} = f(u)$ has the eigenvalues $\lambda_{1,2} = \pm i\omega_1$ and $\lambda_{3,4} = \pm i\omega_2$ with $\omega_2 > \omega_1 > 0$ (for the other λ_j, $\mathrm{Re}\,\lambda_j < 0$). To reduce the system to the normal form, we should know whether $\omega_2 = 2\omega_1$ or $\omega_2 = 3\omega_1$. In the first case there remain quadratic ("resonant") terms in the equations for z (see (3a)). In the second case the normal form will have additional cubic terms. Thus we require that $\omega_2 \neq 2\omega_1$ and $\omega_2 \neq 3\omega_1$. By normalizing the original system up to third order inclusively, we obtain the following model system (see Section 4)

$$\dot{z}_1 = i\omega_1 z_1 + z_1(A_{11}r_1 + A_{12}r_2),\ \dot{z}_2 = i\omega_2 z_2 + z_2(A_{21}r_1 + A_{22}r_2),\ r_k = |z_k|^2 \quad (5)$$

from where the equations for r_k are easily derived:

$$\dot{r}_1 = r_1(a_{11}r_1 + a_{12}r_2),\quad \dot{r}_2 = r_2(a_{21}r_1 + a_{22}r_2);\quad a_{ik} = 2\mathrm{Re}\,A_{ik} \quad (6)$$

The stability of the equilibrium $z_1 = z_2 = 0$ in system (5) (with real dimension 4) is equivalent to that of the equilibrium $r_1 = r_2 = 0$ in system (6) if the system is considered only in the quadrant $r_1 \geq 0$, $r_2 \geq 0$. This problem is not difficult and has the following solution.

Kamenkov criterion (1936) (see [4], [12]). System (6) is asymptotically stable if

1) $a_{11} < 0$, 2) $a_{22} < 0$ and 3) when $a_{12} > 0$ and $a_{21} > 0$, then $\Delta = a_{11}a_{22} - a_{12}a_{21} > 0$.

The system is unstable if $a_{11} > 0$, or $a_{22} > 0$, or $\Delta < 0$ for $a_{12} > 0$ and $a_{21} > 0$.

8. COMMENTARY TO SECTION 7

1. The Kamenkov criterion is also valid for the original system $\dot{u} = f(u)$. Recall that system (5) can strictly speaking be called "model" only after that is proved.

2. Geometric formulation. System (6) has two rays $r_1 = 0$ and $r_2 = 0$ as the trajectories (more precisely, the invariant lines). It can have the third invariant ray $r_2 = cr_1$ with $c > 0$. By (6), $\dot{\alpha} = b\alpha^2$ on the invariant ray $r_1 = c_1\alpha$, $r_2 = c_2\alpha$. For system (6) to be asymptotically stable, it is obviously necessary that there should be no rays with $b \geq 0$. The criterion implies that this condition should also be sufficient.

3. When one of the inequalities of the criterion turns into equality

($a_{11} = 0$, or $a_{22} = 0$, or $\Delta = 0$ for $a_{12} > 0$ and $a_{21} > 0$), the fourth and fifth order terms are involved and system (5) becomes not model.

4. The Kamenkov criterion admits the following "mechanical" interpretation. System (5) describes two connected nonlinear oscillators. Either of them has some "friction": when $r_2 \equiv 0$, $\dot{r}_1 = a_{11}r_1^2 < 0$, and when $r_1 \equiv 0$, $\dot{r}_2 < 0$. If $a_{12} < 0$, then the oscillator 2 oppresses the oscillator 1 ($\dot{r}_1 < a_{11}r_1^2$), and if $a_{12} > 0$, it acts as exciter ($\dot{r}_1 > a_{11}r_1^2$), the action of the oscillator 1 is analogous. The common equilibrium $z_1 = z_2 = 0$ is stable, when at least one of the oscillators does not excite the other. If they both excite each other, then the product of "coupling" coefficients must not be too large (there should be $a_{12}a_{21} < a_{11}a_{22}$).

5. The Kamenkov problem gives birth to five problems of the degeneracy level 3. Three of them correspond to the resonances between the frequencies ($\omega_2 = \omega_1$, $\omega_2 = 2\omega_1$ and $\omega_2 = 3\omega_1$) and the two others to the inequalities of this criterion turning into equalities (see 3).

9. THE MOST INTERESTING CASES OF THE DEGENERACY LEVEL $\nu = 3$

1) Two pairs of pure imaginary λ for $\omega_2 = 2\omega_1$ (see [8], [7]). This resonance almost always leads to instability. In contrast to Section 7, here the cubic terms of the original equations can be neglected: only the resonant quadratic terms "work". By normalizing up to second order, we obtain the following model system

$$\dot{z}_1 = i\omega z_1 + B_1 z_1^* z_2, \qquad \dot{z}_2 = 2i\omega z_2 + B_2 z_1^2 \tag{7}$$

Instability appears when $B_1 \neq 0$, $B_2 \neq 0$ and $B_2 \neq cB_1$, $c < 0$.

2) Two zero λ (and a Jordan cell) with additional degeneracy: $a_0 = 0$ [11]. The normalization is here only useful to give two "critical" variables x and y. For $a_0 = 0$ (see Section 6) we have

$$\begin{cases} \dot{x} = y + d_0x^2 + dxy + gy^2 + d_1x^3 + \dots \\ \dot{y} = bxy + cy^2 + ax^3 + b_1x^2y + g_1xy^2 + a_1x^4 + \dots \end{cases} \tag{8}$$

By changing variables, we can make the coefficients d_0 and d_1 vanish.

Ljapunov criterion. The equilibrium $x = y = 0$ of system (8) is asymptotically stable at $d_0 = d_1 = 0$ when the following inequalities hold

$$\text{a) } b^2 + 8a < 0 \quad , \qquad \text{b) } b\,[\, a(c - 2d) - 3a_1 \,] + 5ab_1 > 0 \tag{9}$$

The equilibrium is unstable when any of the inequalities reverses sign.

Remark. The simplifying changes of variables can be chosen otherwise by making zero the other coefficients of system (8). Formulas (9) in this case assume another form.

3) Three pairs of pure imaginary λ in the generic state (without frequency resonances, see [13],[6]). In this case (and for any $n_2 > 3$) the geometric formulation of Section 8 is valid: it suffices for asymptotic stability that the model system should have no "growing" (and "neutral") rays.

4) Two pairs of pure imaginary λ for $\omega_2 = 3\,\omega_1$. Here exist sufficient and (separately!) necessary conditions of asymptotic stability in the form of inequalities on the coefficients of the normalized system. Finding conditions close to necessary and sufficient ones (of the type of Kamenkov and Ljapunov criteria) means in a sense finding all the limit cycles of the auxiliary system of two (real) differential equations (see [1] ,[5]). This problem has no analytical (formular) solution, therefore at the resonance $\omega_2 = 3\,\omega_1$, in contrast to all the other 19 cases with $\nu \leq 3$, there is no stability criterion defined explicitly. This circumstance, not so important for applications, becomes principal in the theory. Having no inequalities defining the stability criterion for even one case with $\nu = 3$, one cannot enumerate <u>all</u> the cases with $\nu = 4$. Therefore (and also because of great number of cases with $\nu = 4$) the degeneracy level $\nu = 3$ seems to be a natural boundary for the consideration in the general theory[1)].

LITERATURE

[1] Arnold, V.I.: Algebraic unsolvability in problems of Ljapunov stability and of the topologic classification of singular points in analytical systems of differential equations. Funktsional. Analiz. <u>4</u> (1970) 3, 1-9; English transl. in Functional Anal.Appl. <u>4</u>(1970).

[2] Arnold, V.I.: Geometric methods of the theory of ordinary differential equations. Springer-Verlag, Berlin 1983.

[3] Hassard, B.D., Kazarinoff, N.D. and Wan, Y. -H.: Theory and applications of Hopf bifurcation. Cambridge University Press, Cambridge 1981.

[4] Kamenkov, G.V.: Stability and oscillations of nonlinear systems. Nauka, Moscow 1972 (Russian).

[5] Khazin, L.G. and Shnol, E.E.: The simplest cases of algebraic unsolvability in problems of asymptotic stability. Dokl.Akad.Nauk SSSR <u>240</u>(1978)6, 1309-1311; English transl. in Soviet Math.Docl. <u>19</u>(1978) 3, 773-776.

[6] Khazin, L.G. and Shnol, E.E.: Stability of critical equilibrium states. Scientific Centre for Biological Research, USSR Academy of Sciences, Pushchino 1985 (Russian).

[7] Khazina, G.G.: Certain stability questions in the presence of resonance. PMM <u>38</u>(1974)1, 56-65; English transl. in J. of Appl. Math.and Mechanics <u>38</u>(1974)1, 43-51.

1) See also Remarks in Section 3.

[8] Kunitsyn, A.L.: On the stability in a critical case of pure imaginary roots under resonance. Differ.Uravnen. 7(1971)9, 1704-1706(Russian).

[9] Kunitsyn, A.L. and Markeev, A.P.: Stability in resonance cases. Itogi nauki i tekhniki, ser. Obshchaya mekhanika.(ed. VINITI, Moscow) 4(1979), 58-139 (Russian).

[10] Ljapunov, A.M.: Stability of motion. Academic press, New York 1966.

[11] Ljapunov, A.M.: Investigation of a particular case of the problem of stability of motion. Collected works, USSR Academy of Sciences, Moscow-Leningrad(1956) 2, 272-331 (Russian).

[12] Malkin, I.G.: The theory of stability of motion. Nauka, Moscow 1966 (Russian).

[13] Molchanov, A.M.: Stability in the case of a neutral linear approximation. Dokl.Akad.Nauk SSSR 141(1961)1, 24-27 (Russian).

PERMANENCE IN POPULATION DYNAMICS

Amann Erwin [1]

1. Introduction

Population dynamics are frequently described by differential equations of the Lotka-Volterra type or the replicator-equation. These are dynamics of the form

$$\dot{x}_i = x_i \, f_i(x) \quad \text{on} \quad R^n_+ \qquad \text{respectively} \tag{1}$$

$$\dot{x}_i = x_i \, (f_i(x) - \bar{f}) \tag{2}$$

on the simplex $S_n = x \in R^n_+ \quad \sum_i x_i = 1$ and $\bar{f} = \sum_i x_i f_i(x)$ an adjustment term to keep the orbits on this simplex.

But for the most examples the terms f_i are asumed to be linear, as for example in the classical Fisher-Haldane-Wright selection equation, the famous hypercycle of Eigen and Schuster but also other differential equations in modern ecology.

This lecture considers primarily the linear replicator equation

$$\dot{x}_i = x_i \, ((Ax)_i - x.Ax) \qquad i=1,\ldots,n. \tag{3}$$

But since, as J.Hofbauer (4) has shown, the dynamics of replicator equation and Lotka-Volterra equation, the later of dimension n-1, are equivalent, we can use some helpful results on Lotka-Volterra eouations as well.

Dynamical systems often show chaotic behaviour, strange attractors and the orbits are thus very sensitive to initial conditions. Thus it is often hopeless to make predictions other than on a statistical basis. But nevertheless it is less important to find the precise assymptotic behaviour than the fact, that all orbits converge to some bounded respectively compact set.

Especially in evolutionary theory it is an important question, wether or not some species (respectively some information) gets lost forever.

[1] Institut for economics, Liechtensteinstr.13, Vienna.

Thus the concept of permanence has been introduced (as well as some weaker concepts as strong or weak persistence) the last few years.
A dynamical system is called <u>permanent</u> if there exists a compact set K in interior S_n (R^n_+) such that all orbits end up in K.
This means that the boundary is a repellor. An equivalent description is, that there must exist a k>0 such that

$$k < \liminf_{t\to\infty} x_i(t)$$

In case of Lotka-Volterra equations we have to require additionaly that all orbits are bounded.
Notice that k does not depend on the initial conditions of the orbit. Thus for every species that is initially present, selection will not lead to extinction.

I want to present here some necessary and sufficient conditions for permanence of the replicator equation. But first some general introduction to degree and index theory:

2. Brower degree and Poincaré index

The Brower degree of a vector field with respect to a set K is defined by

$$\deg (f,y) = \sum_{f(x)=y} \text{sign} \det Df(x)$$

if y is a regular point of the vector field, where Df is of course the Jacobian of f at x. If y is irregular then the degree is defined as the

$$\deg (f,y) = \lim \deg (f,y_n)$$

with (y_n) is a serie of regular points converging to y.
J.Hofbauer showed this year that the degree of the vector field with respect to any bounded open set K containing all interior ω-limits is $(-1)^{\dim K}$ if the replicator equation (resp. Lotka Volterra equation) is permanent.

For the Lotka Volterra equation the dimension is equal to n. For the replicator equation however the dimension is n-1 since it is restricted by $\sum x_i = 1$.

The poincaré index of a rest point $\hat{x}$ is then defined as

$$i(\hat{x}) = \deg (f/_B, 0)$$

where $f/_B$ stands for f restricted to a closed ball B which contains no other fixed point.

If $\hat{x}$ is regular then $i(\hat{x}) = \text{sign} \det Df(\hat{x})$. That is equal to $(-1)^o$ where o is the number of negative eigenvalues of $Df(\hat{x})$.

For n=2 thus the index of a sink or a source is +1, wheras the index of a saddle is (-1).

The famous Poincaré-Hopf index theorem states that if M is a bounded open set (or a manifold) and if f is a vector field that points outwards at the boundary and has O as a regular value, then

$$i = \chi(M) .$$

where the sum goes over all equilibria in M and the right hand side is the so called Euler characteristic. The sum over the indices, thus, does not depend on the vector field.

3. Sufficient conditions

Theorem 1: A replicator equation (2) on S_n is permanent if there exists a continous map $P > 0$ in interior S_n, $P=0$ on the boundary, and a continous function G which coincides with $\dot{P}/P$ in interior S_n such that

$$(4) \qquad 1/T \cdot \int_0^T G(x(t))\, dt > 0 \qquad \text{for a } T > 1 .$$

P is something like a measure for the distance from the boundary and condition (4) tells us that it is an average Ljapunow function.
This theorem has been proved by J.Hofbauer (4) in 1981. Last year Jansen showed that this can be facilitated by choosing $P = \prod x_i^{p_i}$ $p_i > 0$.

This P fulfilles the requirements in theorem 1 with

$$(5) \qquad G(x) = p.Ax - x.Ax$$

This leads to the following

Theorem 1': A replicator equation is permanent if for all x rest point in the boundary S_n

$$(6) \qquad G(x) > 0 \qquad \text{holds.}$$

This theorem corresponds in a certain degree to the definition of saturated fixed points:

Definition: A rest point is called saturated if

$$f_i(z) > \bar{f}(z)$$

holds for all i.

Indeed saturatedness is an eigenvalue condition, considering the external eigenvalues of boundary rest points:

Let $D(z)$ be the Jacobian at the rest point z.

Then $D(z)$ is determined by

$$\partial \dot{x}_i / \partial x_j = \delta_{ij} (f_i - \bar{f}) + z_i \, \partial / \partial x_j \, (f_i - \bar{f}) .$$

Since z is on the boundary, $z_i = 0$ for some i.

Thus

$$D(z) = \begin{pmatrix} 0 & . & . & 0 & T_z^{\,i} & 0 & . & . & 0 \end{pmatrix}$$

with $T_z^{\,i} = f_i - \bar{f}$ is just the external eigenvalue to the eigenvector $(0,\ldots,0,1,0,\ldots,0)$.

Thus we can write condition (6)

$$G(z) = \sum p_i T_z^{\,i} > 0$$

for some $p > 0$.

Saturated rest points are those rest points that have only nonpositive external eigenvalues.

Interior rest points are trivially saturated since $f_i = \bar{f}$ for all i.

4. Necessary conditions

One of the important necessary conditions for arbitrary permanent systems is the existence of an interior rest point. This was first proved by Hutson and Vickers (7) respectively Sieveking (10) in 1983. You can see this applying Browers fixed point theorem on the positive invariant flow of a permanent system.

A later result for linear replicator systems goes a bit further:

Theorem 2: Let (3) be a permanent replicator equation, then there exists a unique rest point $\hat{x}$ in interior S_n and

$$\lim 1/T \int_0^T x_i(t)\, dt = \hat{x}_i$$

The proof of Schuster et al. (9) proceeds as follows:

$$(\log x_i)^{\cdot} = \dot{x}_i/x_i = (Ax)_i - x.Ax$$

If we integrate from 0 to T and divide by T we get

$$(\log x_i(T) - \log x_i(0))/T = \sum_j a_{ij}\, z_j - H$$

with z_i is the time mean of $x_i(t)$ and H the mean of $\bar{f}$.

Taking the limit we get since $0 < x_i(t) < 1$ and log x_i is thus bounded:

$$0 = \sum_j a_{ij}\, \bar{z}_j - \bar{H}$$

This is just the condition for an interior fixed point.

If there are two fixed points in interior then we get from linearity a manifold of fixed points intersecting the boundary, a contradiction to permanence, since there are rest points arbitraryly close to the boundary.

Existence holds for any permanent system, as was first proved by Hutson and Vickers respectively by Sieveking.

Theorem 3: If the Lotka Volterra equation is permanent, then the following conditions hold:

a) $(-1)^n \det D > 0$

b) $\operatorname{trace} D < 0$

c) $\det A > 0$

Proof: Remember the form of the Lotka Volterra equation:

$$\dot{x}_i = x_i\,(r_i - (Ax)_i)$$

Since $\hat{x}$ is the unique rest point in interior R^n_+ and therefor the unique solution of the linear system $Ax = r$, A is non singular. Hofbauers theorem on the degree of a vector field then implies:

$$i(\hat{x}) = (-1)^n$$

$$d_{ij} = -\hat{x}_i\, a_{ij} \tag{7}$$

now implies that the Jacobian D is also non singular. Thus $i(\hat{x})$ is just the sign of det D and together with (7) also $\det A > 0$.

The result for trace D has been already presented in a paper of J.Hofbauer and me last meeting here at the Wartburg. The proof uses the fact, that for permanent systems a certain set K must shrink in Volume. Since trace D is proportional to divergence of the vector field, this implies together with Liouvilles formula that the trace must be negative.

A similar result holds for the replicator equation with $a_{ii} = 0$ for all i

a) $(-1)^{n-1} \det D > 0$

b) $-x.Ax = \operatorname{trace} D < 0$

c) $(-1)^{n-1} \det A > 0$

$a_{ii} = 0$ is no restriction however, since the dynamics are perfectly the same if we add any scalar to a whole column of A.

The proof is quite simple since we easily can change to a Lotka-Volterra system, as J.Hofbauer showed, by the simple change A to A':

$$a'_{ij} = a_{nj} - a_{ij} \qquad i,j = 1,\dots,n-1$$

a) holds since dimension of the replicator equation is n-1.

The trace of the Jacobian is easily to compute: tr D = $a_{ii}\hat{x} - \hat{x}.A\hat{x}$ since we have to add the characteristic value corresponding to the characteristic direction $\underline{1}$ wich is transversal to S_n.

Thus also $\hat{x}.A\hat{x} > 0$.

The interior rest point is determined by

$$(A\hat{x})_i = \hat{x}.A\hat{x} \qquad \text{for all } i$$

We get since $\hat{x}.A\hat{x} = \bar{f} > 0$

$$(A\,\hat{x}/\bar{f})_i = 1$$

by Cramers rule:

$$\hat{x}_i / \bar{f} = \det A_i / \det A \qquad 1/\bar{f} = \sum \hat{x}_i/\bar{f} = \sum \det A_i/\det A$$

where we get A_i by replacing the i-th column by $\underline{1}$.

Thus $\hat{x}_i = \det A_i / \sum \det A_i$

$$\det A = \begin{pmatrix} a_{11} & \cdots & a_{1\,n-1} & 1 \\ \\ a_{n1} & \cdots & a_{n\,n-1} & 1 \end{pmatrix} = \begin{pmatrix} a_{11}-a_{n1} & \cdots & a_{1n-1} & 0 \\ a_{n-11}-a_{n1} & \cdots & & 0 \\ a_{n1} & \cdots & a_{nn-1} & 1 \end{pmatrix} =$$

$$= \begin{pmatrix} -a_{11} & \cdots & -a_{1n-1} \\ \\ -a'_{n-11} & \cdots & -a'_{n-1n-1} \end{pmatrix} = (-1)^{n-1} \det A'$$

Det $A' > 0$ and $\hat{x}.A\hat{x} > 0$ then implies that c) holds.

5. Catalytic networks

Last I want to show the correlation between permanent replicator systems and the corresponding catalytic network.

Let $a_{ij} \geq 0$, then we define the graph of the catalytic network by assign assigning to each species one vertex i, and arrows from j to i if and only if $a_{ij} > 0$ holds, that is iff j catalizes i.

Theorem 4: The graph of a permanent catalytic network is irreducible (that is, for any pair of species (i,j) there exists a sequence of arrows, starting at i and ending at j)

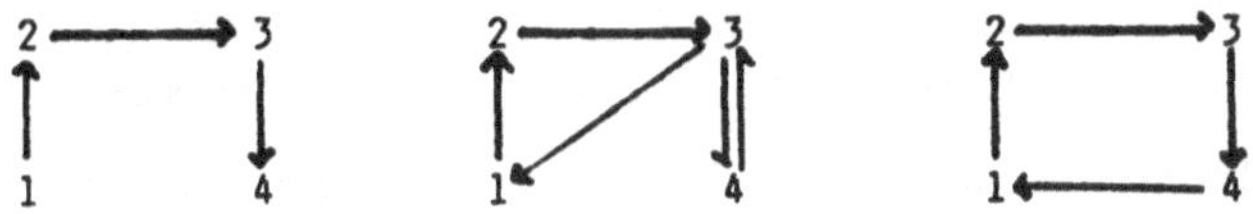

reducible irreducible hamiltonian

Figure

This result has been proved by Schuster and Sigmund in 1980.
In a IIASA paper, going to be published, I show some further results: For $n \leq 5$ the graph of a permanent replicator equation must even be Hamiltonian (i.e. contains a closed path visiting each vertex exactly once, as for example the famous hypercycle of Eigen and Schuster). For $n \geq 6$ however this is not true any more. Again there is a general counterexample to find in this IIASA working paper.

At the moment I am trying to generalize Schusters result on arbitrary replicator equations, i.e. equations where the actions a_{lm} can also become negative, but the diagonal entries stay zero.
Then the conjecture says, that the positive actions must form an irreducible graph.
For let the graph be reducible then there exists a set of numbers say

$$I = \{i_1, i_2, \ldots, i_k\} \quad \text{and} \quad I^c = \{1, \ldots, n\} - I$$

such that $a_{ij} \leq 0$ for all $i \in I^c$ $j \in I$.

In the simplex S_k ($x_j > 0$ $j \in I$) the restricted replicator equation has

(i) either any saturated rest point $\dot{z}$ with $z.Az \geq 0$ or

(ii) there exists an attractor on the boundary trespassing fixed points each of them with nonnegative $z^j.Az^j$.

In both cases the transversal characteristic values T_z^i $i \in I^c$ are all nonpositive. Thus the rest points respectively the ω-limes on the boundary S_k are attractor for the dynamics of the primarily replicator equation, a contradiction to permanence.

REFERENCES

(1) AMANN,E. AND HOFBAUER,J. Permanence in Lotka Volterra and replicator equations; in: Lotka Volterra Approach on Dynamical Systems. Proc. Conf. Wartburg, GDR (1984)

(2) HOFBAUER,J. On the occurence of limit cycles in the Lotka-Volt. equations. Nonlinear Analysis. TMA 5 (1981), 1003-1007

(3) HOFBAUER,J. A general cooperation theorem for hypercycles. Monatshefte Math. 91 (1981) 233-240

(4) HOFBAUER,J. AND SIGMUND,K. Evolutionstheorie und dynamische Systeme Mathematische Aspekte der Selektion. Berlin Hamburg: Paul Parey (1984)

(5) HUTSON,V. AND VICKERS,G.T. A criterion for permanent coexistence of species, with an application to a two-prey one-predator system. Math.Biosciences 63 (1983) 253-269

(6) SCHUSTER,P. SIGMUND,K. AND WOLFF,R. Dynamical systems under constant organization. Part 3: Cooperative and competitive behaviour of hypercycles. J.Diff.Equ. 32 (1979) 357-368

(7) SIVEKING, Vorlesungsmanuskript: Frankfurt am Main (1982/83)

(8) ZEEMAN,E.C. Populationdynamics from game theory. In: Global Theory of Dynamical Systems, Lecture Notes in Mathematics 819. Springer (1980).

MODELS OF EVOLUTION PROCESSES INCLUDING AGE STRUCTURE

W. Ebeling, A. Engel, R. Feistel and V.G. Mazenko[1)]

1. Introduction

In the early days of life on earth each living system consisted of one simple cell only. These simple cells were the units of life. Like little machines they were able to consume free energy, to move in space, to react on certain external factors as light, chemical substances ect. [1-4]. In principle, simple cell organisms have infinite lifetime; the reason for their death is always due to bad living conditions as no food or attacks of prey species. Further simple cell organisms are completed immediately after their birth, they do not develope and do not show aging. With increasing complexity the cell organisms show a finite lifetime consisting of several periods as e.g. youth, period of learning, self-reproduction, period of death. For example yeast cells have a life cycle which has to be modelled in order to reflect their growth properties 5 .Aging is typical especially for the multi-cell organism.We may imagine that in the early days of life on earth sometimes by stochastic events and under the influence of attracting forces between cells, aggregates of cells were formed. These cell aggregates could eventually show evolutionary advantage if the cells located outside developed special defense and exchange properties and the cells inside were specialized to other functions. The developement of many-cell organisms, the differentiation of cells and the division of labour between cells in one organism was a great achievement of evolution of life on earth. However in no case a many-cell organism can be the result of a single act of reproduction (copying of a given template) but more likely the result of a developmental process or in other connections morphogenesis or epigenesis. These terms denote the developement of a single organism through the course of its life cycle. The life span of complicated cell organisms is principally limited even at favourate conditions. However it should be underlined that developement and aging is not a disadvantage in the struggle on life during evolution but a great achievement. Developement makes possible complex structuring, learning, teaching etc.. Therefore the modelling of the individual developement is an important goal of evolution theory.

[1)] Sektion Physik, Humboldt-Universität, 1040 Berlin, Invalidenstr. 42. DDR (V.G. Mazenko leaves in absence from the Chernovtsy State University, Fakulty of Mathematics, 374001 Chernovetsy, USSR)

2. Competition and Selection on Finite Sets of Species with Age Structure

The standard linear model for the age structure of one species due to Mc Kendric [6] and von Foerster [7] reads

$$\frac{\partial}{\partial t} x_i(\tau, t) + \frac{\partial}{\partial \tau} x_i(\tau, t) = -d_i(\tau)\, x_i(\tau, t) \ , \qquad \tau, t > 0,$$

$$x_i(0, t) = \int_0^\infty b_i(\tau)\, x_i(\tau, t)\, d\tau \ , \qquad t \geq 0 , \tag{2.1}$$

$$x_i(\tau, 0) = \varphi_i(\tau) \ , \quad \tau \geq 0 \ , \quad \varphi_i(0) = \int_0^\infty b_i(\tau)\, \varphi_i(\tau)\, d\tau \neq 0.$$

Here t denotes the time, τ the individual age, $x_i(\tau, t)$ the population density of individuals of species i of age τ at time t, $d_i(\tau)$ the death rate at age τ and $b_i(\tau)$ the birth rate at age τ where

$$d_i(\tau),\ b_i(\tau) \geq 0 \in C(R^+) \ , \ \int_0^\infty d_i(\tau)\, d\tau = \infty \ , \ R^+ = [0, \infty).$$

Further $\varphi_i(\tau) \geq 0 \in L(R^+)$ is the initial age distribution. Models of this type where analyzed in many papers (see e.g. [7-11]). Here we shall generalize the model (2.1) to a set of competing species i = 1,2 ... n. For simplicity we shall consider here only the simplest situation of competition [1-4]

$$\sum_{i=1}^{n} \int_0^\infty x_i(\tau, t)\, d\tau = C = \text{const.} \tag{2.2}$$

This condition of constant overall-number of species may be satiesfied in many ways. Let us consider here 3 different situations:

(i) External regulation of the decay rate
We make in (2.1) the replacement

$$d_i(\tau) \longrightarrow d_i(\tau) + \phi(t) \ . \tag{2.3}$$

From eq. (2.2) follows

$$\phi(t) = \langle b(\tau) \rangle - \langle d(\tau) \rangle \tag{2.4}$$

where we have used the notation

$$\langle a(\tau) \rangle = \frac{\sum_{i=1}^{n} \int_0^\infty a_i(\tau)\, x_i(\tau, t)\, d\tau}{\sum_{i=1}^{n} \int_0^\infty x_i(\tau, t)\, d\tau} \ .$$

For the mathematical analysis of our model we introduce the notations ($i = 1, \ldots, n$):

$$K_i(\tau) = b_i(\tau)\Lambda_i(\tau), \quad \Lambda_i(\tau) = \exp\left[-\int_0^\tau d_i(s)\,ds\right]. \tag{2.5}$$

For simplicity we consinder only such species which fulfil

$$\int_0^\infty K_i(\tau)\,d\tau \geq 1 \tag{2.6}$$

i.e., these species would survive without the existence of competition [7,8]. Now we introduce relative densities (frequencies) ρ_i and total densities n_i by

$$x_i(\tau,t) = n_i(t)\,\rho_i(\tau,t), \quad n_i(t) = \int_0^\infty x_i(\tau,t)\,d\tau. \tag{2.7}$$

This representation is called "ecological picture" [2] . Introducing this into eqs. (2.1), (2.3) two systems of equations follow. For the frequencies we obtain

$$\begin{aligned} &\frac{\partial}{\partial t}\rho_i(\tau,t) + \frac{\partial}{\partial \tau}\rho_i(\tau,t) = -[d_i(\tau) + (B_i(t) - D_i(t))]\,\rho_i(\tau,t), \quad \tau, t > 0, \\ &\rho_i(0,t) = \int_0^\infty b_i(\tau)\rho_i(\tau,t)\,d\tau \doteq B_i(t), \quad t \geq 0, \\ &\rho_i(\tau,0) = \varphi_i(\tau) \Big/ \int_0^\infty \varphi_i(\tau)\,d\tau, \quad \tau \geq 0. \end{aligned} \tag{2.8}$$

The population numbers have to satisfy

$$\frac{\partial}{\partial t} n_i = [(B_i - D_i) - \phi(t)]\,n_i, \tag{2.9}$$

where

$$D_i(t) = \int_0^\infty d_i(\tau)\,\rho_i(\tau,t)\,d\tau \tag{2.10}$$

and from (2.4)

$$\phi(t) = \sum_{i=1}^n (B_i - D_i)\,n_i \Big/ \sum_{i=1}^n n_i.$$

By (2.8) the dynamics of $\rho_i(\tau,t)$ does not depend on the selection process, hence by means of the transformation

$$\rho_i(\tau,t) = u_i(\tau,t) \Big/ \int_0^\infty u_i(\tau,t)\,d\tau$$

we get an equ. of type (2.1) for the n_i, the solution of which is well known [8] .
A nessecary condition is that the initial condition for all species satisfies

$$\int_0^\infty [b_i(\tau) - d_i(\tau)]\, \varphi_i(\tau)\, d\tau \geq 0 . \tag{2.11}$$

The stationary solution of the system (2.8) reads

$$\rho_i^0(\tau) = \Lambda_i(\tau) \exp(-\lambda_i \tau) \Big/ \int_0^\infty \Lambda_i(\tau) \exp(-\lambda_i \tau)\, d\tau \tag{2.12}$$

with

$$\lambda_i = \lim_{t\to\infty} [B_i(t) - D_i(t)] .$$

Here the eigenvalue λ_i follows from

$$1 = \int_0^\infty K_i(\tau) \exp(-\lambda_i \tau)\, d\tau . \tag{2.13}$$

In the limit of long time after reaching the stationary distribution of the internal age distribution the population numbers evolve according to

$$\frac{d}{dt} n_i = (\lambda_i - \langle\lambda\rangle)\, n_i . \tag{2.14}$$

In this way follows the asymptotic distribution

$$x_m(\tau, t) \to C\, \Lambda_m(\tau) \exp(-\lambda_m \tau) \Big/ \int_0^\infty \Lambda_m(\tau) \exp(-\lambda_m \tau)\, d\tau \tag{2.15}$$

if $t \to \infty$ and $\lambda_m = \max_i \lambda_i$. Furthermore

$$x_i(\tau, t) \to 0 \quad \text{if} \quad t \to \infty ,\ i \neq m . \tag{2.16}$$

The winner of the competition is the species with maximal eigenvalue
Let us consider an example

$$b_i(\tau) = q_i \exp(-\beta_i \tau) , \quad d_i(\tau) = d_i ,$$
$$q_i, \beta_i, d_i > 0 , \quad \beta_i + d_i < q_i . \tag{2.17}$$

The eigenvalue equation (2.13) reduces to

$$q_i (\beta_i + d_i + \lambda_i)^{-1} = 1 , \quad \lambda_i = q_i - (\beta_i + d_i) \tag{2.18}$$

The winner of the competition is characterized by the maximal eigenvalue

$$\lambda_m = \max_i \left(q_i - \beta_i - d_i \right) \tag{2.19}$$

and the asymptotic distributions are

$$\begin{aligned} x_m(\tau,t) &\to C\,(q_m-\beta_m)\exp\left[-(q_m-\beta_m)\tau\right], \\ x_i(\tau,t) &\to 0\,, \quad i \neq m\,, \quad t \to \infty. \end{aligned} \tag{2.20}$$

The special case $\beta_i = 0$, $i = 1, \dots n$ corresponds to constant birth rates. Let us consider now a second case of interest

(ii) Factorial regulation of the birth rates
In this case we make in (2.1) the replacement

$$b_i(\tau) \longrightarrow A(t)\, b_i(\tau). \tag{2.21}$$

We may imagine that A(t) is the concentration of food which is regulated by a control mechanism. One sees immediately that the condition of constant overall organization (2.2) leads to

$$A(t) = \langle d(\tau) \rangle / \langle b(\tau) \rangle . \tag{2.22}$$

The analysis of eqs. (2.1), (2.21), (2.22) is similar to that given above. The result is the following: The winner of the competition is the species m with maximal integral value

$$\int_0^\infty K_m(\tau)\,d\tau = \max_i \int_0^\infty K_i(\tau)\,d\tau . \tag{2.23}$$

The corresponding asymptotic distributions are for $t \to \infty$

$$\begin{aligned} x_m(\tau,t) &\to C\,\Lambda_m(\tau) \Big/ \int_0^\infty \Lambda_m(\tau)\,d\tau\,, \\ x_i(\tau,t) &\to 0\,, \quad i \neq m . \end{aligned} \tag{2.24}$$

This asymptotic distribution is stationary and stable. Let us consider now again the example given by (2.18). The winner of the competition is the species m with

$$\frac{q_m}{\beta_m + d_m} = \max_i \frac{q_i}{\beta_i + d_i} . \tag{2.25}$$

The corresponding asymptotic distribution is for $t \to \infty$

$$x_m(\tau,t) \to C\, d_m \exp(-d_m \tau),$$

$$x_i(\tau,t) \to 0 \quad , \quad i \neq m . \tag{2.26}$$

The third situation of interest is:

(iii) External regulation of birth and death rates

$$d_i(\tau) \longrightarrow d_i(\tau) + \Phi_1(t) \; , \; b_i(\tau) \longrightarrow b_i(\tau) + \Phi_2(t) . \tag{2.27}$$

The condition of constant overall number (2.2) gives

$$\Phi_1 = \langle b(\tau) \rangle \quad , \quad \Phi_2 = \langle d(\tau) \rangle . \tag{2.28}$$

The virtue of (2.1), (2.27) is that its rhs are positiv for arbitrary initial conditions, Negative values of the rhs would correspond to offspring with age $\tau > 0$ and "negative" offsping respectively thereby complicating the interpretation of the regulation terms $\Phi_1(t)$ and $\Phi_2(t)$. Moreover in (2.27) the inner age structure of the species is coupled to the competition process, so the influence of a selection pressure an the age structure can be studied.

3. Phenotype Space Including Mutations Models

According to earlier papers by Rozonoer and Sedykh [13] as well as by these authors [2,14] we introduce now a phenotype space Q with the elements

$$q = (q_1, q_2, \dots, q_d) \in Q \subset R^d$$

where the components $q_1, q_2, \dots q_d$ denote a set of properties. In difference to our earlier approach we add now the age of the individuum $\tau = q_{d+1}$ as a special property and introduce the extended phenotype space

$$\tilde{Q} \subset R^{d+1} \quad , \quad \hat{q} = (q_1, \dots, q_d, \tau) = (q, \tau) .$$

Further we assume the dynamics

$$\partial_t x(\tilde{q},t) = -\partial_{\tilde{q}} \left(f(\tilde{q})\, x(\tilde{q},t) \right) + \int A(\tilde{q},\tilde{q}')\, x(\tilde{q}',t)\, d\tilde{q}'$$

$$- d(\tilde{q})\, x(\tilde{q},t) - \Phi(t)\, x(\tilde{q},t). \tag{3.1}$$

Here the drift functions are defined by the growth properties of the phenotype which are for individual growth given by

$$\frac{dq_K}{d\tau} = f_K(q) \quad , \quad K = 1, 2, \dots, d. \tag{3.2}$$

For the age component holds

$$f_{d+1} = 1. \tag{3.3}$$

Since the newborns have always the age $\tau = 0$ the birth process is formally a resetting along a characteristic of eq. (3.2) to the plane $\tau = 0$ and the transition function should have the form

$$A(\tilde{q},\tilde{q}') = \delta(\tau)\, B(q(\tau'), q', \tau') \tag{3.4}$$

where $q(\tau)$ is the solution of (3.2) with $q(0) = q$.

In a more simple approximation we may assume

$$A(q,\tilde{q}') = \delta(\tau)\; b(q',\tau') \prod_{K=1}^{d} b_K\left(q_K(\tau') - q'_K\right) \tag{3.5}$$

where $b_K(q_K)$ is a Gaussian-like distribution .
For positive ages $\tau > 0$ we get from eqs. (3.1-3.4)

$$\left(\frac{\partial}{\partial t} + \frac{\partial}{\partial \tau} \right) x(q,\tau,t) = -\left[d(q,\tau) + \Phi(t) \right] x - \frac{\partial}{\partial q} \left[f(q)\, x(q,\tau,t) \right]. \tag{3.6}$$

In order to get the boundary condition to eq. (3.6) on the $\tau = 0$ plane we integrate eq. (3.1) from $\tau = -0$ where $x(q,\tau,t) = 0$ to $\tau = +0$ where by definition $x(q,\tau,t) = x(q,0,t)$. This gives

$$x(q,0,t) = \int dq' \int B(q(\tau'), q', \tau')\, x(q,\tau',t)\, d\tau'. \tag{3.7}$$

This is the continuous analoge to the second part of eq. (2.3). In the case that no mutations may happen the transition function is simply

$$B(q(\tau'), q', \tau') = b(q',\tau') \prod_{K=1}^{d} \delta\left(q_K(\tau') - q_K\right)$$

and we get

$$x(q,0,t) = \int_0^\infty b(q(\tau'),\tau')\; x(q(\tau'),\tau',t)\, d\tau' \tag{3.8}$$

The occurence of mutations means a spreading around the origin of the growth trajectory (characteristic) in the plane $\tau = 0$.

4. Conclusions

The interaction between competition between species and age processes inside every species gives rise to several interesting aspects of evolutionary processes so far not discussed in theoretical models. In this paper simple mathematical models were introduced and discussed which by combining features of the Eigen model of selection processes and the von Foerster - Mc Kendric model of age dynamics allow to investigate this interaction. The discussion of the long time behaviour of the solutions reveals how the species surviving in the selection process could be determined from its age-dependent birth and death rates.

References

[1] Eigen, M., P. Schuster: Naturwiss. 64 (1977) 541; 65 (1978) 341.

[2] Ebeling, W., R. Feistel: Physik der Selbstorganisation und Evolution. Akademie-Verlag Berlin 1982.

[3] Küppers, B.O.: Molecular theory of evolution. Springer-Verlag, Berlin-Heidelberg-New York 1983.

[4] Hofbauer, J., K. Sigmund: Evolutionstheorie und dynamische Systeme. Paray Verlag, Hamburg 1984.

[5] Bley, Th., B. Heinritz: State models - an appropriate tool for improving the efficiency of biotechnical processes In: Lotka-Volterra approach to cooperation and competition in dynamic systems. Math. Res. 23(1985) 241-244.

[6] Mc Kendrick, A.G.: Applications of mathematics to medical problems.-Proc.Edinburgh Math.Soc. 51 (1926) 2, 98-130

[7] Von Foerster, H.: Some remarks on changing populations. - In: The kinetics of cellular proliferations. Brune &Stratton New York 1959 382-407.

[8] Dynamical theory of biological populations, Nauka Moscow 1974 (in Russian).

[9] Matzenko, V.G., V.N. Rubanovski: Application of the direct Liapunov method to analysis of the age structure of biopopulation. Zh. rych. matem. i mate. phys. 23 (1983) 2, 326-332 (in Russian).

[10] Hoppensteadt, F.C.: An age-dependent epidemic model. J. Franklin Inst. 297 (1974) 325-333.

[11] Hadeler, K.P.: Models for parasitic diseases. In: Lotka-Volterra - Approach to cooperation and Competition in dynamic systems. Math. Res. 23 (1985) 162-169.

[12] Gould, S.J.: Ontogeny and Phylogeny. Harward Univ. Press. Cambridge 1977.

[13] Rozonoer, L.I., Ya.I. Sedykh: Avtomatika i Telemekhanika (Moscow) 3 (1979) 119, 5 (1979) 137.

[14] Ebeling, W., A. Engel, B. Esser, R. Feistel: Diffusion and Reaction in Random Media and Models of Evolution Processes. J. Stat. Phys. 37 (1984) 3/4, 369-384.

PATTERN FORMATION IN AN IONIC REACTION-DIFFUSION SYSTEM: A CONTRIBUTION TO THE MORPHOGENE PREPATTERN THEORY OF MITOSIS

By F. Felber[1)], A. Hunding[2)] and H. Malchow[1)]

1. Introduction

The important role of reaction-diffusion processes in morphogenesis, i.e. "the chemical basis of morphogenesis" [1] , has been accepted more and more recently. The theoretical results [2-15] have been supported by experiments now, e.g. the isolation of morphogens in Hydra [16] is well-known where stable gradients of head and foot activators or inhibitors regulate the differentiation of cells. Because biochemical reactions are reactions of charged particles in most cases the study of electrical dissipative structures is of growing interest [10,15,17-23] . In this paper we consider a system with one neutral and two charged reactants which is very similar to a simplified Sel'kov-mechanism [8-10,14] after applying the concept of ambipolar diffusion by Nernst and Schottky as an approximation [15,20-27] . After brief explanation of this concept in chap. 2 the reaction scheme is introduced. Its behaviour is described for the well-stirred case as well as for the unstirred case, where we look for spontaneous pattern formation in a sphere after diffusive instability of a homogeneous substance distribution. The resulting patterns should serve as prepatterns governing mitosis [10] , an idea which is in good agreement with experimental results for nuclear division of the radiolarian Aulacantha scolymantha. The direct-current voltages resulting from the stationary concentration gradients yield the electric forces necessary for the process of mitosis.

2. The concept of ambipolar diffusion

The three fundamental laws for the diffusion of ions in electrolytes are the two Fickian laws

$$X_i(\underline{r},t)\underline{v}_i(\underline{r},t) = -D_i\underline{\nabla}X_i(\underline{r},t) + M_iX_i(\underline{r},t)\underline{\underline{E}}(\underline{r},t), \quad (1)$$

$$\frac{\partial}{\partial t}X_i(\underline{r},t) + \underline{\nabla}\left[X_i(\underline{r},t)\underline{v}_i(\underline{r},t)\right] = 0 \quad (2)$$

---

1) Humboldt-Universität, Sektion Physik, Bereich 04, DDR-1040 Berlin, Invalidenstr. 42

2) Panum Institute, Department of Chemistry 6-2-30, University of Copenhagen, Blegdamsvej 3, DK-2200 Copenhagen N, Denmark

and the Einstein relation

$$M_i = z_i e D_i / kT. \tag{3}$$

Here $X_i(\underline{r},t)$ is the concentration of the ions of kind i at position $\underline{r}$ at time t, $\underline{V}_i(\underline{r},t)$ their mean velocity, z_i their charge number, D_i their diffusion coefficient and M_i their mobility. $\underline{E}(\underline{r},t)$ is the macroscopic electric field satisfying the Poisson equation

$$\underline{\nabla}\,\underline{E}(\underline{r},t) = \frac{F}{\varepsilon}\sum_i z_i X_i(\underline{r},t) = -\Delta\varphi(\underline{r},t) \tag{4}$$

with the potential (r,t).

Nernst and Schottky started from the fact that quasineutrality is a general property of systems of charged particles, because charge separation is prevented by reversal electric fields. In the case of two kinds of ions the relation

$$z_1 X_1(\underline{r},t) + z_2 X_2(\underline{r},t) \approx 0 \tag{5}$$

must hold. This condition can be satisfied only if both kinds of ions diffuse with equal mean velocity

$$\underline{V}_1(\underline{r},t) = \underline{V}_2(\underline{r},t) = \underline{V}(r,t). \tag{6}$$

Therefore we get

$$-D_1\frac{\underline{\nabla} X_1}{X_1} + M_1\underline{E} = -D_2\frac{\underline{\nabla} X_2}{X_2} + M_2\underline{E}\,;. \tag{7}$$

in the case of

$$|z_1| = |z_2| = z \tag{8}$$

it follows

$$D_1\left(\frac{\underline{\nabla} X_1}{X_1} + \frac{ze}{kT}\underline{\nabla}\varphi\right) = D_2\left(\frac{\underline{\nabla} X_2}{X_2} - \frac{ze}{kT}\underline{\nabla}\varphi\right). \tag{9}$$

Because of

$$X_1 \approx X_2 = X \tag{10}$$

we can introduce the approximation

$$\frac{ze}{kT}(D_1 + D_2)\underline{\nabla}\varphi \approx (D_2 - D_1)\frac{\underline{\nabla} X}{X}. \tag{11}$$

Hence we obtain for the potential gradient

$$\underline{\nabla}\varphi = \frac{D_2 - D_1}{D_2 + D_1} \cdot \frac{kT}{ze} \underline{\nabla} \ln X. \tag{12}$$

For the mean concentration X we find

$$X(\underline{r},t)\,\underline{v}(\underline{r},t) = -D_A \underline{\nabla} X(\underline{r},t). \tag{13}$$

Here is D_A the so-called ambipolar diffusion coefficient introduced by Nernst and Schottky

$$D_A = \frac{2D_1 D_2}{D_1 + D_2}. \tag{14}$$

We will use this concept of ambipolar diffusion in the next chapter to investigate a special reaction scheme.

3. A special reaction model

Now we consider a special set of reactions with an uncharged substance X_1 and two charged reactants X_2^- and X_3^+, i.e. $z_1 = 0$, $z_2 = -1$ and $z_3 = 1$:

$$A \xrightarrow{k_1} X_1, \tag{15}$$

$$X_1 + (\gamma - 1)X_2^- + X_3^+ \xrightarrow{k_2} \gamma X_2^- + 2X_3^+ + B, \tag{16}$$

$$X_2^- + X_3^+ \xrightarrow{k_3} C \tag{17}$$

with A, B and C constant. With the scaling

$$c_i = X_i / X^*, \quad \tau = t/t^* \tag{18}$$

the kinetic equations with dimensionless quantities are

$$\dot{c}_1 = 1 - c_1 c_2^{\gamma-1} c_3 \qquad = f_1, \tag{19}$$

$$\dot{c}_2 = c_1 c_2^{\gamma-1} c_3 - \alpha c_2 c_3 = f_2, \tag{20}$$

$$\dot{c}_3 = c_1 c_2^{\gamma-1} c_3 - \alpha c_2 c_3 = f_3 \tag{21}$$

with time unit $\quad t^* = (k_1^\gamma k_2 A^\gamma)^{-\frac{1}{\gamma+1}}$,

concentration unit $\quad X^* = k_1 A\,(k_1^\gamma k_2 A^\gamma)^{-\frac{1}{\gamma+1}}$,

and $\quad \alpha = k_1 k_3 A\,(k_1^\gamma k_2 A^\gamma)^{-\frac{2}{\gamma+1}}$.

The relation $f_2 = f_3$ must hold due to charge conservation. Including flux terms and using the approximation of ambipolar diffusion, i.e. $c_2 \approx c_3 = c$, one obtains

$$\dot{c}_1 = 1 - c_1 c^\gamma + d_1 \Delta c_1 \ , \tag{22}$$

$$\dot{c} = c_1 c^\gamma - \alpha c^2 + d_A \Delta c \tag{23}$$

with the ambipolar diffusion coefficient $d_A = 2\, d_2 d_3 / (d_2 + d_3)$. The diffusion coefficients are scaled by $d_1 = D_1 / D^*$ with the unit $D^* = z^{*2}/t$, where z^* is a reference length. Obviously the system (22,23) is very similar to a simplified Sel'kov-mechanism [8-10, 14] . The only difference is the quadratic decay term of the second component. The only homogeneous stationary solution is

$$c_{10} = \alpha^{\gamma/2} \ , \qquad c_0 = \alpha^{-1/2} \ . \tag{24}$$

A normal-mode stability analysis shows that (c_{10}, c_0) is with $\gamma = 3$

I. a stable node for $0 < \alpha \leq 0{,}318$,

II. a stable focus for $0{,}318 < \alpha \leq 1$,

III. an unstable focus for $1 < \alpha \leq 3{,}15$ and

IV. an unstable node for $3{,}15 < \alpha$.

It can be proved by standard methods [28,29] that $\alpha = 1$ is a Hopf-bifurcation point i.e. limit cycles can be observed for $\alpha > 1$ in the vicinity of the bifurcation point (comp. Figs. 1,2) .

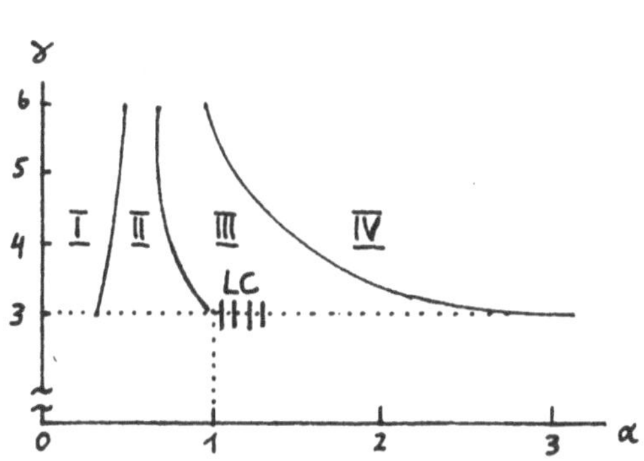

Fig. 1: Stability of (c_{10}, c_0) in the (α, γ)-plane

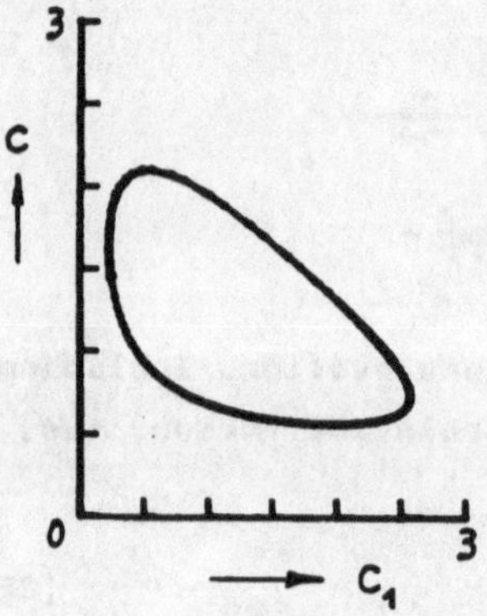

Fig. 2: Limit cycle in the (c_1, c)-plane for $\alpha = 1{,}1$; $\gamma = 3$

4. Pattern formation in the distributed system

The homogeneous solution may break down spontaneously and new solutions may be stable, inhomogeneous concentration distributions in the system (primary bifurcations), giving rise to spontaneous gradient and other pattern formation. It was anticipated early [2] that the arising patterns are closely connected to the well-known eigenfunctions of Laplace's equation. Such a stable pattern may change over in a predictable way to a new stable pattern (secondary bifurcation) by changing simple parameters in the system, such as its size (e.g. radius or other reference length) or a rate constant such as α in (23) (e.g. by increasing inflow of X_1). Here we deal with spherical reaction-diffusion systems, where the qualitative space dependence of the patterns, which arise, and their succession by changing parameters are known to be independent of the particular form of $f(C_1, C)$.
The system (22,23) has to be integrated numerically. Therefore the sphere is divided into 1000 boxes by dividing the radius into 5, the polar angle φ into 20 and the acimutal angle into 10 equal parts, applying then a difference scheme technique for systems of stiff, ordinary differential equations by Gear with initial conditions $(c_1, c) = (c_{10}, c_o)$ + noise. The other parameters have been found by linear stability analysis of (c_{10}, c_o):
The matrix determining stability of small fluctuations with wavelength k (satisfying $\Delta\Phi = -k^2\Phi$) is

$$\begin{bmatrix} -\alpha^{-\gamma/2} - k^2 d_1 & -\gamma\alpha^{1/2} \\ \alpha^{-\gamma/2} & (\gamma-2)\alpha^{1/2} - k^2 d_A \end{bmatrix} \qquad (25)$$

with characteristic polynomial

$$\lambda^2 + \lambda\{-(\gamma-2)\alpha^{1/2} + k^2(d_1 + d_A) + \alpha^{-\gamma/2}\} + \\ + \gamma\alpha^{(1-\gamma)/2} - [(\gamma-2)\alpha^{1/2} - k^2 d_A]\cdot[\alpha^{-\gamma/2} + k^2 d_1]. \tag{26}$$

The necessary and sufficient conditions for $\mathrm{Re}\,\lambda < 0$ follow from Hurwitz's criterion. They are

$$D_1 = \alpha^{-\gamma/2} - (\gamma-2)\alpha^{1/2} + k^2(d_1 + d_A) > 0 \tag{27}$$

and

$$D_1 \cdot \{\gamma\alpha^{(1-\gamma)/2} - [(\gamma-2)\alpha^{1/2} - k^2 d_A]\cdot \\ \cdot[\alpha^{-\gamma/2} + k^2 d_1] > 0. \tag{28}$$

Examine the case $k = 0$ (the spatial homogeneous case) first. We want to make sure that no bulk oscillations are possible. Thus, we must require $\mathrm{Re}\,\lambda < 0$ in this case, i.e. from (27) :

$$\alpha^{-(\gamma+1)/2} < \gamma - 2. \tag{29}$$

Thus, it is necessary that $\gamma > 2$. Inequality (28) is always satisfied then as well.

The case k 0 only adds a positive number to (27). Therefore it is always satisfied, when bulk oscillations are excluded. The only hope for emergence of stable structure is then that (28) is violated. Thus $\{\,\}$ must become negative. Evaluating $\{\,\}$ we have

$$\{\,\} = 2\alpha^{(1-\gamma)/2} + k^2[\alpha^{-\gamma/2} d_A - (\gamma-2)\alpha^{1/2} d_1] + k^4 d_1 d_A < 0. \tag{30}$$

A necessary condition is that $[\,]$ is negative, thus (with $d_A/d_1 = \theta$):

$$\theta\alpha^{-(\gamma+1)/2} - (\gamma-2) < 0. \tag{31}$$

Combining this with (29), we have

$$(\gamma-2)/\theta > \alpha^{-(\gamma+1)/2} > \gamma - 2 \tag{32}$$

which yields the necessary condition

$$\theta = d_A / d_1 < 1. \tag{33}$$

This condition, however is not sufficient.

We would like explicit values of α, d_1, d_A and k to satisfy (30). Let us choose $\gamma = 3$. Thus, (30) becomes:

$$2/\alpha + (k^2 d_1)[\Theta\alpha^{-3/2}] + \Theta(k^2 d_1)^2 < 0. \tag{34}$$

This is a polynomial in $x = k^2 d_1$ with a minimum (which should be negative). For $Ax^2 + Bx + C$ the minimum $C-B^2/4A$ obtains for $x = -B/2A$. Parameters may be chosen accordingly:
The minimum of l.h.s. of (34) is

$$2{,}5\alpha^{-1} - \alpha^2/4\Theta - \Theta/4\alpha^2 < 0$$

or, with $y = \alpha^2/\Theta$:

$$\frac{1}{y} + y > 10. \tag{35}$$

We have already from (29) $1 > \alpha^2$, thus approximately

$$1 > \alpha^2 > 10 \cdot \Theta \tag{36}$$

which means that Θ must be smaller than approximately 1/10.
We have chosen $d_1 : d_A = 1/30$ accordingly. Further parameters may be obtained from (36):

$$1 > \alpha^2 > \frac{1}{3}.$$

Now, the minimum in (34) obtains for

$$x = k^2 d_1 = \frac{1}{2\Theta}\left[(\gamma-2)\alpha^{1/2} - \Theta\alpha^{-\gamma/2}\right] = 11{,}696. \tag{37}$$

Stationary solutions appear as a linear combination of the eigenfunctions Ψ_{nlm} of the Laplacian

$$\Psi_{nlm}(r,\vartheta,\varphi) = j_n(k_{nl} r)\, Y_{nm}(\vartheta,\varphi) \tag{38}$$

i.e. the spherical Bessel functions multiplied by the spherical harmonics.
From [30] we have $k \sim 1.0638 \cdot \pi$ if we want the emergence of $j_2 Y_{2m}$ to arise. This yields the following choises: $d_1 = 1.047$, $d_A = \Theta \cdot d_1$.
Computer calculation were performed with

$\gamma = 3$, $\Theta = 1/30$, $\alpha = 0.70$, $d_1 = 1{,}096$.

Out of the five possible $j_2 Y_{2m}$'s only the pattern $j_2 Y_{20}$ arises as predicted theoretically. A corresponding analysis for the case $j_1 Y_{1m}$, with $k \simeq 0.6626 \cdot \pi$, gave rise to a gradient prepattern for parameters

$\gamma = 3$, $\Theta = 1/30$, $\alpha = 0.70$, $d_1 = 2.720$.

Fig. 3: Formation of the gradient and bipolar pattern from a homogeneous distribution for decreasing diffusion (increasing radius).

The corresponding electrical potentials can be got from (12) and are of order 10^{-2} Volts which is a realistic size for biological systems.

References

[1] Turing, A.M.: The chemical basis of morphogenesis. Phil. Trans. R. Soc. Lond. 237, 5 (1952).

[2] Gmitro, J.F. and L.E. Scriven: A physicochemical basis for pattern and rhythm. In: K.B. Warren (Ed.): Intracellular transport. Academic Press, New York 1966.

[3] Crick, F.: Diffusion in embryogenesis. Nature 225, 420 (1970).

[4] Gierer, A. and H. Meinhardt: A theory of biological pattern formation. Kybernetic 12, 30 (1972)

[5] Fife, P.C.: On modelling pattern formation by activator-inhibitor systems. J. Math. Biol. 4, 353 (1977)

[6] Erneux, T.; J. Hiernaux and G. Nicolis: Turings theory in morphogenesis. Bull. Math. Biol. 40, 771 (1978)

[7] Haken, H. and H. Olbrich : Analytical treatment of pattern formation in the Gierer-Meinhardt model of morphogenesis. J. Math. Biol. 6, 317 (1978).

[8] Billing, G.D. and A. Hunding: Bifurcation analysis of nonlinear reaction-diffusion systems: Dissipative structures in a sphere. J. Chem. Phys. 69 3603 (1978).

[9] Hunding, A.: Dissipative structures in reaction-diffusion systems: Numerical determination of bifurcations in the sphere. J. Chem. Phys. 72, 5241 (1980)

[10] Hunding, A.: Possible prepatterns overning mitosis: The mechanism of spindle-free chromosome movement in Aulacantha Scolymantha. J. theor. Biol. 89, 353 (1981).

[11] Marek, M. and M. Kubicek: Patterns of spatial organization in growth. Z. Naturforsch. 35a, 556 (1980).

[12] Malchow, H.: Lateral diffusive instabilities leading to dissipative structures on a sphere. Studia biophysica 90, 129(1982)

[13] Malchow, H. and R. Feistel: Polarity, symmetry, and gradients in a reaction-diffusion system. Studia biophysica 88, 125 (1982)

[14] Hunding, A.: Morphogen prepatterns during mitosis and cytokinesis in flattened cells: Three dimensional Turing structures of reaction-diffusion systems in cylindrical coordinates. J. theor. Biol. 114, 571 (1985)

[15] Malchow, H. and L. Schimansky-Geier: Noise and diffusion in bistable nonequilibrium systems. Teubner - Texte zur Physik, Teubner-Verlag, Leipzig 1985

[16] Schaller, H.C. and H. Bodenmüller: Morphogene Substanzen aus Hydra. Naturwissenschaften 68, 252 (1981)

[17] Kondepudi, D.K.: Bifurcation, symmetry and the influence of an electric field on chemical dissipative structures. Phys. Lett. 77A, 203 (1980)

[18] Larter, R. and P. Ortoleva: A theoretical basis for self-electrophoresis. J. theor. Biol. 88, 599 (1981)

[19] Ebeling, W. and U. Feudel: Influence of Coulomb interactions on dissipative structures in reaction-diffusion systems. Ann. Physik (Leipz.) 40, 68 (1983)

[20] Malchow, H.; W. Ebeling and R. Feistel: Polarity, symmetry, and gradients in a reaction-diffusion system. II. Patterns of charged particles. Studia biophysica 97, 231 (1983)

[21] Malchow, H.: Dissipative structures in an ionic Lotka-Volterra type chemical reaction system. In: Ebeling, W. and M. Peschel: Lotka-Volterra-approach to cooperation and competition in dynamic systems. Mathemat. Forschung 23, Akademie-Verlag, Berlin 1985

[22] Malchow, H.: Pattern formation in a bistable ionic reaction -diffusion system. J. Non-Equilib. Thermodyn. 10, 15 (1985)

[23] Rastogi, R.P.; I. Das; A.R. Singh and H. Malchow: Chemical waves and multiple steady states in Iodate-Arsenite systems. J. Non-Equilibrium Thermodyn., in press

[24] Nernst, W.: Z. phys. Chem. 2, 613 (1888)

[25] Schottky, W.: Phys. Z. 25, 635 (1924)

[26] Ebeling, W.: Beitr. Plasmaphysik 7, 11 (1967)

[27] Falkenhagen, H.: Theorie der Elektrolyte. Hirzel, Leipzig 1971

[28] Marsden, J.E. and M. Mc Cracken: The Hopf bifurcation and its applications. Appl. Math. Sci. 19, Springer-Verlag, New York 1976

[29] Schneider, K.R.: Andronov-Hopf-Bifurkation im Falle stetiger Parameterabhängigkeit. ZAMM 60, 19 (1980)

[30] Morse, P.M. and H. Feshbach: Methods of theoretical physics. Part II, McGraw-Hill Book Comp.,Inc. New York etc.1953p.1576

CHAOTIC DYNAMICS AND FLUCTUATIONS IN A BIOCHEMICAL SYSTEM

H. Herzel[1)], Th. Schulmeister[2)]

1. Introduction

Our attention is devoted to a biochemical model which describes regulating mechanisms of the living cell [19] . As a result of enzyme regulation a rich variety of dynamic behaviour occurs. We found self-sustained oscillations, coexistence of different attractors, period-doubling and deterministic chaos [8, 10, 19] . Such temporal selforganization is relevant with respect to biological rhythms and arises in certain metabolic pathways [2, 12, 15, 18] .
The paper is organized as follows: Section 2 gives an introduction to the chaotic dynamics in our model. The orbital instability is characterized by a positive Lyapunov exponent, whereas the strangeness of the attractor is described by a non-integer dimension. However, it turns out that these averaged quantities give only a rough description of the chaotic properties. We emphasize that chemical chaos is characterized by large deviations from average behaviour.
Therefore we discuss in section 3 the concept of "nonuniform dynamics". Quantities are defined which measure the deviations from a purely exponential divergence of nearby trajectories.
Grassberger pointed out that nonuniform dynamics and "inhomogeneous attractors" are related [5] . Section 4 is devoted to this problem.
It is widely recognized that real systems are inherently noisy [1, 8, 9].
Thus, it would seem useful to study the effects of random fluctuations. Section 5 contains our results on the influence of noise on Lyapunov exponents and "next-amplitude maps". In this way we may answer the question whether or not properties of the system persist in a "noisy" setting.

2. The model

The following three-variable system contains fundamental regulation mechanisms: the autocatalytic reaction of a substrate x into product y and a reversible deposition of x into an inactive form z

[1)] Department of Physics, Humboldt-University Berlin, 1040 Berlin, Invalidenstr. 42, GDR

[2)] Central Institute of Molecular Biology, Academy of Sciences of the GDR, 1115 Berlin, Robert-Rössle-Str. 10, GDR

$$\frac{dx}{dt} = 1 - Bx - xy^2 - Exy + z$$

$$\frac{dy}{dt} = A(xy^2 - y + D) \qquad (1)$$

$$\frac{dz}{dt} = F(Exy - z)$$

The routes to chaos and a bifurcation diagram are discussed in ref. [19] - In the following we fix some parameters (A = 4; B = 0.35; D = 0.1; F = 0.2) and vary E.

In 1978 the nonperiodic long-term behaviour of solutions of eqs. (1) was taken as indicator of chaos [19] . Nowadays Lyapunov exponents which describe the stability properties of trajectories are widely used as quantitative measures [3,8] . A positive Lyapunov exponent related to orbital instability may be used as a definition of chaos. It can be computed from the linearized equations:

$$\frac{dq_1}{dt} = (-B - y^2 - Ey)\,q_1 - (2xy + Ex)\,q_2 + q_3$$

$$\frac{dq_2}{dt} = Ay^2 q_1 + A\,(2xy - 1)\,q_2 \qquad (2)$$

$$\frac{dq_3}{dt} = FEy\,q_1 + FEx\,q_2 - F q_3$$

These equations govern the behaviour of an infinitesimal deviation $\underline{q}$ from a solution of eqs. (1). The growth of the Euclidean norm of $\underline{q}$ leads for almost all initial vectors $\underline{q}$ (0) to the maximum Lyapunov exponent λ_1 [3] :

$$\lambda_1 = \lim_{t \to \infty} \frac{1}{t} \ln \frac{\| \underline{q}(t) \|}{\| \underline{q}(0) \|} \qquad (3)$$

The sum of all Lyapunov exponents is given by the mean divergence:

$$\lambda_1 + \lambda_2 + \lambda_3 = \lim_{t \to \infty} \frac{1}{t} \int_0^t (-B - y^2 - Ey + 2Axy - A - F)\, dt' \qquad (4)$$

Since one Lyapunov exponent vanishes [3] , we can compute the whole Lyapunov spectrum from eqs. (1)-(4). Using an integration time of 50000 time units we obtained the following estimations (E = 1.5):

$$\lambda_1 = 0.0285 \pm 0.0005$$

$$\lambda_2 = 0 \qquad (5)$$

$$\lambda_3 = -0.248 \pm 0.001$$

Lyapunov exponents give an upper bound of the information dimension D_1 (see section 4) :

$$D_1 \leq 2 + \frac{\lambda_1}{|\lambda_3|} \approx 2.115 \qquad (6)$$

Since this value is slightly above two, we can expect that Poincaré sections or next-amplitude maps look similar to one-dimensional curves. In order to obtain next-amplitude maps as in fig. 1, we store during the integration of eqs. (1) sequences of maximum y-values. Due to the two-band structure in fig. 1a, it appears desirable to use every second maximum only. In this way, the curve in fig. 1b is derived. The similarity with the logistic map is no surprise because period-doubling was observed at adjacent parameters

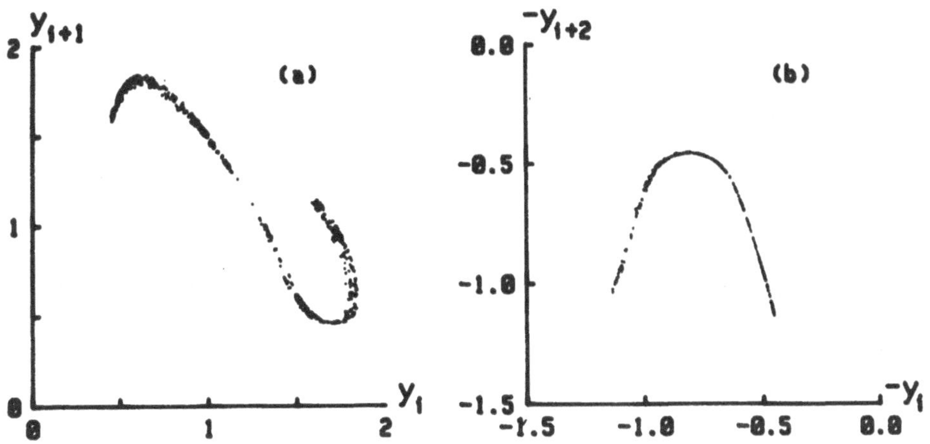

Fig. 1. Next-amplitude maps of chaos in eqs. (1) (E = 1.5)

Up to now we got by standard techniques (Lyapunov exponents, next-amplitude maps) some insight into the chaotic properties. In the next sections it is demonstrated that Lyapunov exponents and the dimension D_1 can be interpreted as averaged quantities. It turns out that the deviations from an average behaviour are very important, at least in chemical systems [10]

3. Nonuniform dynamics

In order to produce temporal self-organization in chemical systems, several reactions are necessary. The reaction rates may vary widely and, therefore, the dynamics is usually very nonuniform in the concentration space. This means that the phase space velocities and the stability properties of trajectories fluctuate strongly.

A positive Lyapunov exponent indicates exponential growth of infinitesimal perturbation in the long-term average (see eq. 3). However, on short time scales drastic deviations from a purely exponential growth may occur. It was found that expansion and contraction change rapidly in our model. In order to treat this problem, we return to eqs. (2) and discuss the growth of the Euclidean norm $\| \underline{q}(t) \|$ in more detail. A time dependent growth rate L(t) can be derived from eqs. (1) and (2):

$$\frac{d}{dt} \ln \| \underline{q} \| = L(t) \tag{7}$$

Eq. (3) shows that the long-term average of L(t) is equal to λ_1. However, fig. 2 displays that the fluctuations of L(t) drastically exceed the mean value ($\lambda_1 = 0.0285$).

In order to quantify this nonuniformity, we analyse the statistics of the growth rates:

$$l_i(\tau) = \int_{t_i}^{t_i+\tau} L(t)\, dt = \ln \frac{\| \underline{q}(t_i+\tau) \|}{\| \underline{q}(t_i) \|} \tag{8}$$

These $l_i(\tau)$ measure the exponential change of the length of $\underline{q}$ during the time interval τ when the system moves along the trajectory. The deviation from average behaviour can be measured on different time scales by the ratio of the standard deviation to the mean value:

$$\begin{aligned} \Delta l(\tau) &= \left(\langle l^2(\tau) \rangle - \langle l(\tau) \rangle^2 \right)^{\frac{1}{2}} \\ \langle l(\tau) \rangle &= \lambda_1 \tau \end{aligned} \tag{9}$$

For small and large time scales previously defined quantities can be recovered: the "Non-Uniformity Factor" (NUF) [6, 14] and a "diffusion constant" Q_{11} [4, 5] :

$$\begin{aligned} \frac{NUF}{\lambda_1} &= \lim_{\tau \to 0} \frac{\Delta l(\tau)}{\langle l(\tau) \rangle} \\ Q_{11} &= \lim_{\tau \to \infty} \frac{\Delta l^2(\tau)}{\tau} \end{aligned} \tag{10}$$

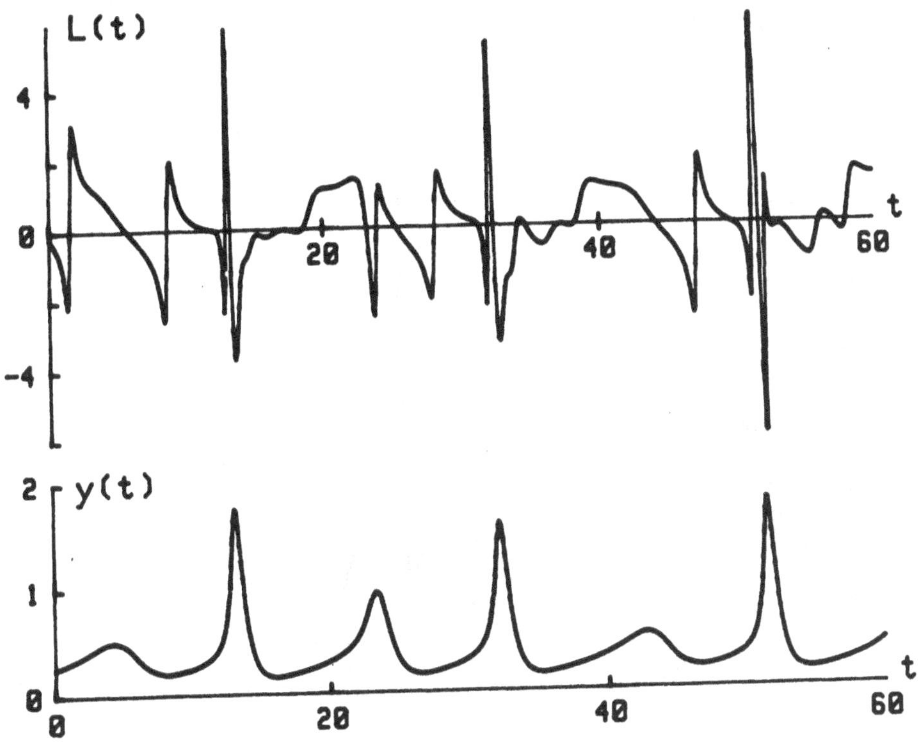

Fig. 2 Time dependence of the expansion rate L(t) and the corresponding concentration y(t) (E = 1.5)

The following table demonstrates the nonuniformity in our model (E =1,5; integration of eqs. (1) and (2) over 10 000 time units).

τ	$\Delta l(\tau)$	$\Delta l(\tau) / \langle l(\tau) \rangle$
0.1	0.128 ± 0.002	46 ± 2
0.2	0.247 ± 0.002	44 ± 2
1	0.88 ± 0.002	31 ± 2
10	1.96 ± 0.1	7 ± 0.5
100	2.14 ± 0.5	0.75 ± 0.2

It can be concluded that only on time scales of more than ten oscillations (mean "period" $\langle T \rangle \approx 9.6$) the growth of perturbations is actually governed by the maximum Lyapunov exponent. Therefore, for scales comparable with the period the behaviour of perturbations cannot be described

iii) harmonic mean: fractal dimension D_0

$$\left(\sum_{i=1}^{M} \frac{1}{n_i(r)}\right)^{-1} \sim r^{D_0} \tag{14}$$

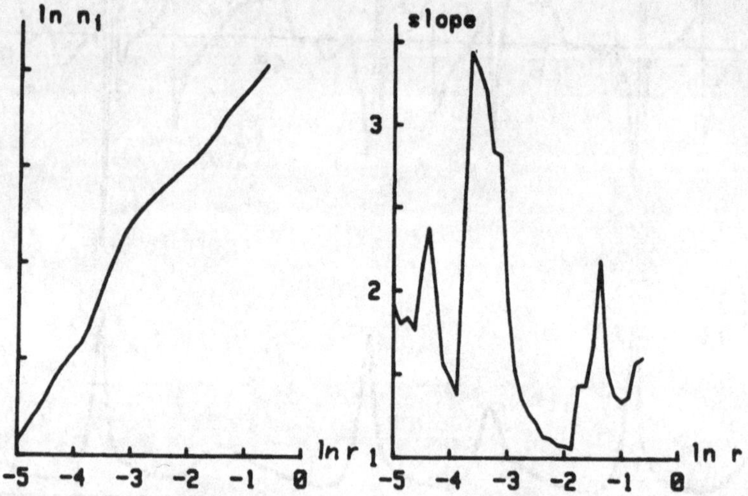

Fig. 3 Number of points $n_i(r)$ in a sphere of radius r around an attractor point versus r and the slope of the log-log plot ($N=5 \cdot 10^5$)

It is well known that differences between the means reflect the distribution of the ensemble. Thus the differences of the dimensions can be taken as a measure of the inhomogeneity of the attractor. Further it turns out that all these dimensions can be estimated from a given time series [13]. However, the correlation exponent D_2 can be obtained with much higher accuracy [5].
It is demonstrated in ref. [7] that the attractor in our biochemical system is indeed very inhomogeneous. The estimated correlation exponent

$$D_2 = 2.02 \pm 0.06 \tag{15}$$

differs considerably from the value in eq. (6).
Until now two kinds of deviations from an average behaviour were discussed separately: "nonuniform dynamics" and "inhomogeneous attractors".

appropriately by the averaged quantity λ_1.
Nonuniform dynamics is intimately related to the effects of noise [8, 9] and to the computational effort which is necessary to estimate Lyapunov exponents. Further, the important question of state prediction depends on the degree of nonuniformity [17]. The close connection between nonuniform dynamics and inhomogeneous attractors is discussed in the next section.

4. Inhomogeneous attractors

In this section the solutions of eqs. (1) are considered in phase space, which is spanned by the concentration variables. After a transient time the motion is governed by attractors. Periodic motion corresponds to an attracting limit cycle whereas chaotic motion takes place on "strange attractors". These can be characterized by non-integer dimensions. Such dimensions are related to the scaling properties of statistical quantities and display, therefore, some kind of self-similarity.
Attractor dimensions can be estimated from a time series of N attractor points ($\Delta t = 0.1$), which is generated using eqs. (1). The fraction of points which are contained in a sphere of radius r and centre $\underline{x}_i$ is ($\underline{x}$ corresponds to the triple (x,y,z)):

$$n_i(r) = \frac{1}{N} \sum_{j=1}^{N} \Theta(r - \| \underline{x}_i - \underline{x}_j \|) \tag{11}$$

$$\Theta(s) = \begin{cases} 0 & \text{if } s < 0 \\ 1 & \text{if } s \geq 0 \end{cases}$$

For a homogeneous fractal one may expect that these "local densities" scale as r^D for small r [16]. However, fig. 3 demonstrates large deviations from such a power law.
In this way a complicated local structure of the attractor is indicated. Nevertheless, averaging over many local density functions leads to estimations of attractor dimensions [5, 16]:

i) arithmetic mean: correlation exponent D_2

$$\frac{1}{M} \sum_{i=1}^{M} n_i(r) \sim r^{D_2} \tag{12}$$

ii) geometric mean: information dimension D_1

$$\left(\prod_{i=1}^{M} n_i(r) \right)^{1/M} \sim r^{D_1} \tag{13}$$

However, from ref. [5] follows that both concepts are intimately related. Thus "diffusion constants" (as, e.g., Q_{11} introduced in eq. (10)) are closely connected with the differences of dimensions.
We emphasize that especially chemical systems exhibit extremely nonuniform dynamics and inhomogeneous attractors. Therefore a proper description of chemical chaos should include the statistics of growth rates (see eqs. (9)) and the statistics of local density functions (see eqs. (11) - (14)).

5. The effects of noise on chaotic dynamics

In order to cross the interface between mathematical models and experimental evidence of chaos, it is appropriate to incorporate the effects of fluctuations [1, 8, 10]. The wide class of additive short-correlated noise is simulated by adding every 0.1 time units Gaussian pseudorandom numbers (standard deviations Q) to the solutions x(t) and y(t) computed from eqs. (1). In the presence of noise the same properties and techniques as in the preceding sections are used: the instability of orbits, measured by Lyapunov exponents, and the structure of attractors, displayed by next-amplitude maps.
The full line in fig. 4 shows the maximum Lyapunov exponent λ_1 versus the parameter E in the absence of noise, whereas the dashed line visualizes the smoothing effect of noise. These estimations of Lyapunov exponents were obtained via linearization along deterministic or noisy trajectories, respectively (see eqs. (1) - (3)).
One may expect that chaos exceedingly sensitiv depends on fluctuations since trajectories are unstable. However, statistical properties as the Lyapunov exponent λ_1 turn out to be robust. Moreover, noise enhances the chaotic behaviour, while destroying periodic orbits. In the deterministic case so-called periodic windows are interspersed among chaotic parameters (see fig. 4). These periodic orbits are associated with chaotic transient and therefore noisy trajectories resemble nearby (in parameter space) chaotic orbits. This kind of "noise-induced chaos" is demonstrated in fig. 5.
Summarizing we conclude that in agreement with other studies essential properties of chaos persist in the presence of noise. Thus chaos is observable in nature and numerical errors have no dramatic conquences.

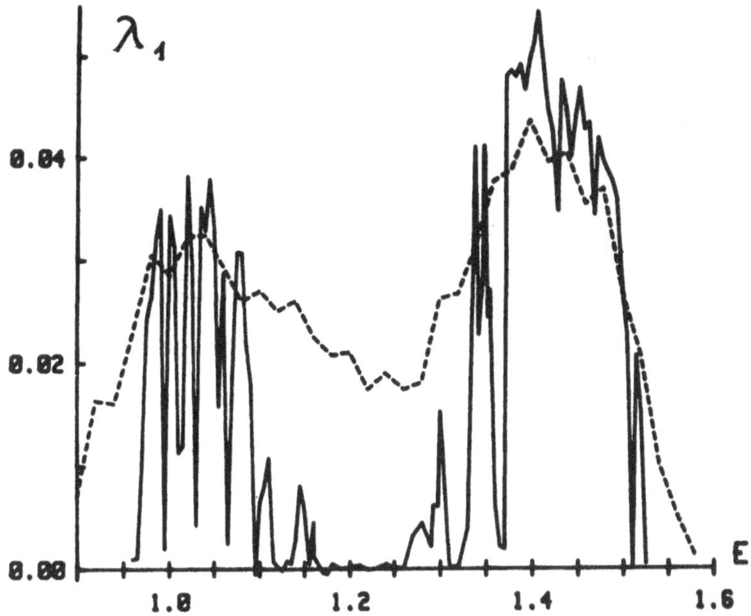

Fig. 4 Maximum Lyapunov exponents versus parameter E. Neighboring points have been joined by a straight line to guide the eye. Full line: Q = 0; ΔE = 0.005 ; Dashed line: Q = 0.0005; ΔE = 0.02

6. Summary and remarks

Deterministic chaos in a rather simple biochemical system was characterized by Lyapunov exponents and non-integer dimensions. However, it turns out that these averaged quantities only partially describe what is actually going on. Therefore we suggest to take local Lyapunov exponents as well as local density functions into account. We emphasize that nonuniform dynamics found in our model is a characteristic property of chemical systems. It can be measured by the statistics of growth rates (see eqs. (8) - (10)) on arbitrary time scales.

In the preceding section the interaction of two kinds of stochasticity, deterministic chaos and noise, was analysed. The enlargment of the parameter region with seemingly chaotic behaviour due to noise should be a remarkable contribution with consequences for the observability of chaos in real systems.

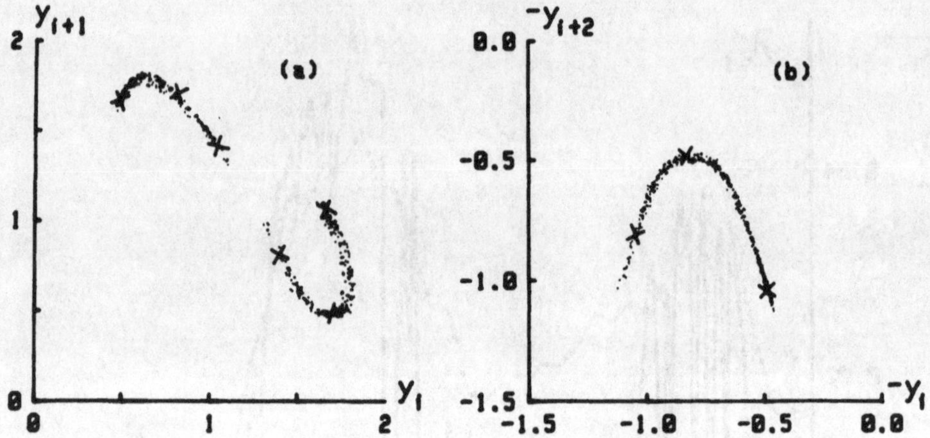

Fig. 5 Obliteration of a periodic window (E = 1.5095; Q = 0.0004). The deterministic cycle is visualized by crosses (compare fig. 1).

References

[1] Crutchfield, J.P., J.D. Farmer, B.A. Huberman, Phys. Reports 92 (1982) 2, 45-82

[2] Decroly, O., A. Goldbeter, Phys. Lett. 105A (1984) 4/5, 259-262

[3] Eckmann, J.P., D. Ruelle, Rev. Mod.Phys. 57 (1985)3,617-656

[4] Fujisaka, H., Progr. Theor. Phys. 70 (1983) 1264-1275

[5] Grassberger, P., I. Proccachia, Physica 13D (1984) 34-54

[6] Haubs, G., H. Haken, Z. Phys. B 59 (1985) 459-468

[7] Herzel, H. Th. Schulmeister, Syst. Anal.-Mod.-Sim., to appear

[8] Herzel, H., thesis, Humboldt Universität Berlin 1986

[9] Herzel, H., W. Ebeling, Phys. Lett. 111A (1985) 1/2, 1-4

[10] Herzel, H., W. Ebeling, Th. Schulmeister, submitted to Z. Naturfo.

[11] Herzel, H., B. Pompe, submitted to Phys. Lett. A

[12] Hess, B., M. Markus, Ber. Bunsenges. Phys. Chem. 89 (1985) 642-651

[13] Kurths, J., H. Herzel, Physica D, to appear

[14] Nicolis, J.S., G. Mayer-Kress, G. Haubs, Z. Naturfo. 38a (1983) 1157-1169

[15] Olsen, L.F., Chaos in biological systems, preprint

[16] Paladin, G., A. Vulpiani, Lett. Nuev. C. 41 (1984) 82-86

[17] Pompe, B., J. Kruscha, R.W. Leven, Z. Naturfo., to appear

[18] Rapp, P.E., in: Chaos (ed. A.V. Holden), Manchester University Press 1986

[19] Schulmeister, Th., stud. biophys. 72 (1978) 3, 205 - 206 and Microfiches 3/14-28

PERIODIC, QUASIPERIODIC AND CHAOTIC MOTION IN A FORCED PREDATOR - PREY ECOSYSTEM

R.W. Leven[1)], B.P. Koch[1)], and G.S. Markman[2)]

1. Introduction

The possibility that dynamical models can give rise to chaotic attractors has now been generally accepted and increasing attention has turned to the question of characterizing such structures and their associated chaotic motion. All the equations describing such models have the common feature of being nonlinear and deterministic while for some range of parameters their solutions are pseudo-random or chaotic. Many of the features of the transition from regular to chaotic dynamics have been shown to be independent of the details of the mathematical model, provided certain topological conditions are obeyed [1, 2].

Experimental studies of hydrodynamic systems [3, 4], periodically forced mechanical or electronic systems [5-7], chemical oscillators [8, 9] and periodically forced biological oscillators [10, 11] have demonstrated transitions from regular to irregular dynamics displaying these theoretically predicted features.

The possible existence of chaotic dynamics in ecological systems first was discussed by May who discovered that the logistic difference equation gives a period doubling bifurcation cascade and a chaotic motion [12]. Although current interest in these problems was stimulated, especially, by the review of May [13] the number of papers which discuss population dynamics of ecosystems from the viewpoint of chaos is very limited and there are only a few such studies on the predator-prey ecosystem [14-17].

As a consequence of the Poincaré-Bendixson theorem one can get only equilibrium points and periodic orbits for a two species autonomous Lotka-Volterra system. However, dynamical complexity can result from the Lotka-Volterra equations in dimensionality equal to or greater than three [14, 15] or in the case of a two-species ecosystem with a time dependent external force [16, 17].

1) Sektion Physik/Elektronik, E.-M.-Arndt-Universität Greifswald, DDR-2200 Greifswald, GDR

2) Rostov State University, Rostov, USSR

This paper presents results on chaotic dynamics of a modified Lotka-Volterra model which describes a two species predator-prey system under the assumption that the environment of the prey changes periodically. In section 2 the model equations are introduced and the characteristic behaviour of the corresponding autonomous system is discussed. In section 3 the existence of different types of motion and the scaling behaviour of periodic horns in the parameter plane are investigated. In section 4 the problem of scenarios leading to chaos is discussed.

2. The model and its equations of motion

The original Lotka-Volterra model is described by the following equations (see e.g. [18])

$$\frac{dN_1}{dt} = \alpha_1 N_1 - \beta_1 N_1 N_2 \qquad (1)$$

$$\frac{dN_2}{dt} = -\alpha_2 N_2 + \beta_2 N_1 N_2 \qquad (2)$$

Here N_1 and N_2 measure the size of the prey and predator populations, respectively, α_1 is the net inherent birth rate of prey, α_2 is the die out rate of predator. The terms $-\beta_1 N_1 N_2$ and $\beta_2 N_1 N_2$ represent the effects of species interaction.
The first step in modifying the system (1),(2) is that we replace the term $\alpha_1 N_1$ which is linear increasing with N_1 by a more complicated function

$$F(N_1) = \alpha_1 N_1 (K - N_1)(N_1 - L) \qquad (3)$$

with $L < K$. As a result in the absence of predators the prey population decreases to zero if the initial value $N_1^{(o)}$ is smaller than the critical value L. This autocatalytic effect (Allee effect) is appropriate when the survival of the prey depends on group behaviour in such areas as foraging, mating, and defense. For $N_1^{(o)} > L$ the population approaches a stationary regime with $N_1 = K$ in the absence of predators.
The second step is that we take into account periodical (seasonal) changes in the environment of the prey, assuming that the value K varies periodically according to

$$K = K_o(1 + A \sin \omega t) \qquad (4)$$

where $K_o A$ and ω are the amplitude and the angular frequency of the oscillatory part of K, respectively, which may be interpreted, for

example, as a seasonal change in the capacity of food of the prey population.
We introduce variables

$$x = \frac{\beta_2}{\alpha_2} N_1 , \quad y = \frac{\beta_1}{\alpha_2} N_2 , \quad \tau = \alpha_2 t$$

and abbreviations

$$\gamma = \frac{\alpha_1}{\beta_2} K_o , \quad a = \frac{\beta_2}{\alpha_2} L , \quad b = \frac{\alpha_2}{\beta_2 K_o}$$

Thus our model equations can be written as

$$\frac{dx}{d\tau} = \gamma x(1 + A \sin \omega\tau - bx)(x-a) - xy \tag{5}$$

$$\frac{dy}{d\tau} = -y + xy \tag{6}$$

The corresponding autonomous system (A = 0) has four equilibrium points (for simplicity we choose $\gamma = 1$):

$$x^{(o)} : (0,0), \; x^{(1)} : (a,0), \; x^{(2)} : (1/b, 0), \; x^{(3)} : (1,(1-a)(1-b)).$$

The fixed point $x^{(o)}$ is a stable node. $x^{(1)}$ is a saddle if $ab < 1$, $a < 1$. For $ab > 1$, $a < 1$ the fixed point $x^{(1)}$ is a stable node, for $ab > 1$, $a > 1$ a saddle point. It should be mentioned that only parameter values $a < 1$, $b < 1$ have a biological meaning. The fixed point $x^{(2)}$ is a stable node for $b > 1$ and a saddle for $b < 1$. The fixed point $x^{(3)}$ is in the first quadrant for $b < 1$, $a < 1$ and stable for $b > b_o = 1/(2-a)$. For $b \approx 1$ we have a stable node which changes into a stable focus with decreasing b. For $b = b_o$ a Hopf bifurcation takes place and a stable limit cycle around $x^{(3)}$ arises. With a further decrease of b the limit cycle widens and the period T of the motion along this cycle increases. At last, for a certain $b = b_* < b_o$ the separatrix of the saddle $x^{(2)}$ approaches the saddle $x^{(1)}$ forming a saddle-saddle connection which attracts the limit cycle around $x^{(3)}$. For $b < b_*$ we have only one stable fixed point $x^{(o)} : (0,0)$.

3. The behaviour of the nonautonomous system in the parameter plane(b,A)

In this section we present numerical results on the motion of the parametrically forced system (5),(6) for a fixed a = 0.0241 and the period of the external force $T_{ext} = 2$ (thus $\omega = \pi$). The parameter b is varied within the interval $b_* < b < b_o = 0.5061$ where the limit cycle exists for the autonomous system ($A = 0$).

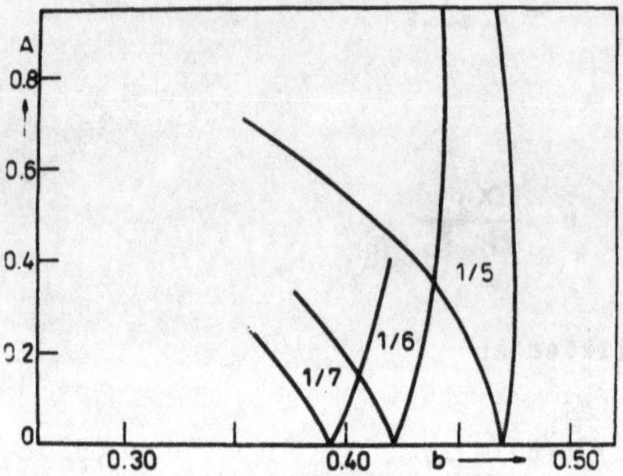

Fig.1 Arnold horns in the (b,A)-plane.

For small values of the "driving" amplitude one typically finds quasi-periodic motion. For certain intervals of b the limit cycle is entrained by the periodic external force because its period T is an integer multiple of the period T_{ext}. With increasing A the intervals of entrainment widen forming typical horns in the b,A-plane similar to those described by Aronson et al. for a two-dimensional map [19]. Fig.1 shows the horns corresponding to periodic motions with periods $5T_{ext}$, $6T_{ext}$ and $7T_{ext}$.

In a fine phase diagram, there are many frequency locking states and to relate all the different periods, we use the winding number $W = M/P$ (for periodic orbits), where P is the period of the orbit in units T_{ext} and M the number of revolutions of the phase point (in a stroboscopic phase representation) under P applications of the Poincaré map. The dependence of W on the parameter b for horns located near to the left boundary of the 1/5-horn for a given $A = 0.195$ is shown in Fig.2. The steps between the different frequency locking states compose a typical devil's staircase.

The periodic states which exist within these horns are characterized by winding numbers

$$W_{n,m} = n/(5n + m), \qquad n = 5, 6, \ldots, \quad m = 1, 2, \ldots$$

For any fixed m the sequences of horns with $n \to \infty$ obey a scaling law

$$b_\infty - b_{n,m} = \beta(m)\, n^{-\alpha} \quad , \qquad \alpha = 2.2 \pm 0.2$$

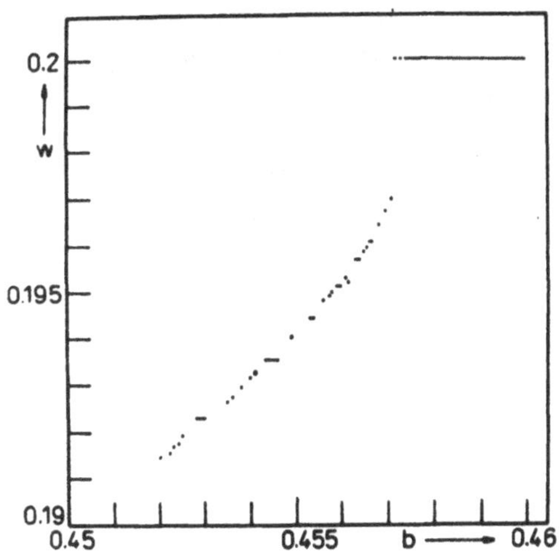

Fig.2 Winding numbers of the Arnold horns near to the large 1/5 - horn as a function of b ($A = 0.195$)

Here b_{∞} is the onset value of the 1/5 - horn and the $b_{n,m}$ are the values of b for which the horns with numbers n,m exist. This result is consistent with the scaling law found by Kaneko [20] and Inoue and Kamifukumoto [17]. Additionally we find that the coefficient $\beta(m)$ scales according to $m^{\alpha'}$ with $\alpha' \approx 2$. We thus conjecture that for the sequences of horns with winding numbers $W_{n,m} = n/(5n + m)$ there exists a generalized scaling law

$$b_{\infty} - b_{n,m} \sim m^{\alpha'}/n^{\alpha} \qquad \text{with} \quad \alpha, \alpha' \approx 2.$$

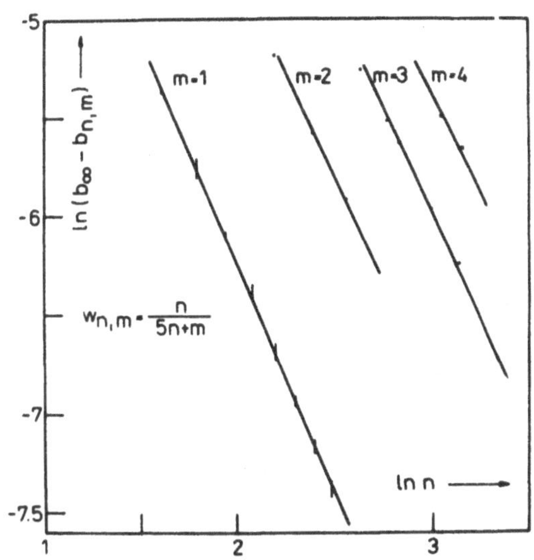

Fig.3 $\ln(b_{\infty} - b_{n,m})$ over ln n plot for horns with winding numbers $W_{n,m} = n/(5n + m)$

4. Scenarios leading to chaos

We observe two routes to chaos, the period doubling one which has been well studied in recent years and the transition from two-frequency quasiperiodicity to chaos. In the interiour of the Arnold horns the motion is characterized by a cascade of period doubling bifurcations leading to an obviously chaotic motion for sufficient large values of A. Fig.4 shows an enlarged picture of the horn with period $T_5 = 5\ T_{ext}$.

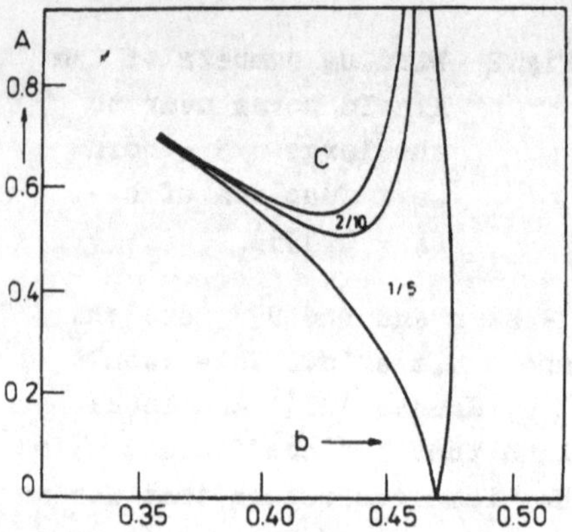

Fig.4 Period doubling route to chaos in the 1/5-horn. In the region C a 5-piece chaotic attractor exists.

Table I gives the Feigenbaum ratio $\delta_n = (A_{n-1} - A_n)/(A_n - A_{n+1})$ for some n-values, where A_n are the values of the driving amplitude at the bifurcation points from period $5 \cdot 2^n$ to period $5 \cdot 2^{n+1}$.
It is seen that already the values δ_2 and δ_3 are very close to the Feigenbaum constant $\delta = 4.669...$

Table I

n	A_n	n
0	0.571	
1	0.6869	4.533
2	0.71247	4.685
3	0.717928	4.667
4	0.7190976	

In the chaotic regime the stroboscopic phase portrait of the motion is characterized by five islands. To ensure that the motion is really chaotic we calculated the largest Lyapunov exponent λ_1 for some parameter values which proved to be strictly positive ($\lambda_1 \approx 0.01$) in this region.
Like in other periodically driven systems (see e.g. [21],[22]) the 5-piece attractor with increasing A merges with those corresponding

to the chaotic motion within other horns forming a "large" chaotic attractor which is characterized by values of λ_1 which are much larger ($\approx$ 0.1).

An interesting problem is the transition from quasiperiodicity to chaos which seems to be less well understood as the period doubling route up to now (see e.g. [23],[24]). In order to study this transition we fixed the parameter b at 0.452 and continuously increased A between 0.19 and 0.25. This, of course, is not the "right way" to obtain a continuous transition from quasiperiodicity to chaos (with increasing A the parameter b should be varied in such a way that the winding number is maintained at a fixed irrational value, what is practically impossible to realize).

Nevertheless, with increasing amplitude A of the external forcing one finds that the torus corresponding to the quasiperiodic motion is more and more deformed. We show this by construction next amplitude maps. Three of these maps for different values of A are shown in Figs.5-7. A comparison with the computed largest Lyapunov exponent shows that the latter becomes positive for values of A for which an additional maximum appears in the next-amplitude map. Despite of the fact that this process is interrupted by the appearance of periodic horns the simultaneousness of the destruction of the torus and the appearance of a positive Lyapunov exponent point to the transition from quasiperiodicity to chaos.

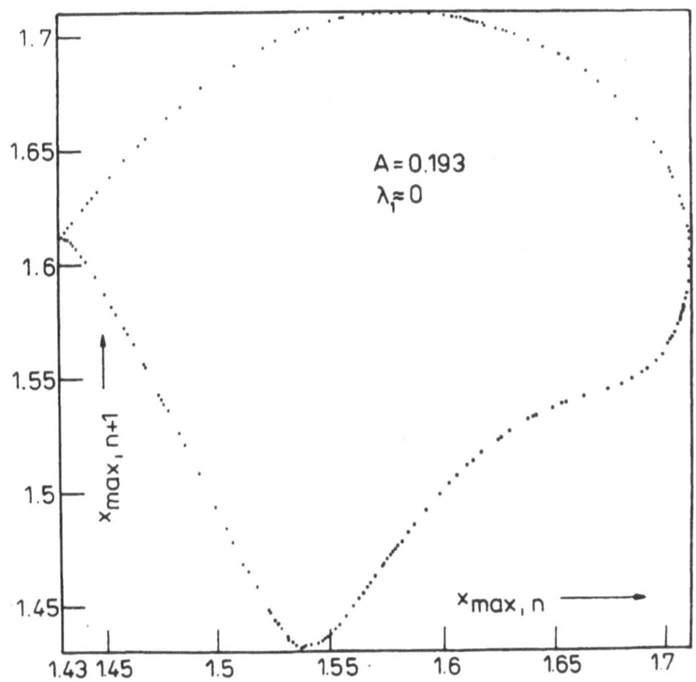

Fig.5 Next-amplitude map $x_{max,n+1}$ over $x_{max,n}$. a = 0.0241
b = 0.452 , A control parameter

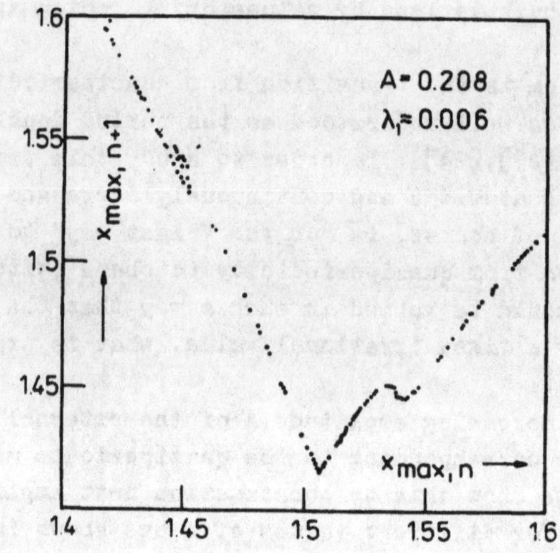

Fig.6 Next-amplitude map $x_{max,n+1}$ over $x_{max,n}$
a = 0.0241 , b = 0.452 , A control parameter

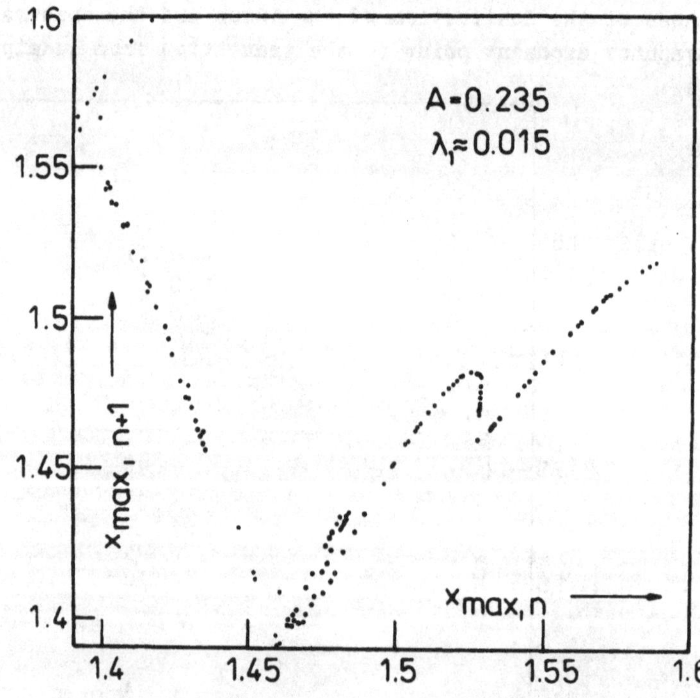

Fig.7 Next-amplitude map $x_{max,n+1}$ over $x_{max,n}$
a = 0.0241 , b = 0.452 , A control parameter

5. Concluding remarks

We have studied a modified Lotka-Volterra model for a predator-prey system including autocatalytic effects as well as periodical changes in the environment of the prey. The solutions of the corresponding non-autonomous two variable system exhibit periodic, quasiperiodic and chaotic behaviour, respectively, in dependence upon the parameter values.

With the amplitude of the periodical excitation as a control parameter we find transitions to chaos via period doubling as well as from a quasiperiodic state. In connection with this it would be worthwhile in a further study to address such interesting problems as, for example, the question whether there exists a critical line in parameter space separating the domains of quasiperiodicity from chaos and a possible scaling of the Lyapunov exponent above such a line, and also the problem of interiour crises which seems to occur when chaotic attractors belonging to the different large horns with winding numbers 1/5, 1/6,... merge to chaotic attractors of larger size with larger values of the Lyapunov exponent.

References

[1] Feigenbaum, M.J. : J.Stat.Phys. 21 (1979) 669.

[2] Eckmann, J.P. : Rev.Mod.Phys. 53 (1981) 643 .

[3] Fenstermacher, P.R., H.L. Swinney, S.V. Benson, and I.P. Gollub: Ann. N.Y. Acad. Sci. 316 (1979) 652 .

[4] Libchaber, A., C. Laroche, and S. Fauve: J.Phys.Lett. 43 (1982) L211 .

[5] Koch, B.P., R.W. Leven, B. Pompe, and C. Wilke: Phys. Lett. 96A (1983) 219 .

[6] Leven, R.W., B. Pompe, C. Wilke, and B.P. Koch: Physica 16D (1985) 371 .

[7] Testa, J., J. Perez, and C. Jeffries: Phys.Rev.Lett. 48 (1982)714.

[8] Pikovski, A.S. : Phys.Lett. 85A (1981) 13 .

[9] Simoyi, R., A. Wolf, andH.L.Swinney: Phys.Rev.Lett. 49 (1982)245 .

[10] Guevara, M.R., L. Glass, and A. Shrier: Science 214 (1981) 1350 .

[11] Hayashi, M. Nakao, and K. Hirakawa: Phys.Lett. 88A (1982) 265 .

[12] May, R.M. : Science 186 (1974) 645 .

[13] May, R.M. : Nature 261 (1976) 459.

[14] Smale, S.: J. Math. Biol. 3 (1976) 5 .

[15] Arneodo, A., P. Coullet, and C. Tresser: Phys.Lett. 79A (1980)259.

[16] Markman, G.S. : "Dynamics of Biological Populations", Inter-Univ. Collection, Gorky State University (1982) p.48 .

[17] Inoue, M. , and H. Kamifukumoto : Progress Theor. Phys. 71 (1984) 930 .

[18] Gole, N.S., S.C. Maitra, and E.W. Montroll: Rev.Mod.Phys. 43 (1971) 231.

[19] Aronson, D.G., M.A. Chory, G.R. Hall, and R.P. Mc Gehee: Commun.Math.Phys. 83 (1982) 303.

[20] Kaneko, K.: Progr.Theor.Phys. 68 (1982) 669.

[21] Leven, R.W., and B.P. Koch: Phys.Lett 86A (1981) 71.

[22] Leven, R.W., and B. Pompe: Proc. of the "3rd Symposium on irreversible processes and dissipative structures", Kühlungsborn 1985.

[23] Abraham, N.B., J.P. Gollub, and H.L. Swinney: Physica 11D (1984) 252.

[24] Ostlund, S., D. Rand, J. Sethna, and E. Siggia: Ann.N.Y.Acad.Sci. 357 (1980) 473.

CHAOTIC† DYNAMICS OF LOGICAL PARADOXES

*John S. Nicolis**

1. Introduction

One may think that it is easy to dismiss a logical (linguistic) paradox of the type "I can't say this" as mere nonsense. Such dismissal implies however the existence of a "higher" cognitive level in the information processing device involved, at which level this inconsistent sentence is classified accordingly.

Think though for a moment of those unfortunate organisms in which say part of a given chromosome codes for proteins which act as antigens to proteins produced by other parts of the *same* chromosome; think in other words of those organisms prone to autoimmune diseases. This is a real biological contradiction. Think also of organisms in which cell differentiation either stops at a premature level or is epigenetically upset (cancer prone organisms, cells deficient in i.e. LDL or Insulin receptors). These organisms are deprived of a number of stable attractors ensuring information processing and global coherence.

Think finally of persons who upon beingordered: "Be Spontaneous"! go onto a vicious, nonconverging reasoning cycle with the concomitant appearance of "Schizophrenic" symptoms. In such cognitive devices the "program" never halts - resulting into "deadlock".

In all above cases we are witnessing a number of inconsistencies or unresolved cognitive difficulties (as a result of which an external stimulus instead of being "categorized" and disposed of, is "endlessly" circulating in the organism); we may partition them into two basic categories namely as due to,

(1) Inability to distinguish the ongoings at two different hierarchical levels. ("Chicago is a trisyllable").

(2) Inability to display stable attention (or cognitive receptors) as eigenfunctions of the map or dynamical flow which map or flow involves the operation that the processor applies by way of conjugating externally impinging stimuli and internal dynamical activity. ("This sentence is false"). An example involving both (1) and (2) cases is: "This sentence contains one error" (see also Section 4).

To begin with, let us classify all cognitive tasks into two broad categories:

(a) "Exercises" or "deductive" problems - whose outcomes always exist, are unique and are implicitly contained in the initial premises - provided that these premises form a complete and self consistent set, and

(b) "*Complex*" problems, of the "Inductive" type whose outcomes may or may not exist, they may not be unique (i.e. multiple) and are not necessarily nested in the initial

* Department of Electrical Engineering, University of Patras, Greece

† Biochemical receptors (under normal conditions) and "Cognitive" receptors (under abnormal conditions) are extraordinarily selective (i.e. sensitive) to their messages (the initial conditions). Hence, the justification for the present "Chaotic" model. Such receptors, upon triggered, cause a parametric amplification of a certain signal, thereby initiating a cascade of metabolic steps or mappings (which may or may not possess hyper-attractors).

conditions but are in part generated by the evolution of an iteractive and recursive procedure or the "flow" itself. Under such circumstances we say that if there is evolutionary (asymptotic) stability in the dynamical flow in state space then there is a ***discriminable pattern*** or a "solution".

In dealing with problems of category (a) one goes through regressive steps on a sequence which converges, eventually "halts" and gives the solution. No ***new*** information is produced during this process although admittedly, the agent who deals with the problem may be ignorant of how to turn the "implicit" into "explicit" or how to choose amongst different strategies of approach. We bypass this issue here.

In problems of category (b), convergence cannot be guaranteed. Such problems are "dynamical" and their outcome can only be understood in a historical that is an evolutionary context. When we say that the information required for the solution of an "inductive" problem is not "hidden" exclusively in the initial conditions - so there is ***no*** a priori truth ("$\alpha-\lambda\eta\vartheta\varepsilon\iota\alpha$") to be revealed - we mean that at certain moments the evolving scenario that is the trajectory of the system in state space ***branches-off*** or bifurcates into multiple asymptotically stable regimes (stationary, periodic, quasi-periodic or aperiodic) in ways which depend on critical values of a number of control parameters.

These control parameters stand for the coupling among the variables, and their numerical values are shifted by ever-present (internal and/or external) fluctuations. When these parameters acquire their "critical" values the coupling between the variables concerned changes in ways which precipitate instabilities - meaning that infinitesimally small changes in the parameters lead to finite changes in the variables. Beyond these instabilities the (new) structure(s) may not retain any trace of memory of the previous regimes - although it (they) evolved from them via quite deterministic processes. Similar "loss of memory" takes place in systems without bifurcations but with a sensitive dependence of their flows to initial conditions. (i.e. in the map $X_{n+1} = 2X_n$ (mod 1) beyond the $\xi_c \sim \tau \log_2 10$ iteration where τ is the accuracy of the initial conditions). It is exactly this mixture of determinism and (long- or short-term) unpredictability which makes the problems of "inductive" type so relevant to information sciences.

Most systems relevant to information sciences (e.g. linguistics) contain multiple, hierarchical feedback-loops, which inevitable make them "Self"-referential. If such systems do not possess "closure" or logical consistency they are "paradoxical". The "truth(s)" if any - revealed by such systems is associated with "Selfconsistency" and this characteristic, implying the existence of stable attractors in the iterative and recursive system we call "self-referential sentence". The search for self-consistency of a self-referential sentence, then leads us to the formulation of a boundary value problem. If so, the suggestion follows immediately that "logic" or "rationality", so far a static concept, can be dynamically implemented and replaced with "evolutionary stability".

All linguistic schemes (Genetic, Natural, Artificial . . . .) are discrete. We may in some simple cases approximate them with Markov Chains of simple or multiple memory.

There are two ways, starting from dissipative non-linear dynamics (possessing a Basin with one or multiple attractors), one can deduce a Markov chain:

(a) Taking a strange attractor in N-dimensional state space we first perform an (arbitrary) analog-to-digital conversion by way of a Poincare return mapping which amounts in cutting the trajectory with a hyperplane, recording the isophasic crossings and deriving a set of discrete coupled nonlinear difference equations in N-1 (or less) dimensions. Further we choose an (arbitrary) partition of the discrete map above and then we derive a Markov chain with number of states the partition number and transitional probabilities determined by the local slopes of the map.

(b) Concerning dynamics with more than one coexisting attractors, external noise can act catalytically and provide a sequential passage from attractor-to-attractor, thereby giving rise to a Markov Chain with number of states equal to the number of coexisting attractors and transitional probabilities determined by the type of superimposed noise. (Noise, acting on the control parameters of the system alters the topology of fractal basin boundaries thereby shifting some initial conditions from one basin to another).

2. What is a cognitive channel?

The human mind possesses the unique capability of "mapping" the external (as well as part of the organism's internal) world - that is, it "compresses" long and complex strings of impinging environmental stimuli ("observations") - and then uses these "minimal length algorithms" in order to simulate physical phenomena - thereby revealing the "laws of nature". We may theorize that this process of "self"-organization and category formation is implemented via a set of coexisting (strange) attractors in the cognizant apparatus - each one of which attracts (and therefore compresses) whole subsets of "initial conditions" the sum - total of which constitute the set of external stimuli. This set of initial conditions forms the "Basin" of the attractors and the processes of partition and category formation in the mind involves the topology of the separatrices among the individual subsets of the Basin.

We propose that the set of messages should be *identified* as a set of initial conditions or the Basin of the attractors in an N-dimensional state space where N stands for the number of variables mediating the dynamical phenomenon under consideration.

The receiver, or, more generally the cognitive apparatus is identified as the set of attractors, of much lower dimensionality embedded as invariant compact subsets in the same N-dimensional state space.

Falling into the attractor via (short or longer) transients in state space corresponds to the act of progressive signal abstraction (or compression).

The percentage of compressible sequences in a string say of 0's and 1's of length N is extremely small. Considering the 2^N a priori equiprobable sequences of length N (the members of the Basin) the number of compressible sequences to say K bits is just,

$$\Sigma = 2^1 + 2^2 + \ldots\ldots + 2^{N-K-1} = 2^{N-K} - 2$$

and the percentage of compressible sequences is

$$\eta = \frac{\Sigma}{2^N} \sim 2^{-K}$$

So, although in principle there is a numerous set of attractors at each cognitive level, the percentage of those which really deserve the name, that is possess an information dimensionality $C = N - K$ bits ($K \sim 0(N)$) is just $\eta = 2^{-K}$, that is insignificantly small.

The moral is that the human mind can simulate a very small subset of natural phenomena. Their overwhelming majority literally "passes through" our heads undetected.

Even so how come that our templates-attractors *really* have as Basin the set of *externally* impinging stimuli? To accept this one has to a priori digest the hypothesis that *there exists* a "homology" between *some* patterns of the external world and the categories of the software of the human brain. [The stimuli *are* solutions of the governing dynamics $\dot{x} = F(x;\mu)$].

But if the human brain is just an unsuspected evolutionary byproduct of the "laws of Nature" how come that its software can simulate the very laws which, acting upon matter, triggered its formation? Be as it may according to our model, the thalamocortical pacemaker whose activity is manifested as the E.E.G is the "Noisy" agent which permutates the attractors on a time division multiplexing basis.

To directly reconstitute the Basin (or part of it) from its broken symmetry-dynamical "projections" (the attractors) is of course impossible. It can only be done by trial and error if one knows the implicit dynamics and starts from arbitrary initial conditions leading to the observed attractors. Even then, the portion of the explored Basin is of the order of 2^{-N+C}. To commute between attractors one has to use a "time-division-multiplexing" device such as the thalamocortical pacemaker of the brain giving as a one-dimensional trace the E.E.G. But in case where one likes to "hold" in his mind the "whole" of the attractors group - all at once, the E.E.G may become literally "epileptic" and its dimensionality falls drastically ($D \sim 2{,}05$); it becomes a "quasi-Rössler-attractor" - and in that matter a ***very poor information processor***. During the execution of mental tasks as well as during R.E.M sleep the E.E.G is ***multidimensional***. If one likes indeed to jump between *two* or more coexisting attractors one has to "tunnel" via a higher dimension than the dimension of the highest order-attractor. In the case for example where the basin is the *Necker cube* and the two (unstable) attractors are the "forward" and the "backward" images, one has to mirror the one attractor via a *four*-dimensional state space in order to get the other. In case where the Basin (in a Hilbert space) is the "*electron*" two attractors are the "corpiscule-picture" and the "wave-picture", complementary and mutually exclusive. The appearance of each attractor depends on how we prepare and start the "experiment" that is, on the initial conditions.

3. Chaotic dynamics of pattern formation - both self-consistent and paradoxical

The question "Where exactly information is stored?" seems to receive a rather original answer from a series of well-known experiments on the perception of random dot interference patterns performed several years ago by Glass and Perez.

The essential conclusions we draw from those experiments is that the information lies neither in the externally impinging patters (which may be just a random dot aggregation on the plane) nor on the dynamically stored pattern-replica of the receiver (which again may be random); it rather emerges from the flow in the neighborhood of the *singularities*, as deduced from the pertinent eigenvalues of the linear or nonlinear, discrete of continuous transformation (mapping or flow) - which transformations the processor imposes between the incoming and the internal pattern.

So one is inclined to consider, at least in a number of cases where such experimental documentation exists, that what we "really" see or cognize in the world is perhaps just the *attractors* (or the repellors) of the maps or the flows which the processor generates by way of *relating* external stimuli with internal activity. Biological processors, possessing identical nervous systems tend to generate more or less identical maps or flows-thereby creating sd it were by "social consensus" universal pattern classification schemes. Not only in Perception but also in Cognition and Linguistics emerging stable patterns may be identified as eigenfunctions of nonlinear operators which operators are involved in the recursive character of these systems. For example, recent work in game theory as applied to biology and animal contests as well as to the process of mutual simulation between competing hierarchical systems - each trying to "break" the code of the opponent first, shows that rationality in such tournaments can be completely replaced by evolutionary stability. This points to the emergence of evolutionary stable strategies (E.S.S) which are point attractors for the probabilities of moves adopted by the contestants: such asymptotically stable strategies lead to a stable population polymorphism or stable behavioral repertoires by a single individual.

In Linguistics the situation is different since, here, unlike animal or finite state automata games, self-reference is involved. We conclude with two examples which show that self-consistency hence logic, is provided also by the existence of stable attractors.

Take as a first example the symbolic self-referential loop: "This sentence contains $x_0$0's, $x_1$1's, $x_2$2's, $x_3$3's, $x_4$4's, $x_5$5's, $x_6$6's, $x_7$7's, $x_8$8's, $x_9$9's". Let the 10-dimensional vector $\mathbf{x}(x_0, x_1, x_2, \ldots, x_9)$ start from a randomly selected initial condition and then provide the next iteration by checking the truth of the previous one. For example, start from

$$\mathbf{x}_a\,(0,1,2,3,4,5,6,7,8,9).$$

The next iteration gives

$$\mathbf{x}_b\,(2,2,2,2,2,2,2,2,2,2,)$$

Next comes

$$\mathbf{x}_c\,(1,1,11,1,1,1,1,1,1,1)$$

followed by

$$\mathbf{x}_d\,(1,12,1,1,1,1,1,1,1,1)$$

and finally

$$\mathbf{x}_e\,(1,11,2,1,1,1,1,1,1,1)$$

which obviously is *the* attractor corresponding to the selected initial condition and which makes the self-referential sentence above, self-consistent that is, true. (through exhaustive search one can find another point attractor as well as a limit cycle of period 2. Here the basin of attraction is partitioned into three subsets). As a second example take the sentence: "This sentence has $\underline{X}$ letters". Start iterating from x_a = "one". Then $F(x_a) = 25$ where F is the number of letters after putting in the sentence to the given initial condition.

The second "seed" then is x_b = "twenty five" which leads to $F(x_b) = 32$. Likewise x_c = "thirty two" leading to $F(_c) = 31$ and finally x_d = "thirty one" lands the system to $F(x_d) = 31$ - which is the attractor corresponding to the initially chosen seed "one". (There is another point attractor corresponding to a different subset of the basin).

Note finally that while the truth assessed by the second sentence is "English bound" (translated e.g. into French it becomes false) the first sentence conveys a "translinguistic" truth.

The sentence now "This sentence is false" by a similar token possesses two "repellors" in cognitive space - hence the perpetual oscillation of the processor between them.

4. Paradoxes due to mixing up ongoings at different hierarchical levels

In many instances hierarchical systems are "almost decomposable" in that one can study the dynamical ongoings at a given level taking what is just underneath as a boundary condition and what takes place above as a constant. This can be done whenever the rate constants differ by orders of magnitude from level to level (lower level characterized by larger rate constants or smaller relaxation times).

In linguistic systems, however, this decoupling of levels cannot be accomplished due to the existence of evolutionary feed forward loops (essentially hierarchical maps) mediating the dynamics between successive levels. Specifically in linguistic systems, one makes use at the same time of two hierarchical levels where statements and meta-statements are interlocked.

Let us now examine in more detail the circumstances of breaking down the "mean-field approximation". Consider a dynamical system reduced to a one-dimensional map. The probability of finding the system at point x at a time t increases due to transitions from other points x' of the interval. It decreases due to transitions leaving this point. So, the "master equation" becomes

$$\frac{dP(x;t)}{dt} = \sum_{x'} w(x,x')P(x';t) - P(x;t) \sum_{x' \neq x} w(x',x)$$

where $w(x,x')$ is the transitional probability for jumping from $x' \rightarrow x$, and $w(x',x)$ the transitional probability for jumping from $x \rightarrow x'$. The "steady state" or "stationary" asymptotic C of this equation is the invariant measure of the discrete dynamics (the map) at the lower level. The shape of such a p.d.f. is in many cases multi-humped so that the mean value far from being the most probable is in quite a few cases the least probable. Now the crucial thing is that the transitional probability rates (generally unknown, but depending on the shape of the map) are in most cases nonlinear functions of x. Multiplying now both members of the M.E. of by, x, x^2..... and integrating or summing in the interval of x we get a series of "phenomenological", higher level "macroscopic" equations with reference to the various moments $< x >, < \delta x^2 >$.... of the p.d.f. $P(x,t)$ namely,

$$\frac{d}{dt} < x > = f_1 \{ < x >, < \delta x^2 >, \ldots \}$$

$$\frac{d}{dt} < \delta x^2 > = f_2 \{ < x >, < \delta x^2 >, \ldots \}$$

....

Because of the nonlinearity of the operators f_1 (due to the nonlinearity of the transitional probabilities) we get a great number (infinite) of coupled nonlinear differential equations, with respect to the moments and cross-moments (cross-correlations). Moreover, in the vicinity of bifurcations, or, for self-sustained chaos filling the whole interval, these moments may become comparable. We then understand very clearly that under such circumstances successive hierarchical levels may get mixed up. Example: "This sentence contains two errors": At one (grammatical) level the sentence is false. At a higher (cognitive) level it is true. But then again it is false . . . . . and so on; we get an interlevel contradiction.

5. A note on the characteristics of fractal basin boundaries

We consider an N dimensional flow $\dot{x} = F(x;\mu)$. If an initial condition (message) specified with a given (however small) error ε is within that distance (ε) from a fractal separatrix with dimensionality d the attractor to which it will be falling cannot be predicted with certainty. So there is a fraction of phase space consisting of "fuzzy" messages. The topology of the fractal basin boundaries (separatrices) depends on the system's (the processor's) parameters and so a shifting of these parameters or an uncertainty, δ about their exact values affect also the ability to predict the attractor from an initial condition - independently of the precision ε with which this initial condition has been chosen. If the system parameters vary, then, as has been shown by Grebogi, Ott and Yorke an attractor may be destroyed in a collision with a fractal separatrix (this is called a "crisis") and beyond the specific value of the control parameter for which the crisis takes place a long transient develops, whereas, moving the control parameter in the opposite direction *creation* of an attractor takes place. Given then ε, d and δ we can investigate the topology of the individual subsets of the Basin and we may calculate the uncertainty (or the information gained thereof) associated with the category to which the given stimulus belongs. Knowing on the other hand the state space dimensionality of the impinging messages *and* the (fractal) dimensionalities of the coexisting attractors one can evaluate in each case the degree of (average) compressibility that a given N-dimensional message is subjected to. Together the *entropy of partitioning* and the *degree of compressibility* give the two essential macroparameters characterizing the cognitive process at a given hierarchical level. (this process may not be accomplished in one single hierarchical step; in a second step the attractors (if numerous at the lower level) will play the role of the members of a hyper-Basin towards a new hierarchy of fewer hyper-attractors and so on).

6. Outlook

In biology the dissipative multi-hierarchical genetic map starting between the levels of nucleic acids and aminoacids, is of course a self-referential system. Yet in a "Healthy Organism" the eigenfunctions of this map (different types of proteins, and specialized cells - as well as species) are *coexisting stable attractors* trapped into a "web" of fractal separatrices (basin boundaries). In case where the stability of some of those attractors is upset *and* the topology of the fractal boundaries changes (due to a shifting of some control parameters by noise) and "attractor crisis" may develop as a "collision" across a fractal boundary and a "functional illness" may erupt as a result.

Such attractor crises may refer either to an individual or to a species-level thereby creating an ecological exstinction or a renewal - depending in which way the control parameters are shifted. [On a different field we may speculate that transmutation and creation of new elements or particles may, by the same token, be characterized as attractor crisis in the eigenvalue spectrum of a dissipative non-linear Schrödinger's - like map].

All these "paradoxes", that is all these dynamical deliberations destroying self-consistency, catalyzed by noise, far from being pathological are triggers in promoting ecological evolution and cognitive progress towards higher, more complex abstractions. Nevertheless, no matter how many attractors we appreciate cognitive, based on compressibility of initial conditions always refers to low dimensional, broken symmetry patterns. Under such circumstances paradox always looms. The only radical way of transcending it would amount to moving to higher dimensionality (an infinite one) and recover there correlations and perfect symmetry.

"Reality" however (the multidimensional sum total of the BASIN's members) constructed "backwards" from the set of attractors seems to be out of cognitive reach. One is only reminded of the illusion of the Platonic Constrained Caveman - who perceives and interpretes himself as his own shadow - thereby "demoting" the Basin to a mere attractor.

7. Summary

Consistent cognitive patterns are taken as stable eigenfunctions of the dynamical flow or the map mediating environmental stimuli and "internal" dynamical activity in the brain. Such maps simulate self-referential sentences - modelled as cognitive (digital) feedback loops. A cognitive dynamical processor a) may or b) may not possess attractors at all - stable or unstable. [In general these attractors are separated by fractal basin boundaries]. In case b) asymptotic stability is not ensured, only transients exist and so self-referential schemes remain "open" that is inconclusive. In case a) that of *coexisting repellors* self-referential schemes become paradoxial. Paradoxes (like the "Necker cube" ambiguity) may also be generated via modifying the topology of the fractal Basin boundary thereby changing the fuzziness over the final disposition of a given external stimulus - namely the uncertainty as to which attractor will "shallow" this initial condition.

Another origin of paradox lies in cases where there is a breakdown of hierarchy between successive cognitive levels - where statements and meta-statements are mixed up.

This breakdown may be due to the P.D.f of the dynamical ongoings at the lower level being multihumped - thereby giving rise (at the higher level) in a number of collective properties (moments or "macrovariables") comparable with the number of the bare variables at the lower level.

In conclusion, we may say that paradoxes emerging in natural language may be perhaps of only academic interest, (mental illness aside). Paradoxes in artificial (computing) languages may give rise if undetected by the programmer, to "deadlock" in the machine.

Paradoxes in genetic languages however may literally spell the difference between life (self-consistency) and death. Hence the importance of providing dynamical criteria and self-corrective procedures - that is controlled perturbations of the parameters of the flow or map - which create or maintain stable attractors and avoid "crises". "Genetical Engineering" and "Biological Control" are *ad hoc* ways ti achieve just that.

Key References

[1] Nicolis, J.S. "Dynamics of Hierarchical Systems - An Evolutionary Approach", *Synergetics*, Vol. 25, Springer-Verlag (1986).

[2] Glass L. and R. Perez, "Perception of Random Dot Interference Patterns", *Nature* 246, p. 360.

[3] McDonald, S.W., Grebogi, C., Ott, E. and J.A. Yorke, "Fractal Basin Boundaries" *Physica* 17 D (1985), p. 125.

CONTROL OF GROWTH PROCESSES

Peschel, M., Academy of Sciences GDR, Centre for Scientific Instruments
Voigt, M., Academy of Sciences GDR, Department Mathematics/Informatics
Mende, W., Academy of Sciences GDR, Institute Geography/Geoecology

Introduction

The recent development of microelectronics offers the possibility, to realise complex functions in a redundant hardware by software. The knowledge won in control theory in the early 70-s can now be used efficiently in microcomputer controllers.

In this paper the principle of an adaptive micro-computer-controller will be proposed mainly based on research work of the 70-s in connection with dissertations on socalled switch-relais-control.

Switching relais control circuits are very simple and can be easily realised, but they have a remarkable disadvantage, namely that we meet in the control variable trajectory permanent oscillations may be with relatively large deviations from the prescribed value.

Therefore, we introduce in this paper an adaptive behaviour into the classical concept of a switching relais controller.

Additionally we have to combine this controller also with a classical PID-controller to overcome the idleness of the new controller at the very beginning of a control process and to give him continuous-like properties for all his actions.

The Main Ideas for the Control Structure

For a control structure with one or more feedback loops we need the following parts of the circuit.

1. Control System

These are multivariable systems either linear

$$dx_i/dt = \sum G_{ij}x_j$$

or nonlinear Lotka-Volterra Systems

$$dx_i/dt = x_i(\sum G_{ij}x_j)$$

in future also combined models

$$dx_i/dt = \sum H_{ij}x_j + x_i(\sum G_{ij}x_j)$$

2. Leading System

The strategy for leading the control circuit depends on the control aims we are interested in. We have already checked the following possibilities

- We prescribe fixed values for all control variables and want to realise a stable equilibrium point by the control
- We prescribe through the leading system a trajectory $x_{iS}(t)$ as reference for the state x(t) of the control system to follow quickly and with a small follow-error
- We use the follower-system principle for the purpose of parameter identification of the control system. Now the control system plays the role of a model and the leading system is the real system to be identified, the state-variables of both systems represent an implicit information on the parameter fit. Now the controller tries to adapt the model parameter values on the base of the corresponding deviation concept of the closed loop.

3. Deviation System

The deviation system realises the deviation concept of the closed loop. Depending on the control aim there are different deviation options possible.

The most simple case is the classical one with

$$\Delta_i = x_i(t) - x_{iS}(t)$$

Buth we should take into account for the deviation model also the driving forces of the control system given by differential equations of the form

$$dx_i/dt = f_i(x,t)$$

The right-hand sides of these equations are the driving forces of the control system.

In our program for the controller we use the following concept

$$\Delta_i^* = -\mathrm{sgn}(f_i(x_S,t)).\ \mathrm{abs}((f_i(x,t) - f_i(x_S,t))^{Exp}$$

and in general a linear superposition with certain weights of both concepts.

4. The Controller System

This is the most important part of the adaptive controller we propose. For this concept we used two classical origins:

- The PID-controller
  Because we need at the output of every feedback channel an integrator

we make use of the PID-controller in the following form

$$\int_0^t (KI\ \Delta_i + KP(\Delta_i - \Delta_i')/D + KD(\Delta_i - 2\Delta_i' + \Delta_i'')/D^2)dt$$

Herein Δ_i designates the deviation in the i.th channel, and Δ_i its value after one tact delay, Δ_i'' its value after two tacts delay.

The Switch-Relais Controller

This controller consists of a relais with tolerance zone and a switcher communicating with a switch-period T_a the actual state of the relais output, followed by an integrator to sum the pulses.

The switch relais controller has together with a linear control system rather interesting properties which were studied in detail in the dissertations /1/, /2/ and /3/.

Properties which are important for our adaptive controller are listed now:

After the switcher are observed socalled complex combination oscillations if the feedback loop for a linear control system is closed. These combination oscillations consist of tracks of pulses with equal sings. Such a track of pulses of the same sign we call a half-period of the combination oscillation. After some time a sequence of such half-periods constitutes a periodic oscillation.

. There may occur an initial transient trajectory which after some time flows into the periodic complex combination oscillation.

. It is very important that the initial value configuration of the linear control system has no influence on the observed combination oscillation, only the initial transient trajectory is implied in different way by the different initial states configuration of the linear control system.

. For a given linear plant with specified parameters a veriety of different oscillations might occur. This spectrum of different combination oscillations depend on the ration y(o)/Gy, where y(o) is the initial value of the integrator and Gy the transfer factor of the integrator.

Despite of the fact that the integrator is responsible for the splitting up of combination oscillations, this dependence is in some sense a weak one, because we meet a remarkable symmetry property, namely the combination oscillations depend only on values of the mentioned ratio in the following interval

$$-0.5 \leq y(o)/Gy \leq 0.5$$

andin other intervals the whole picture repeats periodically, that

means we observe an influence of y(o)/Gy mod 1.
From a large amount of studies of the switch relais controller for linear control systems we found the following empirical properties.

Non-vanishing mean values of the integrator are consistent with the integral-like feedback, that means the control circuit can possess a permanent control variable deviation despite of the integrator. For the length of the period of the complex combination oscillation the ration $\mathcal{X} = T_{large}/T_a$ is very important. Here T_{large} denotes the maximal time constant of the linear control system.

A large value of $\mathcal{X}$ means after our experience a big length of the half-periods of the combination oscillations and as a consequence relative large control errors in connection with the permanent oscillation of the control variable.

From this property we draw the conclusion:
If we fight successfully against large values of $\mathcal{X}$ by measures of adaptive control we should be able to reduce the control error.

5. The Concept of the Adaptive Controller

We want to design the adaptive controller in such a way, that the described controller can be used as a one-dimensional controller in every feedback channel separately without the necessity to apply a decoupling network in the case of a multivariable control system.

That means we shall use the integrator at the output of every feedback channel as the control input in exactly one driving force of the control system.

This control input can be applied to the control system in different way. Classically the additive control was largely spread. But this is not obligatory, especially for the control of growth processes a multiplicative control is more suitable for example a control of the reproduction rates of the different species.

On this line we have studied up to now the following two options. Additive input

$$dx_i/dt = x_i(\sum G_{ij}x_j + Int_i) \text{ or in the linear case}$$

$$dx_i/dt = \sum G_{ij}x_j + Int_i$$

Diagonal control (multiplicative input)

$$dx_i/dt = x_i(Int_i x_i + \sum_{j \neq i} G_{ij}x_j) \text{ or in the linear case}$$

$$dx_i/dt = Int_i x_i + \sum_{j \neq i} G_{ij} x_j$$

The diagonal control was very successful already in connection with the so-called replicator dynamics for the purpose of cluster-building (build up of so-called biotopes in the set of competing species /4/)

After a proposal of W. Mende we also studied other multiplicative control inputs. It should be enough to apply in any driving force exactly one integral variable Int_i, that means we could not only think about controlling the reproduction rates but also interaction coefficients G_{ij} with $j \neq i$.

First simulations show that this also works but usually not so good as the diagonal control, we observe permanent oscillations with relatively large deviations from the prescribed values of the control variables.

Very important is to make use of the so-called principle of reference control.

In the case of referance control we seek for such control values with which the driving forces vanish, namely

additive control

$$\sum G_{ij} x_j + y_i \approx 0$$

diagonal control

$$y_i x_i + \sum_{j \neq i} G_{ij} x_j \approx 0$$

If we apply the corresponding control values we try to force the control system to pass on his trajectory only equilibrium points. Obviously the reference control principle applied to Lotka-Volterra equations lead to the same analytical expressions for the corresponding control variables y_i.

The concrete application of the principle of reference control is not without problems.

What should we substitute into the equations for the determination of the reference control, the actual state variables values x_i or their prescribed values x_{iS}?

We could also think about substituting a linear combination of both control variables values, on the other hand we could compute both types of reference controls y_i respective y_{iS} and combine these reference controls linearly.

We had good success by substituting the actual control variables into the determination equation for the reference control. It is further important in which points of time we decide to apply new values of the reference controls y_i.

One successful possibility was to apply new reference control in the channel in such points of time when the control deviation changes its sign.

In any case it is quite important always when we apply new reference controls y_i to readjust the integrators Int_i from zero, otherwise we would meet unsuitable delay effects of the successful control actions.

Most important for the concrete adaptive controller we propose is the mode how we introduce adaptivity into its behaviour. The adaptivity we used up to now based mainly on the influence of the paprameter $\mathcal{X} = T_{large}/T_a$ on the period of combination osciallations. Here we are able to influence the switching time-period

$$T_a = A.D$$

We assume that A is an integer and D is the elementary time-step of the digital controller.

If we want small half-periods of the combination oscillations we should descrease $\mathcal{X}$ by increasing A with the consequence, that then only small control variable deviations could occur.

The disadvantage of such a strategy would be that at the very beginning of the control process when theh control deviations are relatively large the feedback wouldonly seldom influence the control system. Therefore, a latent combination with a classical PID-controller proved to be quite successful.

We adaptively vary not only the switch number A but also the transfer factor Gy of the integrator. There are quite different schemes to change A, respectively Gy monotonously if larger half-periods are observed. The most simple way which already works is a monotonous dependence on the actual control deviation in every point of time we want to make an adaptive change.

Some Experiences with the Adaptive Controller up to Now

Without realising the controller but using the diagonal control principle for the case of Lotka-Volterra equations in the Replicator form we could determine stable clusters of coexisting species. This can be considered as persistence control of a subset of all species. This approach coulo be successfully applied to solve so-called partitioning or assignment problems.

If we want to realise a fixed valued multivariable control task with the adaptive controller for a growing system under cooperation and competition conditions this aim can be interpreted as persistence by control of all species in a prescribed equilibrium point within the positive cone.

By the approach above described it was possible to realise fixed-value multivariable control for multivariable linear systems and multivariable nonlinear systems (Lotka-Volterra systems). We used the reference control additively as well as multiplicatively (diagonal control).

We also realised by simulation a follower-controller for multivariable linear systems and Lotka-Volterra systems. But this controller worked relatively slow. We had to provide a time reserve for him, that means the leading system changed its states with a reduced velocity, so that the control loop was able to follow with the tact velocity of the micro-computer. For an on-line real-time follower system the actual state of development of the controller is not yet sufficient.

Very interesting seems the application of the adaptive controller for the identification of the parameters of an unknown real system. We checked this possibility successfully for the following option: We used the model for the unknown real system as a control system in the multivariable control circuit above described. The unknown real system took the role of the leading system. We used the state-variables of both systems as the control variables respective their prescribed value with the idea that they contain an implicit information about the parameters. We drove the leading system with a reduced velocity and realised a follower-control circuit. The only difference was that now the parameters of the model (control system) are the state-variables which we influence by the feedback loops.

It could be shown that by this approach the diagonal elements of linear systems and Lotka-Volterra systems could be successfully identified. It was not possible by this way to identify the parameters of a regression model.

Very important for all our research was the fact that the control circuits worked successfully without applying a decoupling network.

References

/1/ Franke, K.: Beitrag zur Ermittlung stationärer Eigenschwingungen und der Stabilität einer Klasse von Abtastrelaisregelungen. Dissertation A. TH Karl-Marx-Stadt, 1968

/2/ Regel, P.: Unte rsuchungen zu Abtastrelaisregelkreisen Dissertation A. TH Karl-Marx-Stadt, 1971

/3/ Haase, H., Helbig, H.: Untersuchung spezieller Eigenschaften einer Klasse von Abtastrelaisregelkreisen und deren Ausnutzung zur Kennwertermittlung
Dissertation A. TH Karl-Marx-Stadt, 1973

/4/ Peschel, M., Mende, W.: The Predator-Prey Model: Do we live in a Volterra World? Akademie Verlag Berlin, Springer Verlag, 1986

THE BELOUSOV-ZHABOTINSKII REACTION: A PARADIGM FOR THEORETICAL STUDIES OF DYNAMICAL SYSTEMS

A. ARNEODO
Laboratoire de Physique Théorique, Université de Nice
Parc Valrose, F-06034 NICE Cédex (France)

F. ARGOUL, P. RICHETTI and J.C. ROUX
Centre de Recherche Paul Pascal, Université de Bordeaux I
Domaine universitaire, F-33405 TALENCE Cédex (France)

1. INTRODUCTION

Nonequilibrium chemical systems have been widely studied as examples of nonlinear systems which display a rich variety of interesting dynamics [1,2]. One of the most common features of these chemical reactions is the alternating sequence of periodic and chaotic states formed from combinations of large and small amplitude oscillations. Such a scenario has been extensively investigated in the best understood oscillating system, namely the Belousov-Zhabotinskii reaction. Experimentally nonperiodic states have been found to occur between periodic states [3-9]. For a long time there has been considerable skepticism regarding the existence of deterministic chaos in well-controlled nonequilibrium reaction [10-12]. This is to be imputed to the misleading aspect of the nonperiodic regimes which look very like "stochastic" mixtures of two adjoining periodic states. The fundamental question is to know whether or not such a chaos results from the nonlinear properties of the B.Z. kinetics or from some experimental artifacts.

Recent progress in the experimental studies and the development of new tools for data analysis, e.g. the reconstruction of phase space trajectories from a single time series, have allowed one to distinguish unambiguously between deterministic chaos and external noise. In particular, from Poincaré sections and 1D maps obtained from these experimental phase portraits, several well-known routes leading to chaos have been recognized: type-I intermittency [13], period-doubling bifurcations [14], fractal torus [15]. The most persuasive experiment has been performed at the Texas University [14,16-19]. When decreasing the flow rate, the following succession of regimes has been recorded:

$$(1) \qquad P_1^0,\ C_1^0,\ P_1^1,\ C_1^1,\ P_1^2, \ldots\ P_1^n,\ C_1^n,\ P_1^{n+1}, \ldots$$

where P and C stand for periodic and chaotic states respectively; the upper index defines the number of small amplitude oscillations in a basic pattern, while the lower index denotes the number of large amplitude relaxation oscillations. The analysis of the first chaotic state C_1^0 is summarized in Fig. 1a. A detailed investigation of the intermediate control-parameter range has shown that the transition from the relaxation oscillations to chaos proceeds via period-doubling bifurcations [14]. The map constructed from the phase portrait of C_1^0 clearly yields a smooth curve and not a scatter of points. This rules out any suspicious stochastic interpretation of these complex oscillations. The dynamics obeys a deterministic law; it is described by the iteration of a 1-D map with a single extremum [20]. Such a characteristic shape of the experimental curve explains the observation of some steps of the reverse chaotic band-merging cascade. These chaotic states are separated by periodic windows. Within the experimental resolution, the order of occurrence of these periodic states and the observed iteration pattern are in quite good agreement with the U (universal) sequence generated by unimodal 1-D maps [17]. The return maps which describe the chaotic dynamics in C_1^1, C_1^2... are more complex. Nevertheless one can suspect that the underlying 1-D map contains more than one hump [21]. The computations of the largest Lyapunov exponent either on the reconstructed 1-D maps [18] or from measurements of the divergence rate of nearby trajectories using appropriate numerical methods [19] converge to the same conclusion: the exponent is definitely positive which attests that there is sensitive dependence on initial conditions, the hallmark of deterministic chaos.

However in spite of this whole set of concordant investigations of the B.Z. reaction, some doubts still remain regarding the actual status of the chaotic behavior recorded in the experiments [10-12]. Our main goal here is to elucidate the origin of this skepticism and to answer several of the interrogations still present in the mind of many chemists.

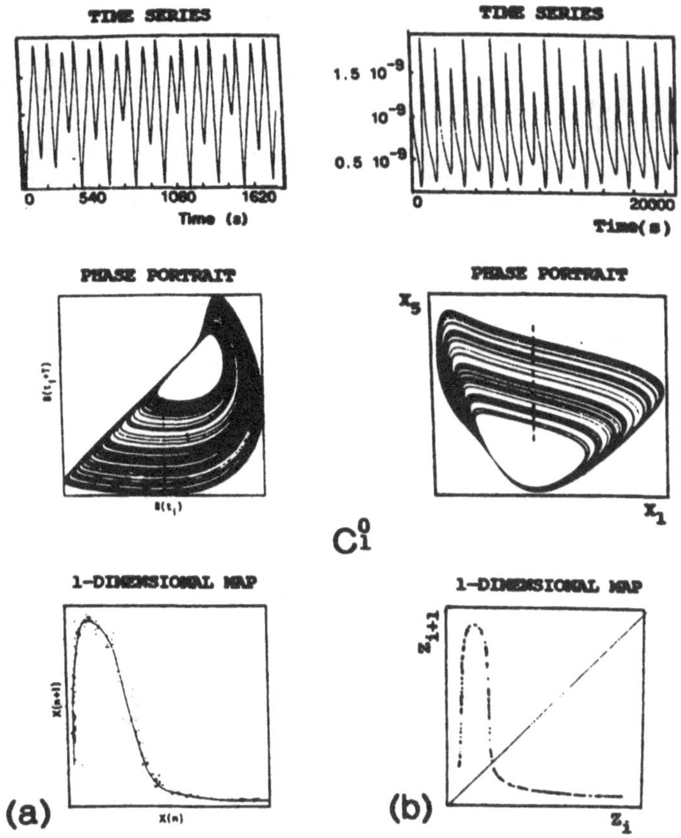

Fig 1: Time series, phase portrait and 1-D map for the chaotic state C_1^0 (a) data from the Texas experiment [14,16-18]; the measurable observable is the bromide ion potential; (b) simulation of the seven-variable model detailed in Table I [32]; the time series corresponds to the time evolution of $X_1 = [Br^-]$.

2. MODELLING THE BELOUSOV-ZHABOTINSKII REACTION

One of the main ingredients of the controversy regarding the existence of chaos in the B.Z. reaction is the considerable number of numerical investigations of "realistic" chemical models which have failed so far to reproduce the scenarios to chaos observed in bench experiments. From the detailed mechanism of the B.Z. reaction elucidated by Field, Koros and Noyes [22,23], a 3-dimensional skeleton model, the Oregonator model [24], was proposed and shown to reproduce the oscillatory behavior of the original model. However any attempt to find chaos in such a model has been yet unsuccessful. A similar conclusion has been reached in many different simulations of modified versions of the Oregonator [10,25-27]; a periodic sequence is generally observed without any chaotic behavior for intermediate control parameter values between two successive periodic states. The only study of an Oregonator model yielding chaos has been the Ringland and Turner's analysis of a seven-variable model [28-29]. However, it should be emphasized that there are drastic discrepancies between the model predictions and the data.

In a preliminary study [30] of a seven-variable model of uncatalyzed reactions of aromatic compounds with bromate, we have succeeded in reproducing an alternating periodic-chaotic sequence strikingly similar to the scenario observed in the Texas experiment [14,16-19]. In a recent work [31,32] we have reformulated this model to account for the chemical mechanism of the B.Z. reaction. It consists of 9 elementary steps as reported in Table I together with the ordinary differential equations which describe the changes with time of the seven concentrations of independent species.

Table I The reaction scheme		The seven-variable differential system
$BrO_3^- + Br^- + 2H^+ \xrightarrow{k_1} HBrO_2 + HOBr$	(R1)	$\partial_t X_1 = -a_1X_1 - a_2X_1X_2 - a_3X_1X_3 + k_8X_5X_6 + k_o(X_1^o - X_1)$
$HBrO_2 + Br^- + H^+ \xrightarrow{k_2} 2HOBr$	(R2)	$\partial_t X_2 = a_1X_1 - a_2X_1X_2 - a_4X_2 + a_5X_4^2 - 2k_5X_2^2 + a_6X_4 - k_oX_2$
$HOBr + Br^- + H^+ \xrightarrow{k_3} Br_2 + H_2O$	(R3)	$\partial_t X_3 = a_1X_1 + 2a_2X_1X_2 - a_3X_1X_3 + k_5X_2^2 - a_7X_3 - k_oX_3$
$BrO_3^- + HBrO_2 + H^+ \underset{k_{-4}}{\overset{k_4}{\rightleftharpoons}} 2BrO_2^{\cdot} + H_2O$	(R4)	$\partial_t X_4 = 2a_4X_2 - 2a_5X_4^2 - a_6X_4 - k_oX_4$
$2HBrO_2 \xrightarrow{k_5} HOBr + BrO_3^- + H^+$	(R5)	$\partial_t X_5 = a_6X_4 - k_8X_5X_6 - k_9X_5X_7 - k_oX_5$
$BrO_2^{\cdot} + Ce(III) + H^+ \xrightarrow{k_6} Ce(IV) + HBrO_2$	(R6)	$\partial_t X_6 = a_7X_3 - k_8X_5X_6 - k_oX_6$
$HOBr + MA \xrightarrow{k_7} BrMA + H_2O$	(R7)	$\partial_t X_7 = k_8X_5X_6 - k_9X_5X_7 - k_oX_7$
$BrMA + Ce(IV) \xrightarrow{k_8} Br^- + R + Ce(III) + H^+$	(R8)	$X_1=[Br^-]$, $X_2=[HBrO_2]$, $X_3=[HOBr]$, $X_4=[BrO_2^{\cdot}]$,
$R + Ce(IV) \xrightarrow{k_9} Ce(III) + P$	(R9)	$X_5=[Ce(IV)]$, $X_6=[BrMA]$, $X_7=[R]$. $X_1^o = [Br^-]_o$.
P is an inert organic product; BrMA is the bromomalonic acid.		$a_1=k_1[BrO_3^-][H^+]^2$, $a_2=k_2[H^+]$, $a_3=k_3[H^+]$, $a_4=k_4[BrO_3^-][H^+]$;
R is an oxydized derivative of malonic acid MA;		$a_5=k_{-4}[H_2O]$, $a_6=k_6[Ce(III)][H^+]$, $a_7=k_7[MA]$,

This model enters the class of systems which exhibit chaotic behavior in nearly homoclinic situations [33-35]. As sketched in Fig. 2 the dynamics in chemical model are generally located on a pleated slow manifold and one can look for the conditions where

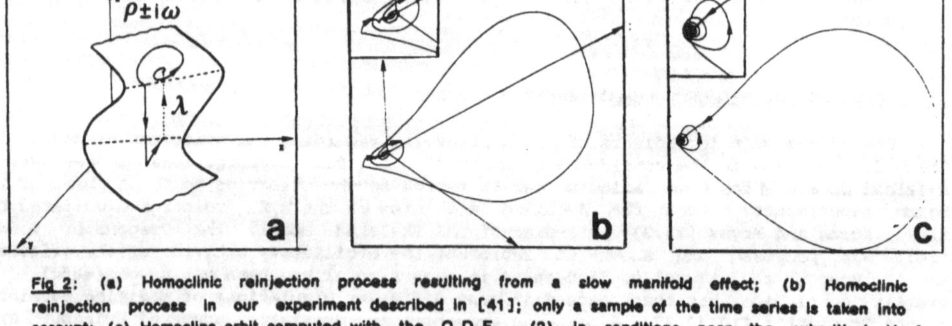

Fig 2: (a) Homoclinic reinjection process resulting from a slow manifold effect; (b) Homoclinic reinjection process in our experiment described in [41]; only a sample of the time series is taken into account; (c) Homoclinc orbit computed with the O.D.E. (2) in conditions near the subcritical Hopf bifurcation of the origin.

the flow on the lower surface of the pleat has the property of returning trajectories to a small neighborhood of the "unstable" focus lying on the upper surface. Then since the contraction along the stable direction is expected to be greater than the dilatation in the spiralling out behavior from the saddle focus ($|\lambda|/\rho>1$), one can refer to a theorem by Shil'nikov [36] which ensures the existence of chaos in a parameter neighborhood of the conditions for homoclinicity. As experienced in [37-40], these conditions favour the existence of spiralling strange attractor like behavior which is well understood through the iteration of a multi-humped (or multi-branched) 1-D map [41].

Following such a strategy, we have reproduced on increasing the flow rate k_o in our seven-variable model [32], a sequence of events in good agreement with the sequence (1) observed in the Texas experiment. In Fig. 1b we have shown our numerical chaotic state C_1^o. The comparison with the experimental attractors C_1^1 and C_1^2 (a second hump appears in the 1-D map) is even more suggestive that the nonperiodic regimes observed along this one-parameter path are very likely to be interpretable in terms of the chaotic orbits which occur on the way to homoclinicity.

3. EXPERIMENTAL EVIDENCE OF HOMOCLINIC CHAOS

In a very recent experiment [41], we have succeeded in locating the B.Z. system in a situation where both a local instability i.e. a subcritical Hopf bifurcation, and a global instability i.e. a homoclinic bifurcation, are competing. On increasing the flow rate from low values we have traced the thermodynamics branch to a critical value where a 2 % variation in the control parameter has induced a discontinuous transition to the chaotic regime depicted in Fig. 3. The homoclinic reinjection process in the neighborhood of a saddle focus (Fig. 2b) is patent either in the time series or in the reconstructed phase portrait. Apparently at random a large amplitude relaxation oscillation occurs and preceeds an episode of small amplitude oscillations the envelop of which increases nearly exponentially (Fig. 3a). This slow manifold reinjection mechanism is manifest when following the trajectory on the phase portrait (Fig. 3b). On the other hand the subcritical nature of the Hopf bifurcation has been identified in a hysteresis phenomenon. The proximity of both this local bifurcation and the homoclinic bifurcation allows us to deduce that the Shil'nikov's condition [36]: $|\lambda|/\rho>1$ is necessarily fulfilled since the Hopf bifurcation value corresponds to $\rho = 0$ [42]. Therefore this concomitant situation brings an experimental evidence of the existence of homoclinic chaos in the B.Z. reaction.

In fig. 3c, we have shown a 2-D Poincaré map obtained from intersecting the chaotic attractor in Fig. 3b with a transverse 2-D plane. The whole set of points lie to a good approximation along a smooth curve which attests that the dynamics is attracted on a two-dimensional fractal surface. Unfortunately preliminary attempts to extract a well-defined single-valued 1-D map from the 140 intersections of the trajectory with the Poincaré plane recorded during our ten hour experiment have yet failed. In order to figure out whether or not such a failure results from fluctuations or some experimental artifacts, we have numerically investigated a third-order O.D.E. which has been shown to arise naturally as the truncated normal form of a triply degenerate problem [35,38]:

$$\dddot{X} + \eta\ddot{X} + \nu\dot{X} + \mu X + k_1X^2 + k_2\dot{X}^2 + k_3X\dot{X} + k_4X\ddot{X} + k_5X^2\dot{X} = 0 \qquad (2)$$

As emphasized in [43], (2) also accounts for the interaction of a subcritical Hopf bifurcation and a homoclinic bifurcation. We have ascertained analytically the conditions for the existence of a subcritical Hopf bifurcation of the origin. We have located the homoclinic conditions using a numerical technique elaborated in [39]. A so-computed homoclinic orbit biasymptotic to the origin is shown in Fig. 2c. Thanks to a simple trial and error method we have traced the homoclinic bifurcation up to a point where it coincides with the subcritical Hopf bifurcation. Then when moving along a one-parameter path one observes a transition from rest to homoclinic chaos (Fig. 4) which displays the same characteristics as the discontinuous transition observed in the bench experiment (Fig. 3). Thanks to this high statistic numerical simulation we have

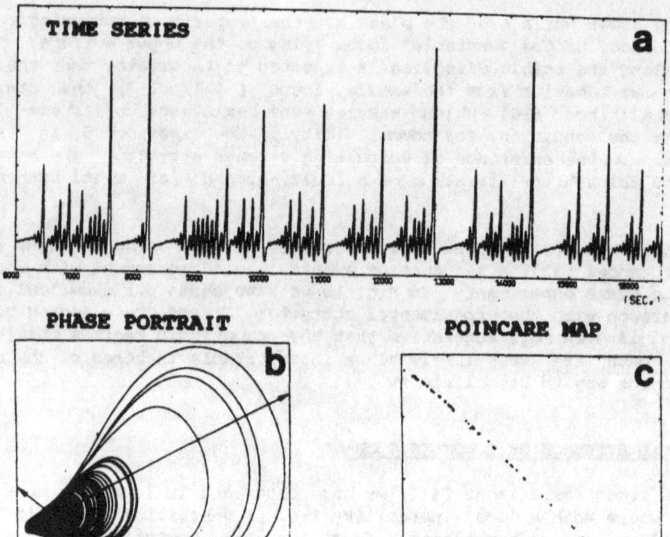

Fig 3: The homoclinic chaos evidenced in our experiment in [41]. (a) the time dependence of [CeIV]; (b) the reconstructed phase portrait (c) the corresponding Poincaré map.

been able to define the multi-branched 1-D map illustrated in Fig. 4c and which contains about 3000 points. As discussed in [41], such a 1-D map is a natural guide to the understanding of homoclinic chaos. Unfortunately it is hopeless to extract this 1-D map from the data since the duration of our experiment cannot exceed a few hours (drift in the experimental conditions) while days are required to evidence the main trends shown in Fig. 4c. Actually it is not illusory to identify experimentally a 1-D map with a few branches in moving slightly the B.Z. system away from the simultaneous occurrence of the Hopf and homoclinic bifurcations.

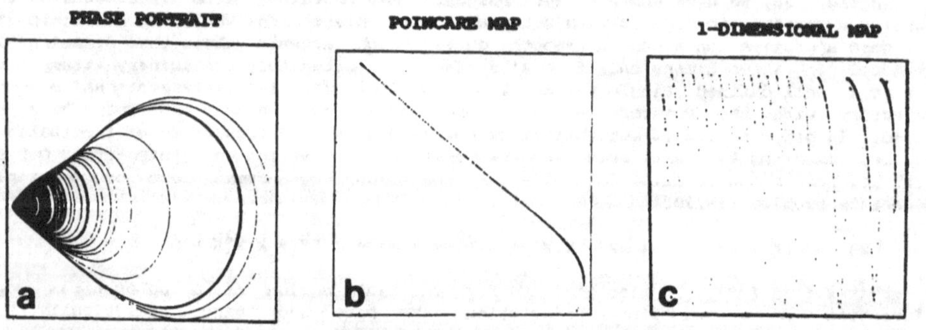

Fig 4: The homoclinic chaos computed with the O.D.E. (2) in conditions near the subcritical Hopf bifurcation of the origin [41], (a) 2-D projection of the phase portrait (b) Poincaré map (c) 1-D map.

4. QUASIPERIODICITY IN THE B.Z. REACTION: A CODIMENSION-THREE EXPERIMENT

The two basic instabilities in the B.Z. system are the Hopf bifurcation [42] (codimension-one bifurcation) at the origin of the oscillating nature of this reaction

[31] and the hysteresis bifurcation [32] (codimension-two bifurcation) which accounts for the phenomenon of bistability [44,45]. On locating the B.Z. system in the parameter space near the "bi"critical surface where both these instabilities occur simultaneously, one can expect the resulting dynamics to be described by the three-dimensional normal form [46,47]:

$$\dot{X} = \lambda + \nu X + aX^3 + b|Z|^2 + qX|Z|^2 + \text{h.o.t.} \tag{3.a}$$
$$\dot{Z} = (\mu + i\Omega)Z + cZX + \text{h.o.t.} \tag{3.b}$$
$$\dot{\bar{Z}} = (\mu - i\Omega)\bar{Z} + \bar{c}\bar{Z}X + \text{h.o.t.} \tag{3.c}$$

where "h.o.t." stands for higher order terms which can be transformed to be axisymmetric about the X-axis up to arbitrarily high order. λ, ν and μ are small parameters which characterize the distance to the polycritical surface. Among the different bifurcation diagrams which account for the dynamics in the neighborhood of the polycritical surface ($\lambda=\nu=\mu=0$), Langford [46] has stressed on the diagram shown in Fig. 5 which presents a secondary Hopf bifurcation leading to quasiperiodicity (soft mode instability).

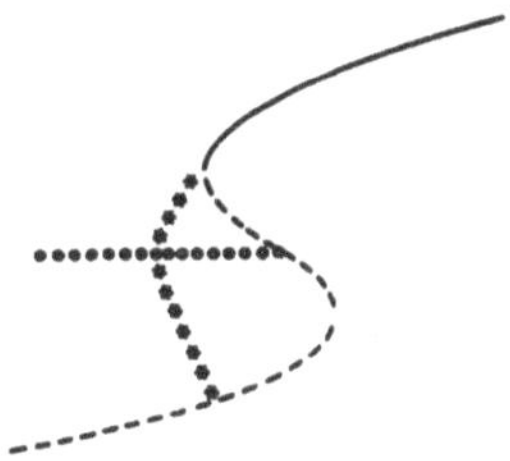

Fig 5: Bifurcation diagram for the Hopf-hysteresis interaction [46], solid (broken) lines indicate stable (unstable) steady states, solid (open) circles indicate stable (unstable) periodic orbits; asterisks denote torus branch.

In a recent experimental study of the B.Z. reaction [31,48], we have reported on the observation of a sequence of events in good agreement with this theoretical bifurcation diagram. Selected steps of this sequence are illustrated in Fig. 6. The corresponding Poincaré maps and circle maps are shown in Figs 7 and 8 respectively. Several observations converge to the conclusion that this experiment was conducted in conditions near the simultaneous onset of the Hopf and hysteresis bifurcations:

- the initial periodic state looks nearly sinusoidal with only a few harmonics of the fundamental frequency f_1 ($\sim$ 40 mHz) above the noise level (Fig. 6a)
- with a very small increase in the flow rate, the system switches from a periodic to a quasiperiodic regime (Fig. 6b). On the time series one can notice the appearance of a slow periodic modulation of the amplitude of the original oscillations. The corresponding phase portrait displays a toroidal attractor while the Poincaré map in Fig. 7b attests of the ergodic covering of the toroidal surface.
- on increasing further the flow rate, the amplitude of the modulation keeps growing (Fig. 6c) which manifests on the phase portrait as a swelling of the global size of the torus at the expense of the size of its inner part which shrinks to a thin tube. Such an evolution comes with a decrease of the low frequency f_2 and thus the appearance of much more harmonics in the power spectrum.
- when one reaches high flow rate values the time series displays larger and larger windows of nearly stationary behavior interrupted from time to time by bursts of large amplitude oscillations (Fig. 6d). With a further increase in the flow rate, the system ultimately transits to a steady state (the flux branch) which belongs to its inner part.

Beyond this rather qualitative description of the entire scenario, additional quantitative results lend strong support to a local interpretation of this experiment. The relative size f_1/f_2 of the two involved frequencies is characteristic of the distance to the bicritical surface; it has been measured $f_1/f_2 \sim 1/30$. The smallness of this ratio may also explain that no frequency lockings between f_1 and f_2 were detected during our experiment. The stiffness of the transition to quasiperiodicity has been studied quantitatively; as predicted by the theory [46,49] such a transition is found to be degenerated in the sense that it does not fit the classical square root law with respect to the control parameter (the amplitude of the modulation behave like $(\lambda-\lambda_c)^\beta$ with $\beta \sim 1/4$). Finally, as seen on the Poincaré maps, there is no evidence of

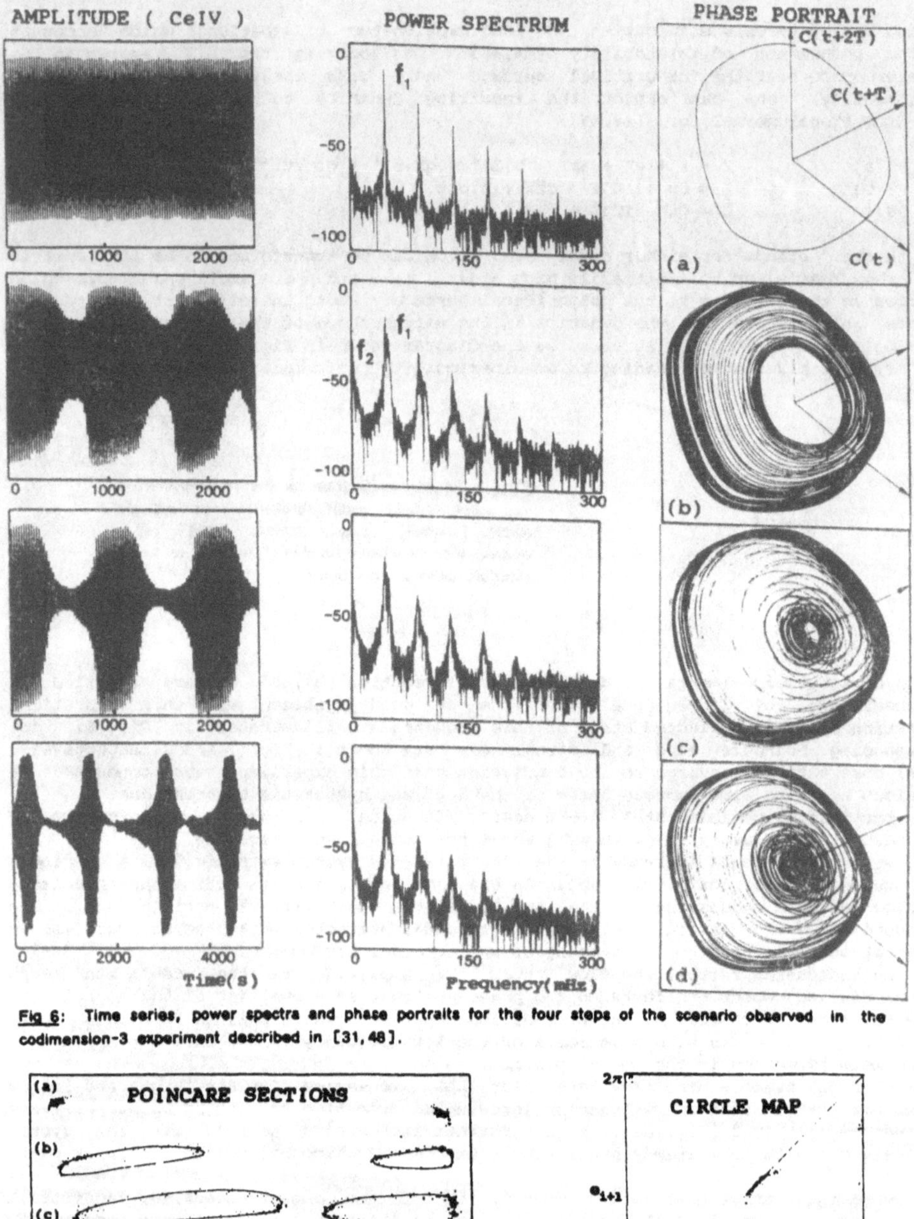

Fig 6: Time series, power spectra and phase portraits for the four steps of the scenario observed in the codimension-3 experiment described in [31,48].

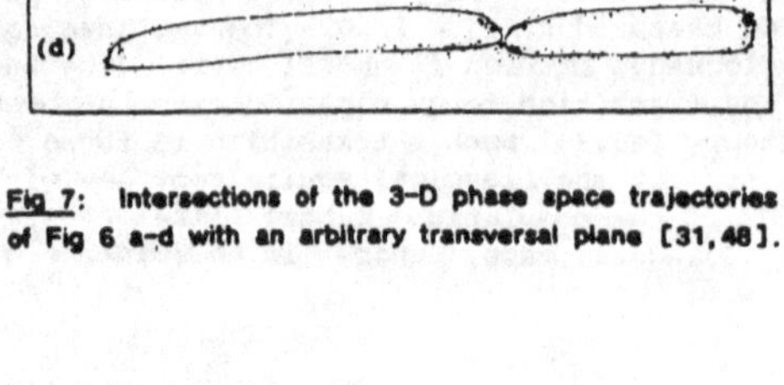

Fig 7: Intersections of the 3-D phase space trajectories of Fig 6 a-d with an arbitrary transversal plane [31,48].

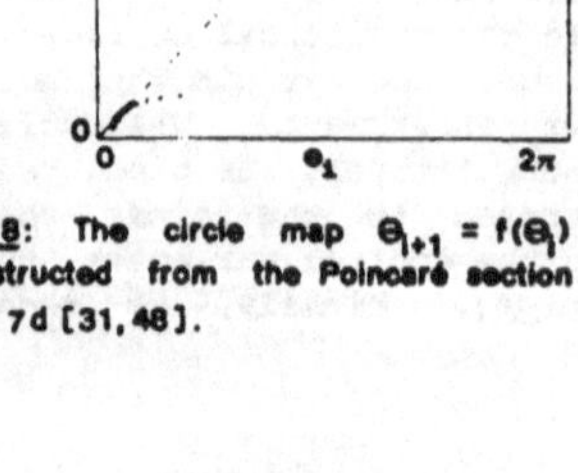

Fig 8: The circle map $\Theta_{i+1} = f(\Theta_i)$ as constructed from the Poincaré section in Fig 7d [31,48].

stretching and folding effects of the torus surface which could lead to a fractal object (Fig. 7). To the precision of the data, the circle maps defined from these Poincaré maps are invertible (Fig. 8). Thus referring to the theory of diffeomorphisms of the circle [50-52], we can assert that no chaos is to be expected in this experiment. This is a strong indication that this experiment has been conducted in nearly axisymmetric conditions as predicted by the normal form (3). Numerical investigations of this normal form in nearly bicritical conditions have corroborated such experimental findings [32].

5. FRACTAL TORUS AND THE ALTERNATING PERIODIC-CHAOTIC SEQUENCES.

Unfortunately the normal form calculation used to derive (3) cannot be carried out in general to infinite order since the series does not usually converge [49,50]. This results from the fact that axisymmetry is a highly nongeneric condition. As studied in [32,47], on introducing nonaxisymmetric perturbations in (3), one witnesses an alternation of phase lockings and quasiperiodic flows up to a critical value of the control parameter where phase locking intervals overlap. This feature comes with the loss of differentiability of the torus and its break up [53,54]. The fractalization of the torus is associated with the creation of Smale horseshoes [55] and the appearance of chaotic dynamics.

When taking the B.Z. system away from the local situation depicted in §4, the breaking of the underlying torus into a fractal objet has been clearly evidenced experimentally [31]. The wrinkles on the toroidal surface detected in this experiment together with the mixing of the reconstructed trajectories are unescapable manifestations of axisymmetry breaking. When further releasing the constraints from bicriticality, such a transition from quasiperiodicity to chaos transforms into alternating periodic-chaotic sequences which look very much like the sequences observed in the pioneering experiments on the B.Z. reactions [3-9]. As mentionned in the introduction, all the arguments about the status of the chaotic behavior in chemical systems have been focussed on such sequences. In [31,32], we have reconsidered these sequences both experimentally and numerically. Among the states recorded in the bench experiment:

$$P_o^1,\ C^1_{n1\ldots ni}\ (n_i \geq 4),\ C^1_{3,4},\ P^1_3,\ C_3^{1,2},\ P_3^2,\ P_3^3 \ldots\ C_3^{4,5},\ P_2^1,\ P_2^{1,2},$$
$$P_2^2,\ P_2^3, \ldots\ C_2^n\ (n \sim 10),\ C^n_{1,2},\ C^n_1; \qquad (4)$$

we have selected a typical chaotic state $C_3^{4,5}$ in Fig. 9. The chaotic aspect of the corresponding signal is mainly contained in the small amplitude oscillations: their number is distributed at random between 4 and 5 while their amplitudes vary from one pattern to the next one. As seen on the reconstructed phase portrait the presence of two underlying saddle foci is patent and the recorded dynamics is very reminiscent of a two-frequency dynamics. However on the contrary to the Texas sequence [14,16-19] which involves only one fundamental frequency and which displays macroscopic chaos (Fig. 1), the microscopic manifestation of chaos in these sequences makes difficult its detection from an experimental point of view [56].

A real understanding of these last sequences has only been possible through a confrontation of experiments and simulations. In [32] we have emphasized the flexibility of the seven-variable Oregonator type model detailed in Table I which also accounts for these two-frequency sequences. In particular we have numerically reproduced the ordering detected in the bench experiments. In Fig. 10 we have shown both a periodic state $P_3^{1,5,6,7}$ which clearly enlightens the underlying toroidal motion and a chaotic state $C_1^{3,4}$. While the chaotic aspect of C_1^3 is hardly perceptible on the time series only (Fig. 10b), the reconstruction of the phase portrait is much more convincing (Fig. 10c). When analyzing the circle map $\Theta_{i+4} = f(\Theta_i)$ corresponding to one of the four pieces of the Poincaré map associated to the last small amplitude oscillations, one recovers a characteristic single humped 1-D map. This numerical finding of a noninvertible circle map [49,50] allows us to conclude definitely to the existence of deterministic chaos in the sequences involving two frequencies. Although such a conclusion was far beyond the experimental resolution, it has been corroborated

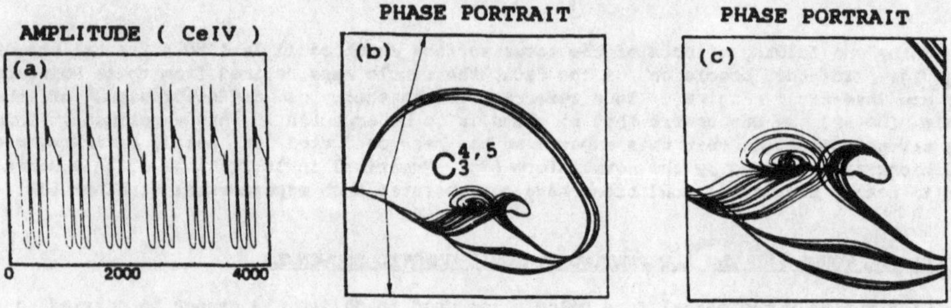

Fig 9: The chaotic state $C_3^{4,5}$ which belongs to the periodic-chaotic sequence detected experimentally in [31]. (a) time series (b) phase portrait (c) enlargement of the phase portrait.

by the observation of frequency-locking overlappings in either numerical and experimental studies of the B.Z. reaction [31,32].

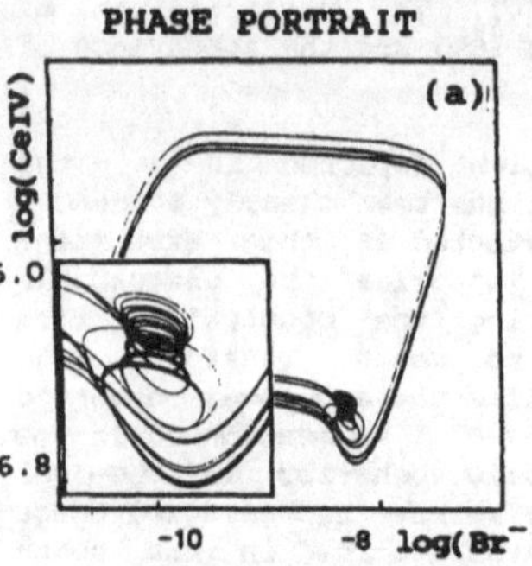

Fig 10 Selected states in the periodic-chaotic sequence computed with the seven-variable model defined in Table I [32] (a) Phase portrait of the periodic state $P_{3,4}^{1,5,6,7}$. (b), (c) and (d) are the time series, phase portrait and circle map of the chaotic state C_1^3.

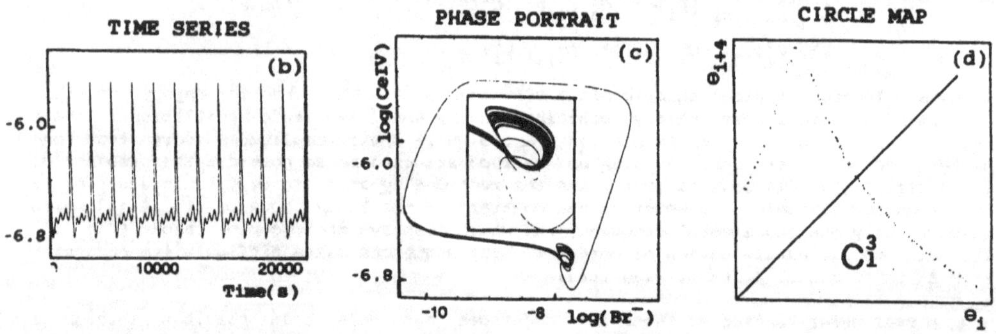

We thank J.L. Hudson for a critical reading of the manuscript. We acknowledge the financial support of a CNRS Grant (ATP "Mathématiques appliquées et méthodes numériques performantes"). One of us (A.A) is very grateful to the warm hospitality of the CRPP where this work has been done.

REFERENCES

[1] Nonlinear Phenomena in Chemical Dynamics, eds. C. Vidal, A. Pacault (Springer, Berlin 1981).
[2] Nonequilibrium Dynamics in Chemical Systems, eds. C. Vidal, A. Pacault (Springer, Berlin 1984).
[3] R.A. SCHMITZ, K.R. GRAZIANI and J.L. HUDSON, J. Chem. Phys. 67, (1977) 3040.
[4] J.L. HUDSON, M. HART and D. MARINKO, J. Chem. Phys. 71 (1979) 1601.
[5] C. VIDAL, J.C. ROUX, A. ROSSI and S. BACHELART, C. R. Acad. Sci. Paris, 289c (1979) 73.
-9-

[6] J.C. ROUX, A. ROSSI, S. BACHELART and C. VIDAL, Physica 2D (1981) 395.
[7] J.C. ROUX, A. ROSSI, S. BACHELART and C. VIDAL, Phys. Lett. 77A (1980) 391.
[8] S. BACHELART, Bordeaux Thesis (1981).
[9] C. VIDAL, S. BACHELART and A. ROSSI, J. Phys. (Paris) 43 (1982) 7.
[10] N. GANAPATHISUBRAMANIAN and R.M. NOYES, J. Chem. Phys. 76 (1982) 1770.
[11] R.M. NOYES, in "Stochastic Phenomena and Chaotic Behavior in Complex Systems", ed. P. Schuster (Springer, Berlin 1984) p. 107.
[12] P. GRAY and S.K. SCOTT, J. Phys. Chem. 89 (1985) 22.
[13] Y. POMEAU, J.C. ROUX, A. ROSSI, S. BACHELART and C. VIDAL, J. Phys. Lett. 42 (1981) L271.
[14] J.C. ROUX and H.L. SWINNEY, in ref. [2] p. 33.
[15] J.C. ROUX and A. ROSSI, in ref. [2] p. 141.
[16] J.C. ROUX, J.S. TURNER, W.D. McCORMICK and H.L. SWINNEY, in "Nonlinear Problems: Present and Future", eds. A.P. Bishop, D.K. Campbell, B. Nicolaenko (North Holland, Amsterdam 1982) p. 409.
[17] R.H. SIMOYI, A. WOLF and H.L. SWINNEY, Phys. Rev. Lett. 49 (1982) 245.
[18] J.C. ROUX, R.H. SIMOYI and H.L. SWINNEY, Physica 8D (1983) 257.
[19] A. WOLF, J.B. SWIFT, H.L. SWINNEY and J.A. VASTANO, Physica 16D (1985) 285.
[20] P. COLLET and J.P. ECKMANN, Iterated Maps of the Interval as Dynamical Systems (Birkhauser, Boston 1980).
[21] K. COFFMAN, W.D. McCORMICK, H.L. SWINNEY and J.C. ROUX, in ref. [2], p. 251.
[22] R.J. FIELD, E. KOROS and R.M. NOYES, J. Am. Chem. Soc. 94 (1972) 8649.
[23] D. EDELSON, R.J. FIELD and R.M. NOYES, Int. J. Chem. Kin. 7 (1975) 417.
[24] R.J. FIELD and R.M. NOYES, J. Chem. Phys. 60 (1974) 1877.
[25] K. SHOWALTER, R.M. NOYES and K. BAR-ELI, J. Chem. Phys. 69 (1978) 2514.
[26] I.B. SCHWARTZ, Phys. Lett. 102A (1984) 25.
[27] J. RINZEL and I.B. SCHWARTZ, J. Chem. Phys. 80 (1984) 5610.
[28] J.S. TURNER, in "Self Organization in Dissipative Structure", eds. W.D. Schieve, P. Allen, University of Texas Press 1982, p. 41.
[29] J. RINGLAND and J.S. TURNER, Phys. Lett. 105A (1984) 93.
[30] P. RICHETTI and A. ARNEODO, Phys. Lett. 109A (1985) 359.
[31] F. ARGOUL, A. ARNEODO, P. RICHETTI and J.C. ROUX, "From quasiperiodicity to chaos in the Belousov-Zhabotinskii reaction. I Experiment", preprint 1986.
[32] P. RICHETTI, J.C. ROUX, F. ARGOUL and A. ARNEODO, II Modelling and Theory", preprint 1986.
[33] O.E. ROSSLER, Phys. Lett. 57A (1976) 397 and 60A (1977) 392; Z. Naturforsch 31A (1976) 259; Bull. Math. Biol. 39 (1977) 275.
[34] J.J. TYSON, J. Chem. Phys. 66 (1977) 905; J. Math. Biol. 5 (1978) 351.
[35] J.C. ROUX, P. RICHETTI, A. ARNEODO and F. ARGOUL, J. Mec. Th. & Appl. (numéro spécial) (1984) 77.
[36] L.P. SHIL'NIKOV, Soc. Math. Dokl. 6 (1965) 163; Math. USSR Sbornik 10 (1970) 91.
[37] A. ARNEODO, P. COULLET and C. TRESSER, Comm. Math. Phys. 79 (1981) 573; J. Stat. Phys. 27 (1982) 171.
[38] A. ARNEODO, P. COULLET, E.A. SPIEGEL and C. TRESSER, Physica 14D (1985) 327.
[39] P. GASPARD and G. NICOLIS, J. Stat. Phys. 31 (1983) 499.
[40] P.A. GLENDENNING and C. SPARROW, J. Stat. Phys. 35 (1984) 645.
[41] F. ARGOUL, A. ARNEODO and P. RICHETTI, "First experimental evidence of homoclinic chaos in the Belousov-Zhabotinskii reaction", preprint 1986.
[42] J.E. MARSDEN and M. McCRAKEN, Hopf Bifurcation and its Applications, Applied Math. Sci. 19 (Springer, N.Y. 1976).
[43] P. RICHETTI, F. ARGOUL and A. ARNEODO, "Type-II intermittency in a periodically driven nonlinear oscillator", preprint 1986.
[44] K. BAR-ELI and R.M. NOYES, J. Phys. Chem. 82 (1978) 1352.
[45] J. BOISSONADE and P. DE KEPPER, J. Phys. Chem. 84 (1980) 501.
[46] W.F. LANGFORD, in "Nonlinear Dynamics and Turbulence", eds. G. Iooss, D.D. Joseph, Pitman Press (1982).
[47] W.F. LANGFORD, in "International Series of Numerical Mathematics" 70, eds. T. Kupper, H.D. Mittelmann, H. Weber (Birkhauser, Verlag, Basel 1984) p. 285.
[48] F. ARGOUL and J.C. ROUX, Phys. Lett. 108A (1985) 426.
[49] J. GUCKENHEIMER and P. HOLMES, Nonlinear Oscillations, Dynamical Systems and Bifurcations of Vector Fields, (Springer, Berlin 1984) and references therein.
[50] V.I. ARNOLD, Supplementary Chapters to the Theory of Differential Equations, (Nauka, Moscow, 1978).
[51] G. IOOSS, Bifurcation of Maps and Applications (North Holland, Amsterdam, 1979).
[52] M.R. HERMAN, Lect. Notes in Math. 597 (1977) 271; Publ. Math. I.H.E.S 49 (1979) 5.
[53] J.H. CURRY and J.A. YORKE, Lect. Notes in Math. 668 (1978) 48.
[54] D.G. ARONSON, M.A. CHORY, G.R. HALL and R.P. McGEHEE, Comm. Math. Phys. 83 (1982) 303.
[55] S. SMALE, Bull. Amer. Math. Soc. 6 (1967) 803.
[56] J. MASELKO and H.L. SWINNEY, Physica Scripta 52 (1984) 269.

THE NITROGEN TRANSFORMATION CYCLE IN WATER

P.Brunovský[1)], T.Kmeť

1. Introduction

This paper is a preliminary report on our analysis of the mathematical model of the nitrogen transformation cycle in water established in [1].It has two parts.The first part /Section 2/ is a parameter study of the equilibria of the transformation cycle and their stability.In Section 3 we give a possible explanation of the change of the uptake rates of different nitrogen compounds by the phytoplankton reported in [2].It comes from the solution of an optimal control problem based on the model of [1].

The model describes the aerobic transformation of nitrogen compounds in water, consisting of

- decomposition of complicated organic compounds into simpler ones with ammonium as the resulting product
- oxidation of ammonium and nitrite
- assimilation of nitrate and ammonium.

Specific kinds of microorganisms take part in this process:

- heterotrophic bacteria /x_1/ assimilate and decompose dissolved organic nitrogen /x_6/ coming from nitrogenius detritus /x_5/ with ammonium /x_7/ being one of the products
- chemoautotrophic bacteria are responsible for the two step oxidation of ammonium into nitrate /x_9/: by Nitrosomonas /x_2/ ammonium oxides into nitrite /x_8/ which is the energy source for Nitrobacter /x_3/
- the cycle is closed by phytoplankton /x_4/ which assimilates ammonium as well as nitrate and dissolved organic nitrogen /Fig.1./.

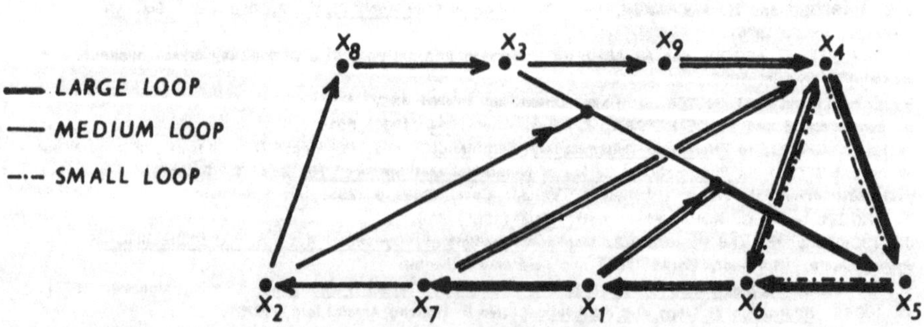

Fig. 1. Schematic diagram of the nitrogen cycle in water.

1) Institute of Applied Mathematics, Comenius University, Mlynská dolina, 842 15 Bratislava, ČSSR

In [I] the following system of differential equations is proposed as a model for the nitrogen transformation cycle:

$$\dot{x}_i = x_i(U_i(x) - E_i(x) - M_i(x)) \qquad \text{for } i=1,2,3,4$$

$$\dot{x}_5 = \sum_{i=1}^{4} M_i(x)x_i - K_5 x_5$$

$$\dot{x}_6 = K_5 x_5 - U_1(x)x_1 - P_6(x)x_4 + E_4(x)x_4$$

$$\dot{x}_7 = E_1(x)x_1 - U_2(x)x_2 - P_7(x)x_4$$

$$\dot{x}_8 = E_2(x)x_2 - U_3(x)x_3$$

$$\dot{x}_9 = E_3(x)x_3 - P_9(x)x_4$$

where

$$U_i(x) = \frac{K_i\ x_{i+5}}{1 + g_i\ x_{i+5}} \qquad \text{for } i=1,2,3$$

$$p = u_6 x_6 + u_7 x_7 + u_9 x_9$$

$$U_4(x) = \frac{K_4\ p}{1 + g_4\ p}$$

$$L_i(x) = \frac{a_{2i-1}\ U_i(x)}{1 + a_{2i}\ U_i(x)} + 1 - \frac{a_{2i-1}}{a_{2i}}$$

$$M_i(x) = g_{2i+3} + g_{2i+4}\ L_i(x)$$

$$E_i(x) = U_i(x)\ L_i(x) \qquad \text{for } i=1,2,3,4$$

$$P_i(x) = \frac{K_4\ u_i\ x_i}{1 + g_4\ p} \qquad \text{for } i=6,7,9.$$

All constants, for the meaning of which we refer to [1], are positive.

In a standard way one can check local existence, uniqueness and global forward existence of solutions starting in the positive orthant as well as forward invariance of the simplex

$$S_c = \left\{x : \sum_{i=1}^{9} x_i = c,\ x_i \geqslant 0 \text{ for all } i\right\}$$

for any c.

Also, we note that in order that x_i /i=1,2,3,4/ grows for at least some positive values of $x \in S_c$ for $c>0$, it is necessary that the constant K_i expressing the effect of temperature and light intensity, is sufficiently large, namely

$$K_i > K_i^{\circ} \qquad /1/$$

where

$$K_i^{\circ} = \frac{g_i(a_{2i}(g_{2i+3}+g_{2i+4})-a_{2i-1}g_{2i+4})}{a_{2i-1} - a_{2i}^2(g_{2i+3}+g_{2i+4})} .$$

Throught the paper we assume /1/ for i=1,2,3.

2. Steady states and their stability

Observe that the first 4 equations representing the nitrogen balance in living organisms have the following structure:

$$\dot{x}_i = x_i\, f_i(p_{i+5}) \qquad i=1,2,3,4 \quad /2/$$

with

$$f_i(p_{i+5})(p_{i+5} - b_{i+5})>0 \quad \text{for } p_{i+5}>0,$$

where $p_{i+5} = x_{i+5}$ for i=1,2,3, $p_9 = u_6x_6 + u_7x_7 + u_9x_9$,

$$b_{i+5} = \frac{a_{2i}(g_{2i+3}+g_{2i+4}) - a_{2i-1}g_{2i+4}}{K_i(a_{2i-1}-a_{2i}^2(g_{2i+3}+g_{2i+4})) + g_i(a_{2i-1}g_{2i+4}-a_{2i}(g_{2i+3}+g_{2i+4}))} .$$

The constants u_6,u_7,u_9 are the uptake proportions for x_6,x_7,x_9 by phytoplankton.

First, assume $K_4 < K_4^{\circ}$, then $\dot{x}_4 < -rx_4$ for some $r>0$, which means that temperature and light intensity is not sufficient for the phytoplankton to assimilate nitrogen in a rate that would allow it to sustain itself.

Denote

$$E_o = \{x \in S_c : x_i=0 \text{ for } i=1,\dots,5,\ x_i \leq b_i \text{ for } i=6,7,8\} .$$

Then, we have the following result:

If $x(0) \in \operatorname{int} S_c$ for $c>0$ then, $x(t)$ tends to a single point of E_o. Let now $K_4>K_4^{\circ}$. An easy argument yields that there are four possible kinds of equilibria

- interior /large loop/ equilibria with $x_i>0$ for all i
- medium loop equilibria with $x_2=x_3=0$, $x_1>0, x_4>0$

- small loop equilibria with $x_1=x_2=x_3=0$, $x_4>0$
- trivial equilibria with $x_1=x_2=x_3=x_4=x_5=0$.

The terminology is motivated by diagram in Fig.1 : if some x_i vanishes for $i\leq 4$ then the corresponding microorganisms are extinct and some link in the diagram is missing.

We now examine the existence and stability of equilibria of the particular loops. It follows from /2/ that an interior equilibrium can exist only if the equations

$$x_6 = b_6\ ,\ x_7 = b_7\ ,\ u_6x_6 + u_7x_7 + u_9x_9 = b_9\ ,$$

have a positive solution, i.e.

$$\check{b}_9 = 1/u_9(b_9 - u_6b_6 - u_7b_7) > 0\ .$$

The total amount of nitrogen c has to satisfy $c>c_1= b_6+b_7+b_8+\check{b}_9$. If this is the case the equations for x_1,x_2,x_3,x_4,x_5 have a unique solution which is positive, provided the condition

$$P_6(b_6) + P_7(b_7) > E_4(b_6,b_7,\check{b}_9)$$

is satisfied. This condition means that equilibrium assimilation of x_6, x_7 exceeds excretion. We shall keep this assumption throughout the paper.

Stability of the equilibrium point has been tested numerically by computing the characteristic exponents /i. e. eigenvalues of the linearization / at the equilibrium numerically. Except of one zero characteristic exponent /due to the invariance of S_c/ the rest of them have been found to have real parts safely negative. For $c\to c_1+$ the interior equilibrium collapses with a trivial one.

The medium loop equilibrium has to satisfy

$$x_9 = 0,\ x_6 = b_6,\ u_6x_6 + u_7x_7 = b_9.$$

This system of equations has a solution with x_7 positive if and only if

$$\check{b}_7 = 1/u_7(b_9 - u_6b_6) > 0.$$

For $c>c_2$ there is a family of medium loop equilibria parametrized by

$$0\leq x_8 < c-c_2,\quad c_2 = b_6 + \check{b}_7.$$

The stability analysis shows two zero characteristic exponents, three ones with negative real parts /one of them can be expressed explicitly/ and the characteristic exponents $f_2(\check{b}_7)$, $f_3(x_8)$. We have

$$f_3(x_8)<0 \text{ if and only if } x_8<b_8 \text{ ;}$$
$$f_2(\grave{b}_7)<0 \text{ if and only if } \grave{b}_7<b_7, \text{ i.e. } \grave{b}_9<0.$$

It follows that the equilibrium family can be stable only if the interior equilibrium does not exist.

If $\grave{b}_9<0$ then for $c>c_2$ we have a stable equilibrium family

$$E_{14}^c = \{x: x_2 = x_3 = x_9 = 0, x_i = x_i^c(x_8), i = 1,4,5, x_6 = b_6,$$
$$x_7 = \grave{b}_7, 0\leqslant x_8< \min\{c - c_2, b_8\}\}.$$

More precisely, there is an open nieghbourhood of E_{14}^c such that if $x(0)\in \text{int } S_c \cap U$ then $x(t)$ tends to a single point of E_{14}^c for $t\to\infty$.For $c\to c_2+$,E_{14}^c collapses into E_o.

The small loop equilibria have to satisfy $x_9 = 0$, $x_7 = 0$, $u_6x_6=b_9$, or, $x_6 = \grave{b}_6 = b_9/u_6$.For $c>c_3 = \grave{b}_6$ one obtains a family of equilibria parametrized by $0\leqslant x_8<c - c_3$.The stability analysis shows 2 zero and 5 characteristic exponents with negative real parts /all of which can be expressed explicitly/ and the characteristic exponents $f_2(x_8)$, $f_1(\grave{b}_6)$. They are negative if and only if $x_8<b_8$, $\grave{b}_7<0$, respectively.

If $\grave{b}_7<0$, we obtain a stable family of equilibria

$$E_4^c = \{x: x_1 = x_2 = x_3 = x_7 = x_9 = 0, x_i = x_i^c(x_8), i = 4,5, x_6 = \grave{b}_6,$$
$$0\leqslant x_8<\min\{c - c_3, b_8\}\}.$$

Again, for $c\to c_3+$, E_4^c collapses into E_o.

Summarizing, we note that

- the small loop always has an equilibrium for sufficiently large c
- the existence of equilibrium of some loop implies existence of equilibria of all smaller loops
- existence of the equilibrium of some loop implies instability of all equilibria of all smaller loops.

Thus, we see that for given values of the parameters of the model at most one of the loop equilibrium families can be stable.This suggests the question whether the only stable loop equilibrium family is globally stable.

We show that this is not always true.In particular, we show that the trivial stable equilibrium family E_o may be non-empty while the interior equilibrium exists, i.e. $\grave{b}_9>0$.

To prove this, we have to find $x_i\in[0, b_i)$, i=6,7,8 such that

$$0<u_6x_6 + u_7x_7 + u_9x_9<\grave{b}_9$$

and
$$x_6 + x_7 + x_8 + x_9>c_1 .$$

This is possible if the plane $u_6x_6 + u_7x_7 + u_9x_9 = b_9$ intersects the plane P: $x_6 + x_7 + x_8 + x_9 = c$ for $c > c_1$ in the region $0 \leq x_i < b_i$, $i = 6,7,8$. For $x \in P$ we have

$$c = x_6 + x_7 + x_8 + x_9 = (1 - u_6/u_9)x_6 + (1 - u_7/u_9)x_7 + x_8 .$$

This equation has a solution with $0 \leq x_i < b_i$ for $i = 6,7,8$ if either $u_6 > u_9$ or $u_7 > u_9$, i.e. the uptake rate of ammonium or that of dissolved organic nitrogen is higher than that of nitrate.

3. Varying uptake rate proportions

Following the model of 1 , so far we assumed that the uptake rate proportions u_6, u_7, u_9, expressing the preference of phytoplankton for different kinds of nitrogen compounds are constant. The observations reported in 2 show that this is not the case: phytoplankton prefers ammonium in case of sufficiency of the latter and assimilates other nitrogen compounds as a supplement only.

In [2] this is explained by inhibition. We propose an alternative explanation: phytoplankton uses light as its only energy source for assimilation, reduction and synthesis of nutrients optimally in order to maximize its biomass.

Mathematically, this can be formulated as the following optimal control problem:

Given $T > 0$, find $u(t) = (u_6(t), u_7(t), u_9(t))$ which maximize

$$\int_0^T x(t)\, dt$$

under the constraints

$$0 \leq u_i \leq u_{i\,max}, \quad i = 6,7,9,$$

$$aU_4(x,u) + bP_6(x,u) + cP_9(x,u) + dE_4(x,u) \leq S.$$

The last inequality expresses the fact that the amount of energy used for synthesis, reduction and excretion of nutrients by phytoplankton cannot exceed a certain value S which depends on light intensity.

We have solved this optimal control problem numerically. The results show that if the concentration of ammonium is sufficiently high, phytoplankton does not assimilate other nitrogen compounds. If, as a result of assimilation, the concentration of ammonium drops below a certain threshold, phytoplankton starts to assimilate dissolved organic nitrogen and assimilates ammonium with maximal possible rate. If the concentrations of both ammonium and organic nitrogen are low, phytoplankton starts to assimilate nitrates as well while assimilating the former nitrogen compounds at maximal possible rate. If the concentration of all three forms of nitrogen is low, all of them are assimilated by phytoplankton at maximal possible rate /Fig.2./.

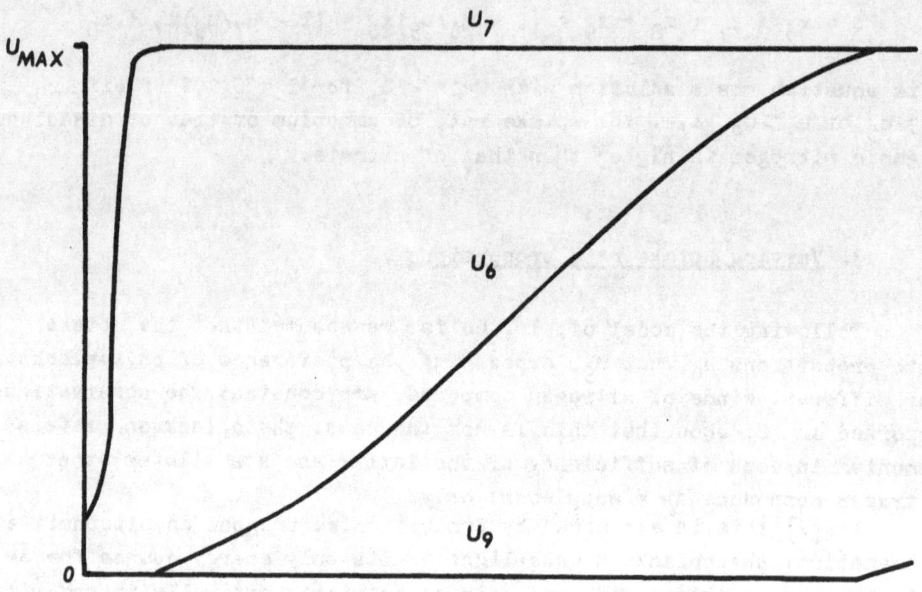

Fig. 2. Optimal uptake rates vs. time

The uptake rates obtained by solving the optimal control problem exhibit remarkable qualitative resemblance to those observed by the authors of [2]. Although we have not been able to prove this rigorously, they are largely insensitive to T and at any instant T they almost precisely coincide with those values of u_i maximizing the instantaneous biomass increase.

The time change of the uptake rates of different forms of nitrogen compounds resulting from the preference pattern of phytoplankton indicates the possibility of oscillations in the nitrogen transformation cycle. To obtain them the uptake rates u_i in the system of nitrogen transformation cycle model equations have to be allowed to vary in time, their values at any moment determined by solving the optimization problem. A numerical study of the resulting system of equations is in the course of work.

References

[1] A. Leonov: The chemical-ecological modelling of aquatic nitrogen compound transformation processes, IIASA Report WP-80,86, Laxenburg 1980

[2] W. Zevenboom, L.C. Mur: Simultaneous short-term uptake of nitrate by oscillatoria agardhii grown in nitrate- of light- limited continuous culture, Journal of General Microbiology 126/1981/, 355 - 363.

MODELLING OF POLYMER GROWTH

G. Czajkowski[1]

1. Introduction

In many problems of statistical physics macroscopical information crosses with microscopical physics of the considered systems. In this situation information-theoretical approach seems to be a proper and natural tool of their theoretical elaboration.

In polymer physics the two levels are : microscopic polymerization process and macroscopic information contained in average molecular weights.

Due to the stochastic character of polymer growth (polymerization) high polymers are polymolecular (or polydisperse), i.e. they consist, in general, of molecular species varrying in degree of polymerization, usually over a wide range. The form of the distribution of the degree of polymerization (or, equivalently, of the polymer weight) is one of the most important characteristics of polymer materials.

The polymerization of a homologous polymer chain can be considered as a birth-and-death process corresponding to the macroscopic kinetic equation

$$\dot{x} = v_+(x) - v_-(x) =: f(x) \tag{1}$$

x is the degree of polymerization (= the number of structural units or, up to a constant, the weight of an x-mer), v_+ is the rate of growth and v_- the rate of destruction of an x-mer. In the stochastic picture x stands for a stochastic variable varrying continously in an intervall (a,b) (e.g. $0, \infty$). We assume that its probability

---

[1] Address: Department of Physics II, Academy of Technology and Agriculture, Al. Prof.S. Kaliskiego 7, 85-790 Bydgoszcz, Poland.

distribution changes in time due to a Fokker-Planck equation (FPE)

$$\dot{p}(x) = (d/dx)\ (fp) + (q/2)\ d^2p/dx^2 \qquad , \qquad (2)$$

q is a certain coefficient ("drift coeff."), see, for example, [9] . We may assume that the stationary solution of the FPE stands for the equilibrium distribution of the degree of polymerization (the so-called "number distribution") which is also the distribution of polymer weight. Thus (as it is clear) the kinetics of polymerization determines stationary statistical characteristics of a polymer sample (cf., e.g., [3] , [5]) - fig.1.

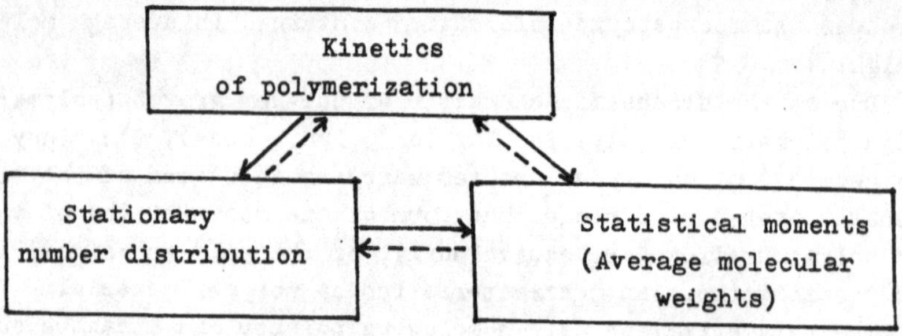

Fig.1

2. Stationary distribution of the degree of polymerization

Macroscopic counterparts of the microscopic number distribution are certain observables related to statistical moments of the distribution, the so-called average molecular weights

$$M_j = (m_j/m_{j-1}) = \int_0^\infty dx\ x^j\ p(x) \ \Big/ \int_0^\infty dx\ x^{j-1} p(x) \qquad (3)$$

which can be determined experimentally, cf., e.g., [1] . In particular M_1 (denoted also by M_n) is the so-called <u>number - average molecular weight</u>, M_2 , denoted usually by M_w , is the <u>weight-average</u> m.w. , M_3 and M_4, denoted by M_z and M_{z+1} are the so-called " z- " and "(z+1)-" average m.w. . For the terminology and notation cf., e.g., [1],[4] . Certain experimental procedures may yield other (non-entire) type of moments such as e.g. the so-called viscosity average m.w. :

$$M_v = (m_{1+a}/m_1)^{1/a} \qquad (4)$$

a being a constant lying between 0.5 and 2.0, see, for example, [1] .

Following problems appear :

1. Suppose that some average weights $M_{j_1}, M_{j_2}, \ldots, M_{j_k}$, can be measured. How to calculate the remaining $M_{j_{k+1}}, \ldots,$? For example by light scattering experiments only the quantities M_w and M_z can be measured quite exactly ; M_n, however, with much smaller accuracy ([1] , Sec.IV). In this case the problem arise how to calculate M_n by given M_w and M_z. More general : what can be said about a probability distribution p(x) if its certain moments are known.
2. Given a set of observed quantities $M_{j_1}, \ldots, M_{j_k}$; what can be concluded about the polymerization kinetics ?

In what follows we discuss the two problems.

1. Several methods were proposed to establish a probability distribution p(x) if its certain statistical moments are given. We use the information-theoretical method which says that the optimal distribution corresponding to the given statistical knowledge realizes the maximum of information S, where

$$S = - \int_X p(x) \ln p(x) \, dx \qquad (5)$$

X is the range of x (cf., e.g.,[6] , [8]). Assume that average m.w. $M_1, \ldots, M_k$ are known i.e. that $m_1, \ldots, m_k$ are known since

$$m_k = M_1 \cdot M_2 \cdots M_k \qquad (6)$$

(without loss of generality we took the integer-order moments as a basis). Maximizing S [p] under the constraints $m_1, m_2, \ldots, m_k$ yields the optimal (the "most probable" or the "representative") distribution [2] [7] in the form

$$p_{\underline{m}}(x) = p_{m_1, m_2, \ldots, m_k}(x) = Z^{-1} \exp(- \sum_{j=1}^{k} \lambda_j x^j) \qquad (7)$$

with

$$Z(\lambda_1, \ldots, \lambda_k) = \int_0^\infty dx \exp(- \sum_{j=1}^{k} \lambda_j x^j) \qquad (7a)$$

The coefficients λ_j's (generalized inverse temperatures of the order j and with respect to the quantity x [6]) are functions of the prescribed moments by the relations

$$m_j = - \partial \ln Z \ / \partial \lambda_j \ , \ j=1,\dots,k \tag{8}$$

and $m_0 = 1$. Averages of other quantities g(x) under assumption that $m_1,\dots,m_k$ are given will be obtained from

$$\langle g \rangle_{\underline{m}-\text{estim}} = \int_0^{\infty} g(x)\, p_{\underline{m}}(x)\, dx \ , \tag{9}$$

($\underline{m}$ - estimated).

2. The relation to polymerization kinetics :

Identyfying the most probable distribution $p_{\underline{m}}(x)$ with the stationary solution of the FP-equation (2)

$$p_{st}(x) = Z^{-1} \exp(- \frac{2V(x)}{q}) \tag{10}$$

with

$$V(x) = - \int_{x_0}^{x} (v_+(y) - v_-(y))dy \ ,$$

$$Z = \int_0^{\infty} dx \exp(- 2V/q)$$

[9] , we have

$$\frac{2\,V(x)}{q} =: \tilde{V}(x) = \sum_{j=1}^{k} \lambda_j x^j \tag{11}$$

and

$$v_+(x) - v_-(x) = - \sum_{j=1}^{k} j\ \lambda_j\ x^{j-1} \tag{12}$$

by an appropriate choice of units. From (12) and (8) the kinetic rates can be calculated in terms of experimentally available average molecular weights. This part of the graph from fig.1 (stationary m.w. $\rightarrow$ polymerization kinetics) will be illustrated on two examples.

3. Information-theoretical m_1, m_2 - distributions

Suppose that for a given polymer the number average m.w. M_n and the weight average m.w. M_w are known (for most materials they are tabulated cf.e.g., [11]). The statistical moments m_1, m_2 are given as

$$m_1 = M_n , \qquad m_2 = M_n M_w \quad . \tag{6a}$$

If

$$1 \leqslant M_w/M_n \leqslant 2 \tag{13}$$

the number distribution has the form [2]

$$p_{m_1,m_2}(x) = Z^{-1}(\lambda_1, \lambda_2) \exp(-\lambda_1 x - \lambda_2 x^2) . \tag{14}$$

The parameters λ_1, λ_2 , Z result from the equations (7-8) which can be put into the form

$$\lambda_1 = F_1(U)/m_1 , \qquad \lambda_2 = F_2(U)/m_1^2 ,$$
$$Z(\lambda_1, \lambda_2) = \lambda_2^{-1/2} Z^*(U) \tag{15}$$

with

$$F_1(U) = \beta(U)\, m_1^*(U) , \quad F_2(U) = [m_1^*(U)]^2 \tag{16}$$

$$m_k(U) = \int_0^\infty dz\, z^k \exp(-\beta z - z^2) \quad / Z^* \tag{17}$$

$$Z^*(U) = \int_0^\infty dz \exp(-\beta z - z^2) \quad .$$

The parameter β follows from the equation

$$U = m_2^*(U) / m_1^{*2}(U) , \tag{18}$$

β is a monotonic increasing function of U ; it is negative for U $< \pi/2$ and positive for U $> \pi/2$ [2] , cf. also Table 1.

Now we can say what type of polymerization kinetics produces the distribution (14). From (12) we have

$$v_+(x) - v_-(x) = c_1 - c_2 x , \quad c_1 = -\lambda_1 , \tag{19}$$

and $c_2 = 2 \lambda_2$.

Table 1

U		$F_1(U)$	$F_2(U)$	$\frac{M_z}{M_w}$ calc	$Z^*(U)$
1.01	-14.142	-100.000	50.000	1.01	9.190 (+21)
1.05	-6.324	-19.998	9.999	1.04	3.901 (+04)
1.10	-4.448	-9.902	4.955	1.08	2.491 (+02)
1.20	-2.932	-4.397	2.249	1.12	1.491 (+01)
1.30	-2.021	-2.265	1.256	1.16	4.546
1.40	-1.267	-1.096	0.749	1.20	2.157
1.50	-0.541	-0.365	0.455	1.24	1.238
1.60	0.238	0.125	0.274	1.29	7.787 (-01)
1.70	1.172	0.465	0.157	1.33	5.089 (-01)
1.80	2.473	0.706	0.082	1.38	3.285 (-01)
1.90	4.915	0.879	0.032	1.43	1.897 (-01)
1.95	7.939	0.945	0.014	1.47	1.223 (-01)

The notation 9.190 (+21) means 9.190×10^{21}.

By the relations (15) the kinetic coefficients c_1, c_2 can be calculated in terms of the average m.w. M_n and M_w. The obtained "representative kinetics" is very simple; a linear birth-and-death process. In order to obtain a sharp distributed molecular weight ($U \rightarrow 1$) the rate of creation c_1 should be much greater than the rate of degradation c_2; the number average M_n equals to the ratio c_1/c_2. For the rate of degradation v_2 much greater than the rate v_1 one obtains an exponential distribution ($U \rightarrow 2$, known in chemistry as Flory distribution). Here a master equation approach would be more appropriate than a FPE approach. For a certain Markovian birth and death process we would obtain $p^{st}(x) \sim \exp(-\lambda x)$ with $-\lambda = \ln v_+/v_-$ i.e. $v_- > v_+$, see also Czajkowski, Z. Phys. B43 (1981), 87.

4. Information-theoretical m_1,m_2,m_3 - distributions

For certain polymers $U = M_w/M_n > 2$, or $V = M_z/M_w > 1.5$ which are limiting values for m_1,m_2- distributions. For example, for polystyren samples $1.3 \leq M_z/M_w \leq 4$ ([1] , Sec.IV). In such cases one can use an information-theoretical distribution generated by three moments:

$$p_{m_1,m_2,m_3}(x) = Z^{-1} \exp(- \lambda_1 x - \lambda_2 x^2 - \lambda_3 x^3) . \quad (20)$$

The examples given in Table 2 show that values $M_w/M_n > 2$, $M_z/M_w > 1.5$ can be achieved by moments of the distribution (20). It may also happen that for given M_n, M_w, M_z, $M_w/M_n < 2$ but (M_z/M_w) calculated differs from that obtained experimentally. It follows, for example, from Table 1 that $(M_z/M_w)_{calc} < M_w/M_n$ for m_1,m_2 - distributions. Examples are known where $M_z/M_w > M_w/M_n$ whereas $M_w/M_n < 2$ (e.g. $M_w/M_n = 1.07$, $M_z/M_w = 1.4$ [10]). In such cases a three-moments distribution is obviously more adequate than a two-moments distribution [2] . The related "representative kinetics" will have the form

$$v_+(x) - v_-(x) = c_1 - c_2 x - c_3 x^2 \quad (21)$$

with

$$c_1 = - \lambda_1 , \quad c_2 = 2 \lambda_2 , \quad c_3 = 3 \lambda_3 . \quad (22)$$

As for the case of p_{m_1,m_2} the coefficients λ_1, λ_2, λ_3 can be calculated from the equations (7-8) with $m_3 = M_z M_w M_n$, $m_2 = M_w M_n$ and $m_1 = M_n$. However, an analytical formula for $Z(\lambda_1, \lambda_2, \lambda_3)$ is not known and numerical calculations are more complicated due to the greater number of unknown quantities which may be reduced by using the relation

$$3 \lambda_3 m_3 + 2 \lambda_2 m_2 + \lambda_1 m_1 = 1 \quad (23)$$

([2]). New variables can be introduced

$$\beta_1 = \lambda_1 (\lambda_3)^{-1/3} , \quad \beta_2 = \lambda_2 (\lambda_3)^{-2/3} \quad (24)$$

The regions (in a space spanned by average m.w) $M_w/M_n > 2$, $M_z/M_w > 1.5$ will be achieved by $\beta_1 > 0$ and $\beta_2 < 0$.

Table 2. The range of U and V for the distribution [1]

$$p(x) = Z^{-1}\exp(-\beta_1 x - \beta_2 x^2 - x^3) \quad (25)$$

β_1	U	V
1. $\beta_2 = -5$		
5	1.0375	1.0244
7	1.6834	1.1102
9	3.5701	1.9170
11	2.4755	1.9695
13	2.2058	1.6937
15	2.1247	1.6068
2. $\beta_2 = -7.5$		
10	1.0058	1.0056
14	1.0751	1.0115
15	3.0209	1.0604
15.5	8.5383	1.2659
16.5	6.8113	4.7434
17.5	2.4710	3.7967
3. $\beta_2 = -9$		
20	1.0257	1.0580
21	2.5358	6.5907
21.7	1.7728 (+1)	4.2859 (+02)
21.9	2.2129 (+01)	8.8768 (+02)
22	2.1968 (+01)	1.0030 (+03)
22.5	8.6769	5.6012 (+02)
23.5	2.2338	2.0223 (+01)

[1] $U = m_2/m_1^2 = M_w/M_n$; $V = m_3 m_1/m_2^2 = M_z/M_w$;
m_j are moments calculated with the distr. (25).

For $\beta_1 \gg 1$ U and V become great and changing very rapidly. It becomes apparent for great values of $|\beta_2|$ (Table 2). Great values of U and V correspond to the transition region between a delta-like distribution (U,V $\rightarrow$ 1) peaked over

$$x = (- \beta_2 + (\beta_2^2 - 3 \beta_1)^{\frac{1}{2}}) /3 \qquad (26)$$

and a smooth exponential distribution $\sim \exp(- \beta_1 x)$.

5. Final remarks

In many cases details of kinetic equations of the polymerization process are not known. In such cases we may consider a simple "representative" kinetics yielding the same stationary values of average molecular weights. The parameters of the "representative" kinetics can be calculated in terms of average m.w. which are measurable macroscopic quantities. The final aim is to control the kinetics by parameters which are at our disposal.

Acknowledgement

The author is grateful to Dr. A. Kociszewski from Higher Pedagogical School Bydgoszcz for performing the numerical calculations and for several discussions.

The work was supported by the Polish Ministry of Science and Higher Education, Project M.R.I.7.

References

[1] Cvetkov, W.N., W.E. Eskin, and S. Ya. Frenkel : Structure of Macromolecules in Solutions, Izd. Nauka, Moscow 1964, in Russian.

[2] Czajkowski, G.: Acta Phys.Polon. A44 (1973), 747.

[3] Emanuel, N.M., and D.G. Knorre : Course of Chemical Kinetics, Vyssaja Skola, Moscow 1974, in Russian.

[4] Flory, J.P.: Principles of Polymer Chemistry, Cornell Univ. Press, Ithaca, New York, 1953.

[5] Frenkel, S.Ya: Introduction to the Statistical Theory of Polymerization, Izd. Nauka, Moscow 1965, in Russian.

[6] Ingarden, R.S.: Fortschr.Phys., 13 (1965), 755.

[7] Jaworski, W., and R.S. Ingarden : Acta Phys.Polon. A59 (1981), 635.

[8] Jaynes, E.T.: Phys.Rev. 106 (1957), 620 , and 108 (1957), 171.

[9] Haken, H.: Synergetics, An Introduction. Springer Verlag, Berlin-Heidelberg - New York 1977.

[10] Koningsveld, R.: Adv.Polymer Sci., 7 (1970), 1.
[11] Nesterov, A.E.: Handbook of Physical Chemistry of Polymers. Nauk. Dumka, Kiev 1984, in Russian.

A MATHEMATICAL MODEL OF GROUP DEFENCE IN PREDATOR-PREY SYSTEMS

H.I. FREEDMAN[1)4)] and G.S.K. WOLKOWICZ[2)3)4)]

1. INTRODUCTION

In this paper we consider a model of a predator-prey system which exhibits group defense in the predator functional response. We model this interaction of populations by a system of autonomous ordinary differential equations, described in the next section, of the standard type as discussed in Freedman [7], Ch. 4.

Group defence is a term used to describe the phenomenon whereby predation is decreased or even prevented altogether due to the increased ability of the prey to better defend or disguise themselves when their numbers are large enough. An example of this phenomenon is described by Tener [4]. Lone musk ox can be successfully attacked by wolves. Small herds of musk ox (2-6 animals) are attacked but with rare success. No successful attacks have been observed in larger herds. A second example described by Holmes and Bethel [10] involves certain insect populations. Apparently, large swarms of the insects make individual identification difficult for their predators. A case of group defence in avian brood parasitism is described in [12].

Related examples of nonmonotone consumption occur at the microbial level where there is considerable evidence (Andrews [1], Aris and Humphrey [2], Boon and Laudelout [3], Bush and Cook [4], Yang and Humphrey [12]), which indicates that when faced with an overabundance of nutrient the effectiveness of the consumer can begin to decline. That is, certain nutrients may be growth-limiting at low concentrations as well as growth-inhibiting at high concentrations. This is often seen when microorganisms are used for waste decomposition or for water purification. This phenomenon is called "inhibition" of the consumer at high densities of the resource.

With the phenomenon of group defence in mind it is therefore of interest to study mathematical models of predator-prey interactions in which the predator response function is not necessarily a monotone increasing function of prey density, but rather is only monotone increasing until some critical density and then becomes monotone decreasing.

In this paper we also reexamine the paradox of enrichment in the light of group defence (Rosenzweig [13]). We show that enrichment can, indeed, lead to extinction in a closed ecosystem as originally postulated by Rosenzweig [13]. This is different from the case where mutual interference among predators occurs, once more precluding extinction (see [8]).

---

1) Department of Mathematics, University of Alberta, Edmonton, Alberta, Canada, T6G 2G1
2) Division of Applied Mathematics Brown University, Providence, R.I., U.S.A. 02912
3) Present address: Department of Mathematical Sciences, McMaster University, Hamilton, Ontario, Canada L8S 4K1
4) Research partially supported by the Natural Science and Engineering Research Council of Canada

2. THE MODEL

We propose the following system of autonomous ordinary differential equations of generalized Gause-type as a model of predator-prey interaction with group defence exhibited by the prey.

$$\begin{aligned} \dot{x} &= xg(x,K) - yp(x) \\ \dot{y} &= y(-s+q(x)) \\ x(0) &\geq 0\,,\ y(0) \geq 0\,,\quad \cdot = \frac{d}{dt} \end{aligned} \tag{2.1}$$

where $x(t)$ and $y(t)$ denote the density of prey and predator, populations, respectively. We assume that the functions g, p and q are continuously differentiable and that s and K are positive constants.

Here, $g(x,K)$ represents the specific growth rate of the prey in the absence of predation and is assumed to satisfy

$$\begin{aligned} &g(0,K) > 0\,,\ g(K,K) = 0\,,\ g_x(K,K) < 0 \\ &g_x(x,K) \leq 0 \ \text{ and } \ g_K(x,K) > 0 \ \text{ for any } \ x > 0\,. \end{aligned} \tag{2.2}$$

The function $p(x)$ denotes the predator response function. We assume $p(x)$ satisfies

$$p(0) = 0\,,\quad p(x) > 0 \ \text{ for } \ x > 0 \tag{2.3}$$

and that there exists $M > 0$ such that

$$p'(x) > 0 \ \text{ for } \ 0 \leq x < M\,,$$

and (2.4)

$$p'(x) < 0 \ \text{ for } \ x > M\,.$$

The rate of conversion of prey to predator is described by $q(x)$. In Gause's model $q(x) = cp(x)$ for some positive constant c. We assume $q(x)$ has properties similar to $p(x)$. In particular

$$q(0) = 0\,,\quad q(x) > 0 \ \text{ for } \ x > 0\,,\quad q(M) > s$$

$$q'(x) > 0 \ \text{ for } \ 0 \leq x < M \tag{2.5}$$

and

$$q'(x) < 0 \ \text{ for } \ x > M\,.$$

Specific examples of $g(x,K)$, $p(x)$ and $q(x)$ can be found in Boon and Laudelout [3], Holling [9], May [11], Rosenzweig [13], and Yang and Humphrey [15].

The existence of $M > 0$ is precisely the assumption which models group defense. It is also only reasonable to assume that the same M holds for both p and q since the conversion of prey to predator should increase and decrease as the consumption of prey increases and decreases. We assume that $q(M) > s$ since otherwise the predator cannot survive on the prey at any density. Therefore, there exists $\lambda < M$ such that $q(\lambda) = s$ and there may exist $\mu > M$ such that $q(\mu) = s$. We assume that $\lambda < K$ or again the predator cannot servive on the prey.

From the above assumptions one can conclude that there is always an interior equilibrium, E_λ, of the form $(\lambda, \lambda q(\lambda,K)/p(\lambda))$ where $q(\lambda) = s$, $p'(\lambda) > 0$ and $q'(\lambda) > 0$. Furthermore, it follows from a standard linear analysis (see Freedman [6], [7]) that this equilibrium is stable or unstable according to whether the slope of the prey isocline at E_λ is negative or positive.

Our first observation follows as in Freedman [6], [7].

<u>Theorem 1.</u> <u>If</u> $\lim_{x\to\infty} q(x) \geq s$ (i.e. <u>no</u> $\mu > M$ <u>exists such that</u> $q(\mu) = s$) <u>or if</u> $q(\mu) = s$ <u>where</u> $\lambda < K < \mu$, <u>then solutions of (2.1) with positive initial conditions either approach</u> E_λ <u>or approach a positive limit cycle surrounding</u> E_λ <u>or are themselves periodic solutions surrounding</u> E_λ.

If on the other hand $\mu < K$, the dynamics differ substantially. In particular, for a significant set of initial conditions the predator can be driven to extinction.

The variational matrix about any equilibrium (x^*, y^*) is:

$$V(x^*,y^*) = \begin{bmatrix} x^* g_x(x^*,K) + g(x^*,K) - y^* p'(x^*) & -p(x^*) \\ y^* q'(x^*) & -s+q(x^*) \end{bmatrix}. \tag{2.6}$$

At $(x^*,y^*) = (K,0)$ this becomes

$$V(K,0) = \begin{bmatrix} Kg_x(K,K) & -p(K) \\ 0 & -s+q(K) \end{bmatrix} \tag{2.7}$$

and so both eigenvalues are negative. Therefore, the equilibrium $E_K = (K,0)$ is locally asymptotically stable which implies that there is a set of initial conditions of positive measure in the interior of the positive cone for which $\lim_{t\to\infty} y(t) = 0$.

Also, in the case $\mu < K$, there exists a second interior equilibrium $E_\mu = (\mu, \mu g(\mu,K)/p(\mu)) = (\mu, y_\mu)$.

$$V(E_\mu) = V(\mu,y_\mu) = \begin{matrix} \mu g_x(\mu,K) + g(\mu,K) - y_\mu p'(\mu) & -p(\mu) \\ y_\mu q'(\mu) & 0 \end{matrix}. \tag{2.8}$$

The constant term in the characteristic equation is therefore equal to $y_\mu q'(\mu)p(\mu) < 0$ since $q'(\mu) < 0$. Therefore E_μ is always a saddle point. Observing the direction the solutions must cross the predator and prey isoclines one observes that there cannot be a periodic orbit surrounding E_μ.

Since E_μ is a saddle point, it has a one dimensional stable manifold, $W^s(E_\mu)$ and hence there are two orbits that approach E_μ asymtotically. Consider the orbit that approaches E_μ from the left and call it Γ. There are at most three possibilities. Case 1: In negative time Γ can leave the strip $0 \leq x \leq \mu$. In this case all solutions with positive initial conditions that start outside the region bounded by $W^s(E_\mu)$ approach E_K asymptotically. Case 2: Γ can be a homoclinic orbit, that is tend to E_μ in negative time. In this case the only solutions with positive initial conditions that do not approach E_K originate inside the region bounded by Γ or on $W^s(E_\mu)$. Case 3: Γ^* can remain in the strip $0 \leq x < \mu$ for all backward time. In this case E_λ is either unstable with no periodic orbit surrounding it or it must be surrounded by one or more periodic orbits and the outermost one must be unstable from the outside. Γ then either approaches E_λ (in the case of no periodic orbit surrounding E_λ) or the outermost periodic orbit if time is followed in reverse. The only solutions that do not approach E_K are those that lie on $W^s(E_\mu)$ or those originating inside the outermost periodic orbit (if one exists).

Since enrichment results in an increase of the carrying capacity K, it is now easy to see that our model predicts that sufficient enrichment could result in extinction of the predator if the prey exhibit group defence. Consider model (2.1)* with $K < \mu$. Then E_K is unstable and at least deterministically both populations persist uniformly (see Butler, Freedman and Waltman [5]). In fact, all solutions approach E_λ or some periodic orbit surrounding E_λ. At $K = \mu$, there is a bifurcation of the critical point E_K. As K increases beyond μ, E_K becomes asymptotically stable and E_μ appears in the interior of the nonnegative cone. Thus there is a set of initial conditions of positive measure for which extinction of the predator results and so Rosensweig's warning applies. It is interesting to note that these outcomes are independent of the local stability of E_λ and hence of which side of the prey isocline E_λ lies on. Rather they depend upon whether or not $\mu < K$ and hence on whether or not E_μ lies in the positive cone.

References

1. Andrews, J.F., A mathematical model for the continuous culture of microorganisms utilizing inhibitory substrates, Biotechnol. Bioeng. 10 (1968) 707-723.

2. Aris, R. and A.E. Humphrey, Dynamics of a chemostat in which two organisms compete for a common substrate, Biotechnol. Bioeng. 10 (1977) 1375-1386.

3. Boon, B. and H. Landelout, Kinetics of nitrite oxydation by Nitrobacter Winogradski, Biochem. J. 85 (1962) 440-447.

4. Bush, A.W. and A.E. Cook, The effect of time delay and growth rate inhibition in the bacterial treatment of wastewater, J. Theor. Biol. 63 (1976) 385-395.

5. Butler, G.J, H.I. Freedman and P.E. Waltman, Uniformly persistent systems, Proc. Amer. Math. Soc., in press.

6. Freedman, H.I., Graphical stability, enrichment, and pest control by a natural enemy, Math. Biosci. 31 (1976) 207-225.

7. Freedman, H.I., Determininistic Mathematical Models in Population Ecology, New York: Marcel Dekker (1980).

8. Freedman, H.I. and G.S.K. Wolkowicz, Predator-prey systems with group defense: The paradox of enrichment revisited (preprint).

9. Holling, C.S., The functional response of predators to prey density and its role in mimicry and population regulation, Mem. Entomolog. Soc. Canada 45 (1965) 3-60.

10. Holmes, J.C. and W.M. Bethel, Modification of intermediate host behaviour by parasites, Zoolog. J. Linnean Soc., Suppl. 1 51, (1972) 123-149.

11. May, R.M., Limit cycles in predator-prey communities, Science 177 (1972) 900-902.

12. May, R.M. and S.K. Robinson, Population dynamics of avian brood parasitism, Am. Nat. 126, 475-494.

13. Rosenzweig, M.L., Paradox of enrichment: Destabilization of exploitation ecosystems in ecological time, Science 171 (1971) 385-387.

14. Tener, J.S., Muskoxen. Ottawa: Queen's Printer (1965).

15. Yang, R.D. and A.E. Humphrey, Dynamics and steady state studies of phenol biodegradation in pure and mixed cultures, Biotechnol. Bioeng. 17 (1975) 1211-1235.

VECTOR-TRANSMITTED DISEASES IN STRUCTURED POPULATIONS

K. P. Hadeler, Tübingen [1)]

In [1] and in several subsequent publications [3], [4] a model for a structured infected population has been investigated. The main features of this model are as follows: The hosts are classified according to chronological age and to the individual parasite load. Whereas age is a continuous parameter, the individual parasite load is a nonnegative entire number, i.e., the number of parasites within the particular host. Whereas the host population is governed by a Lotka-Sharpe equation, the parasites within the host follow a birth and death process with immigration. The immigration rate of parasites into hosts, i.e. the acquisition rate of parasites by hosts, is related to the average parasite load of the total population. This model leads to an infinite system of partial differential equations which, in the special case of linear birth and death process, can be condensed into a single partial differential equation.

This structured model gives more insight into the development of the disease within the population than a simple mass-action law (although simpler approximate equations may be derived a posteriori from the structured models [4], [5]). Furthermore counting individual parasites is very appropriate for modelling helminthic diseases such as onchocerciasis, where the individual host carries few parasites. On the other hand, also for diseases with a large number of infective agents a classification of hosts with respect to individual parasite load or individual infectivity is meaningful. One could use certain scaling processes to obtain such models from discrete models in the same way as one approximates birth and death processes by diffusion processes. The results depend on the type of approximation. Therefore I prefer to write down a general diffusion equation and to obtain the side conditions from conservation laws.

It turns out that it is impossible to obtain a reasonable model for a continuous density of infectives alone. For reasons of dimensionality (hosts of any parasite load produce non-infected offspring) there must be a point mass at the noninfected state. In this way the nucleus of the model becomes a diffusion equation associated with an ordinary differential equation.

In a first step I describe the essential variables and parameters of the model. Let u(t,x) be the density of infected hosts of time t, i.e.

---

1) K. P. Hadeler, Lehrstuhl für Biomathematik, Universität Tübingen
Auf der Morgenstelle 10, D-7400 Tübingen

$$\int_{x_1}^{x_2} u(t,x)\,dx \tag{1}$$

is the number of infected hosts at time t with a level of infection between x_1 and x_2. Let U(t) be the number of noninfected hosts at time t. Then

$$P(t) = U(t) + \int_0^\infty u(t,x)\,dx \tag{2}$$

is the total population size of hosts at time t. In biological terms, individuals counted as infected with x = 0 are distinct from those counted noninfected, i.e. a requirement that U(t) should equal u(t,0) is not justified.

Let the functions k(x) and l(x) describe the stochastic birth and death within the parasite population. Thus k(x) is the diffusion rate and l(x) is the drift of the diffusion process which governs the "diffusion" of hosts between levels of infection. I shall require that k(x) is positive. The function l(x) can assume either sign. In the typical situation of a pure death process l(x) is positive. The function $\tilde{\varphi}(t,x)$ describes the rate at which new parasites are acquired by hosts with infection level x at time t. Of course $\tilde{\varphi}(t,x) \geq 0$. We shall assume that $\tilde{\varphi}$ has a product decomposition $\tilde{\varphi}(t,x) = \varphi(t)\phi(x)$, where the function $\phi(x)$ describes the reaction of the hosts, and the function $\varphi(t)$ depends on the average parasite load of the total population.

The parameter γ is the rate at which infected hosts at low infection levels, i.e. at x = 0, loose all their parasites and become noninfected. The function $\tilde{\varphi}_o(t)$ is the rate at which noninfected hosts acquire parasites. For reasons of symmetry I write $\tilde{\varphi}_o(t) = \varphi(t)\phi_o$ with the same $\varphi(t)$ as before. Here $\phi_o > 0$ is a constant.

All parameters just introduced describe mainly the changes of the parasite population within the hosts. The dynamics of the host population has yet to be specified. The parameter $\mu(x) > 0$ is the mortality of infected hosts at a level of infection x, and $\mu_o > 0$ is the mortality of noninfected hosts. Again, there is no necessity to assume $\mu(0) = \mu_o$.

Let $\nu \geq 0$ give the rate of increase of the number of noninfected hosts due to birth from the whole population. We shall either assume that ν is a constant (as in [1]) or that $\nu = \nu(t)$ is obtained from an equation of Lotka type. In the latter case, which seems more interesting, the parameter $b(x) \geq 0$ is the fertility of hosts at infection level x, and $b_o \geq 0$ is the fertility of noninfected hosts.

Finally, the acquisition function φ is not prescribed a priori, but given implicitely as

$$\varphi = \beta f(w) \tag{3}$$

where f is a given function describing the dynamics of vectors (e.g. insects transmitting the disease), β is a positive contact rate between hosts and vectors and w is the average parasite load or a more general functional of the population.

As described above, the population is governed by two equations of the form

$$u_t = (k(x)u_x)_x + (l(x)u)_x - \varphi(t)(\phi(x)u)_x - \mu(x)u, \tag{4}$$

$$U_t = \nu - \varphi(t)\phi_o U - \mu_o U + \gamma u(t,o). \tag{5}$$

These two equations are coupled by the parameter γ and by a boundary condition of the form

$$a_o u(t,0) + a_1 u_x(t,0) + a_2 U(t) = 0. \tag{6}$$

The coefficients are obtained from a conservation law: If $\mu(x)$, μ_o, and ν vanish, then the conservation equations

$$\frac{d}{dt}\left(U(t) + \int_o^\infty u(t,x)dx\right) = 0 \tag{7}$$

should hold. If $u(t,x)$ converges sufficiently fast for $x \longrightarrow \infty$ then the equation (7) implies

$$\begin{aligned} &- \varphi(t)\phi_o U - \mu_o U + \gamma_o u(t,0) \\ &+ \int_o^\infty [k(x)u_x + l(x)u - \varphi(t)\phi(x)u]_x dx \\ &= -\varphi(t)\phi_o U + \gamma_o u(t,0) \\ &\quad -k(0)u_x - l(0)u + \varphi(t)\phi(0)u \equiv 0. \end{aligned} \tag{8}$$

Thus the desired boundary condition at $x = 0$ is

$$k(0)u_x + l(0)u - \varphi(t)\phi(0)u - \gamma u = -\varphi(t)\phi_o U. \tag{9}$$

Now the full dynamical model consists of the following equations. First a diffusion equation for the density of infected hosts

$$u_t = (k(x)u_x)_x + (l(x)u)_x - \varphi(t)(\phi(x)u)_x - \mu(x)u, \tag{10a}$$

then an ordinary differential equation for the number of noninfected hosts

$$U_t = \nu - \varphi(t)\phi_o U - \mu_o U + \gamma u(t,0) \tag{11a}$$

where

$$\nu = b_o U + \int_o^\infty b(x)u(t,x)dx. \tag{11b}$$

It is a matter of taste whether ν is interpreted as a third state variable or equation (11a) is written in the form

$$U_t = (b_o - \mu_o)U - \varphi(t)\phi_o U + \gamma u(t,0) + \int_o^\infty b(x)u(t,x)dx. \tag{10b}$$

The diffusion equation requires two boundary conditions. The first has just been obtained,

$$k(0)u_x = -\varphi(t)[\phi_o U - \phi(0)u] + [\gamma - l(0)]u \text{ at } x = 0. \tag{10c}$$

The most natural boundary condition at infinity has already been used implicitly, namely that

$$u(t,\cdot) \in L_1(0,\infty). \tag{10d}$$

However, the mathematical treatment of the initial value problem is simpler if another boundary condition is imposed, namely that u is in an appropriate Hilbert space.

The initial condition reads

$$u(0,x) = u_o(x), \qquad U(0) = U_o. \tag{10e}$$

The transmission law is given by

$$\varphi = \beta f(w) \tag{10f}$$

where

$$w = \frac{\int_o^\infty xu(t,x)dx}{U(t) + \int_o^\infty u(t,x)dx} \tag{10g}$$

is the average parasite load of the total host population. In a more general model of Gurtin-McCamy type the parameters b_o, b, μ_o, μ could depend also on the total population size P. In a simplified model the rate ν can be fixed and equation (10b) be replaced by equation (11a).

A final comment on the boundary at $x = 0$ is appropriate. In the extremal case $k(x) \equiv 0$ and $l(x) > 0$ where no boundary condition can be imposed, the conservation law requires $l(0) = \gamma$. But if $k(x) > 0$ then γ may be chosen independently of $l(0)$ without contradicting the conservation law.

It can be shown that the system preserves positivity, i.e. for nonnegative initial data the solution remains nonnegative. Furthermore, for constant φ, the initial value problem (10a,b,c) (10e), has a solution in some Hilbert space $L_2(0,\infty)$ with weight function. The problem of existence of solutions to the full problem is still open.

In this problem we do not expect proper stationary solutions, but rather persistent solutions, i.e. solutions for which $\varphi = \bar{\varphi}$ is constant, but

$$u(t,x) = \bar{u}(x)e^{\lambda t}, \quad U(t) = \bar{U}e^{\lambda t} \tag{12}$$

where $(\bar{u},\bar{U})$ is a quasistationary distribution, and λ is an exponent of growth or decay. Such a persistent solution, for given φ, is obtained from the eigenvalue problem

$$-(k(x)u'(x)' - (l(x)u(x))' + \varphi(\phi(x)u(x))' + \mu(x)u(x) = -\lambda u(x)$$

$$-(b_o - \mu_o)U - \gamma\, u(0) - \int_0^\infty b(x)u(x)dx + \varphi\phi_o U = -\lambda U$$

$$k(0)u'(0) = [\gamma \; - l(0)]u(0) - \varphi[\phi_o U - \phi(0)u(0)],$$

+ boundary condition at infinity.

The property of preserving positivity leads to the existence of a leading eigenvalue λ with a nonnegative eigenfunction $(\bar{u},\bar{U})$ which is the persistent distribution. It is unique up to a positive factor. First, for $\varphi = 0$, there is always the solution describing the uninfected population. Then, for each $\varphi > 0$, there is the solution $(\bar{U},\bar{u})$. If this solution is inserted into the equation (10f) (10e), a well-defined $\beta > 0$ is obtained,

$$\beta = \varphi/f(w).$$

Next we introduce age structure into the model. Then the variables are $U(t,a)$ and $u(t,a,x)$.
First we find out what kind of boundary conditions should be imposed. Introduce a coordinate system in $\mathbb{R}^3$ with t,a,x - axes. Of interest are values $t \geq 0$ of the chronological time, values $a \geq 0$ of the individual age. Along the x-axis, the compartment of non-infected and the level of infectedness $x \geq 0$ is represented,

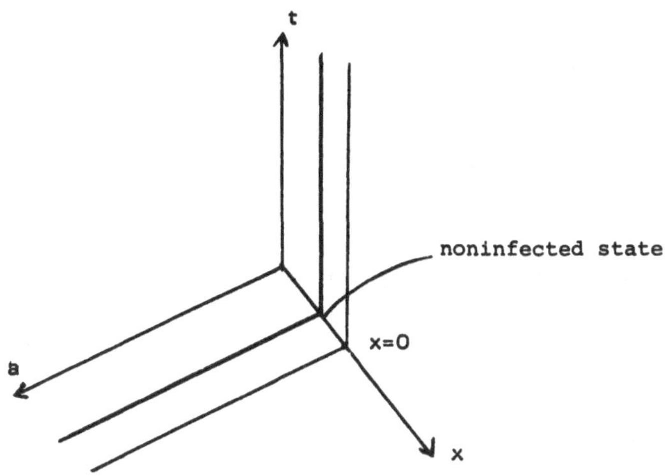

The differential equations read

$$u_t + u_a = (ku_x)_x + (lu)_x - \varphi(\phi u)_x - \mu u, \tag{13}$$

$$U_t + U_a = \gamma u(t,a,0) - \varphi\phi_o(a)U - \mu_o(a)U. \tag{14}$$

At t = 0 there is an initial condition

$$u(0,a,x) = u_o(a,x), \quad a \geq 0 \ ; \quad U(0,a) = U_o(a). \tag{15}$$

At a = 0 the side conditions read

$$u(t,0,x) = 0, \tag{16}$$

$$U(t,0) = \int_0^\infty b_o(a)U(t,a)da + \int_0^\infty\int_0^\infty b(a,x)u(t,a,x)dxda . \tag{17}$$

At x = 0 there is a transition condition

$$\begin{aligned} k(a,0)u_x(t,a,0) &= \varphi\phi(a,0)u(t,a,0) \\ &-l(a,0)u(t,a,0) + \gamma u(t,a,0) - \varphi\phi_o(a)U(t,a). \end{aligned} \tag{18}$$

For given $\varphi \geq 0$ one expects that there is a persistent solution, i.e. a solution of the form

$$\begin{aligned} u(t,a,x) &= e^{\lambda t}\,\bar{u}(a,x), \\ U(t,x) &= e^{\lambda t}\,\bar{U}(a). \end{aligned} \tag{19}$$

The functions $\bar{u},\bar{U}$ satisfy the equations

$$\begin{aligned}
\bar{u}_a &= (k\bar{u}_x)_x + (l\bar{u})_x - \varphi(\phi\bar{u})_x - \mu\bar{u} - \lambda\bar{u} \\
\bar{U}_a &= \gamma\bar{u}(a,0) - \varphi\phi_o\bar{U} - \mu_o\bar{U} - \lambda\bar{U} \\
\bar{U}(a) &= \int_0^\infty b_o(a)\bar{U}(a)da + \int_0^\infty\int_0^\infty b(a,x)\bar{u}(a,x)dx, \\
k_o\bar{u}_x(a,0) &= \varphi\phi(0)\bar{u}(a,0) - l(0)\bar{u}(a,0) + \gamma\bar{u}(a,0) - \varphi\phi_o\bar{U}(a),
\end{aligned} \tag{20}$$

and a boundary condition at $x = \infty$, e.g. $u(t,a,\cdot) \in L_1(0,\infty)$.

In general the coefficient functions depend on the variables x and a.

Next we consider separable problems. It is assumed that the functions k,l,ϕ depend only on x, that the death rate splits additively,

$$\mu(a,x) = \mu_1(a) + \mu_2(x), \quad \mu_o(a) = \mu_1(a) + \mu_2^o, \tag{21}$$

and the birth rate splits multiplicatively according to

$$b(a,x) = b_1(a)b_2(x), \quad b_o(a) = b_1(a)b_2^o. \tag{22}$$

For the solutions we make the separation ansatz

$$u(a,x) = v(a)w(x), \quad U(a) = v(a)W.$$

Then the persistent solution can be found in the following way. First solve the eigenvalue problem depending only on the x-variable,

$$(kw')' + (lw)' - \varphi(\phi w)' - \mu_2 w = \lambda_o w ,$$

$$\gamma w(0) - \varphi\phi W - \mu_2^o W = \lambda_o W$$

with a boundary condition

$$k(0)w'(0) = \varphi\phi(0)w(0) - l(0)w(0) + \gamma w(0) - \varphi\phi_o W$$

and a boundary condition at infinity.

Then the eigenvalue λ_o and the eigenvector w,W is determined (up to a factor). The function v is given by

$$v(a) = e^{-M(a)+ca}, \quad M(a) = \int_0^a \mu_1(s)ds,$$

where the separation constant c is obtained from the characteristic equation

$$\int_0^\infty b_1(a) b_2^o e^{-M(a)+ca} da \cdot W$$

$$+ \int_0^\infty b_2(x) w(x) dx \cdot \int_0^\infty b_1(a) e^{-M(a)+ca} da = W.$$

Finally the growth exponent λ is obtained from the equation $\lambda + c = \lambda_o$.

Hence in the separable case, the problem reduces to a generalized Sturm-Liouville equation and a characteristic equation.

REFERENCES

[1] Hadeler, K. P., Dietz, K.: Nonlinear hyperbolic partial differential equations for the dynamics of parasite populations. Comp. Math. Appl. 9, (Nr. 3) 415-430 (1983)

[2] Hadeler, K. P., Dietz, K.: An integral equation for helminthic infections: Global existence of solutions. In: Conference Proceedings "Recent Trends in Mathematics" Reinhardsbrunn (1982), Teubner-Texte zur Mathematik 50, Teubner Verlag Leipzig (1982/83) p. 153-163

[3] Hadeler, K. P.: Integral equations for infections with discrete parasites: Hosts with Lotka birth law. In: Levin, S., Hallam, T. (eds.) Mathematical Ecology, Trieste. Lecture Notes in Biomathematics. 54, Springer-Verlag, 1984, p. 356-365

[4] Hadeler, K.P., Dietz, K.: Population dynamics of killing parasites which reproduce in the host. J. Math. Biol. 21 (1984), 45-65

[5] May, R. M., Anderson, R. M.: Population dynamics of infectious diseases, II. Nature 280, 455-461 (1979)

NONLINEAR BEHAVIOUR IN MATHEMATICAL MODELS OF ANAEROBIC DIGESTERS

Kubíčková Z. [1], Kubíček M. [2] and Marek M. [3]

1. Introduction

Anaerobic digestion is a microbial process occuring naturally and involving decomposition of organic material (e.g., crop residues, animal manure) in the absence of oxygen. Mixed culture of microorganisms convert organic solids to methane, carbon dioxide and non-biodegradable solids in two principal steps. In the first one fats, carbohydrates and proteins are converted to simple organic acids such as acetic and propionic acid. The "acid forming" process proceeds relatively fast and is not overly sensitive to changes in the environment. In the second step the organic acids are transformed to methane and carbon dioxide or hydrogen. The methanogenic microorganisms have low growth rates and are sensitive to pH, oxygen and other environmental changes. In an anaerobic digester oxygen is excluded and temperature, residence time, feed rate are controlled. Limits of application of anaerobic digesters follow from slow bacterial growth rates, relatively low organic removal efficiencies and the sensitivity of the microorganisms to temperature, pH, toxic materials and ways of operation.

Recent efforts are directed to the increase of the substrate removal rates by an increase of the microorganism concentrations, improving the operation procedures and generally, to better understanding of the complexity of both stationary and dynamic behaviour of digesters which should enable better control. The studies of mathematical models of increasing complexity are useful both in analyzing the details of nonlinear behaviour and also form necessary steps in future introduction of better control strategies.

2. The models

We shall present and analyse three models of increasing complexity, used for description of anaerobic digestion in the flow-through, well mixed anaerobic digester. In the simplest one, only the effects of substrate inhibited growth function will be studied. The second model, proposed by Graef and Andrews [2] the interactions between the liquid and gas phases in the digester are considered. Finally, the third model was introduced by Mosey [6,7] to study the interaction of four groups of methanogenic bacteria.

---

[1] Computer Centre, [2] Dept. of Mathematics, [3] Dept. of Chemical Engineering, Prague Institute of Chemical Technology, Suchbátarova 5, 166 28 Prague 6, Czechoslovakia.

Fig. 1. Flow-through well mixed digester - schematically.
a) One well mixed phase. b) Two well mixed phases.

2.1 A simplified model - Model (S)

We consider one-substrate system with a single group of microorganisms with growth rate function μ including the substrate inhibition in the form

$$\mu = \frac{\hat{\mu}}{1 + K_s/s + s/K_i} , \tag{1}$$

where s is substrate concentration, K_s saturation constant, K_i inhibition constant and $\hat{\mu}$ maximum growth rate. Substrate digestion occurs in the well mixed flow-through digester, cf. Fig. 1a, and we assume that the substrate degradation is linear function of the microorganism concentration x. The balance equations are then in the form

$$\frac{dx}{dt} = \frac{F}{V}(x_0 - x) + \mu x ,$$
$$\frac{ds}{dt} = \frac{F}{V}(s_0 - s) - \frac{\mu x}{Y} . \tag{2}$$

Here F denotes flow-rate, V volume of the digester, x_0 and s_0 are concentrations of microorganisms and substrate in the feed and Y is yield coefficient.

2.2 Andrews - Graef model - Model (A)

In this model [2] two well mixed phases in the digester are considered; the liquid phase - volume V and the gas phase - volume V_G, cf. Fig. 1b. The model is formed by a system of 6 differential equations - mass balances for :

microorganisms concentration

$$\frac{dx}{dt} = \frac{F}{V}(x_0 - x) + \mu x - K_T T , \tag{3}$$

substrate concentration

$$\frac{ds}{dt} = \frac{F}{V}(s_0 - s) - \frac{\mu x}{Y_s} , \tag{4}$$

carbon dioxide concentration in the liquid phase

$$\frac{dc}{dt} = \frac{F}{V}(c_o - c) + \frac{\mu x}{Y_c} + \frac{F}{V}(H_o - H) + \frac{ds}{dt} - \frac{dz}{dt} + K_L a(Kp - c) \quad , \quad (5)$$

carbon dioxide concentration in the gas phase

$$\frac{dp}{dt} = -\bar{p}S_p V K_L a(Kp - c)/V_G - \frac{pS_p V}{V_G}\left(\frac{\mu x}{Y_M} - K_L a(Kp - c)\right) \quad , \quad (6)$$

dissociated substrate concentration

$$\frac{dz}{dt} = \frac{F}{V}(z_o - z) \quad , \quad (7)$$

concentration of toxic substances in the liquid phase

$$\frac{dT}{dt} = \frac{F}{V}(T_o - T) \quad (8)$$

In Eqs (3) - (8) we have used the notation

$$H_o = z_o - s_o \quad , \quad H = z - s \quad , \quad \mu = \hat{\mu} / (1 + K_s/c_{HS} + c_{HS}/K_i) \quad ,$$

$$c_{HS} = \frac{K_1 c s}{K_a(z - s)} \quad . \quad (9)$$

In the above relations we denote (characteristic values of the parameters used in simulations are given in parentheses) :

- F flow rate (1 l/day)
- K Henry's law constant ($0.432\ 10^{-2}$ mol/l Pa)
- K_a acid dissociation constant ($0.316\ 10^{-4}$ mol/l)
- K_i inhibition constant ($1\ 10^{-5}$ mol/l)
- $K_L a$ overall gas transfer film coefficient (30 day^{-1})
- K_1 ionization constant of bicarbonate ($1\ 10^{-6}$ mol/l)
- K_s saturation constant ($0.303\ 10^{-4}$ mol/l)
- K_T toxicity influence rate (2 day^{-1})
- S_p conversion factor (25 l/mol)
- Y_s organism yield (0.02)
- Y_c carbon dioxide yield (0.021)
- Y_M methane yield (0.021)
- V digester liquid volume (10 l)
- x concentration of microorganisms (x_o = 0.001 mol/l)
- s concentration of limiting substrate (s_o = 0.09 mol/l)
- c concentration of dissolved carbon dioxide (c_o = 0.009 mol/l)
- z net cation concentration (z_o = 0.1 mol/l)
- T concentration of toxic components (T_o = 0 mol/l)
- p partial pressure of carbon dioxide ($\bar{p}$ = total pressure = $0.946\ 10^5$ Pa)
- H concentration dissolved bicarbonate
- c_{HS} concentration of unionized volatile acids (mol/l)
- μ specific growth rate (day^{-1})
- $\hat{\mu}$ maximum specific growth rate (0.4 day^{-1})
- V_G volume of gas space in digester (2 l)

2.3 Mosey´s model - Model (M)

Recently Mosey [6,7] has introduced detailed model of the microbial methanation process for a model substrate - glucose. The principle reactions taking place in the liquid phase, where four groups of methanogenic bacteria are present, are schematically shown in Fig. 2. The corresponding stoichiometric equations are given in [6]

Let us assume that the digester is an ideally mixed flow-through reactor, cf. Fig. 2, with a substrate conversion taking place in the liquid phase. When x corresponds to the concentration of the component in the liquid phase, then the balance equations are in the form

$$V \frac{dx}{dt} = F(x_0 - x) + R - T \quad , \tag{10}$$

where R is the term describing the sum of all sources by the reactions considered and T describes the transport of the component from the liquid to the gaseous phase. The list of concentration variables is given in Table 1. The variables 1 - 13 correspond to liquid phase and hence we have 13 mass balances of the type (10).

The mass balance of the component in gas phase is in the form

$$V_G \frac{dy}{dt} = - F_G y + T \tag{11}$$

where y denotes the concentration of the component in the gas phase, cf. variables 14 - 16 in the Table 1, F_G is gas production. We assume that the Henry´s law can be used to determine the relation between gas and liquid phase concentrations $y = H x$, where H denotes modified Henry´s constant. We assume that the gas and liquid phases are essentially in equilibrium.

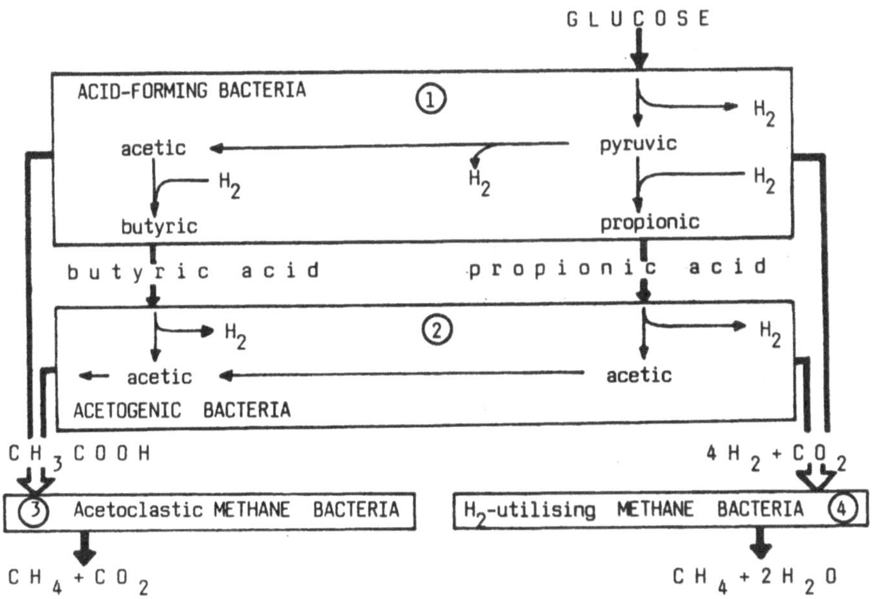

Fig. 2. Anaerobic conversion of glucose by methanogenic bacteria.

Table 1 : State variables

1.	glucose concentration	(mmol/l)
2.	acetic acid concentration	(mmol/l)
3.	propionic acid concentration	(mmol/l)
4.	butyric acid concentration	(mmol/l)
5.	glucose fermenting bacteria concentration	(mg/l)
6.	acetate fermenting bacteria concentration	(mg/l)
7.	propionate fermenting bacteria concentration	(mg/l)
8.	butyrate fermenting bacteria concentration	(mg/l)
9.	hydrogen fermenting bacteria concentration	(mg/l)
10.	hydrogen concentration in the liquid phase	(mmol/l)
11.	carbon dioxide concentration in the liquid phase	(mmol/l)
12.	methane concentration in the liquid phase	(mmol/l)
13.	ammonium hydroxide concentration in the liquid phase	(mmol/l)
14.	hydrogen concentration in the gas phase	(mmol/l)
15.	carbon dioxide concentration in the gas phase	(mmol/l)
16.	methane concentration in the gas phase	(mmol/l)

The entire model is thus formed by a set of 16 ordinary differential equations with relatively complicated right hand sides. The reaction rate equations (included in R in Eq. (10)) for individual reaction steps shown in Fig. 2 are in the form of Monod equations, including the inhibition by hydrogen. The growth rates of microorganisms are described by means of the yield coefficients similarly as in two foregoing models. Control of metabolism of bacteria are formulated on the basis of known redox-potential of the redox-couple $NAD/NADH^+$

3. Nonlinear behaviour of the models

We shall illustrate that above models of anaerobic digestion predict existence of characteristic nonlinear phenomena which can be responsible for inherent control problems in the operation of digesters.

3.1 Multiplicity in the steady state behaviour

The above models exhibit in a certain range of parameters multiplicity of steady states (i.e., the states, where the left hand sides of Eqs (2) - (8), (10), (11) are set to zero). We shall illustrate the dependence of steady state concentrations on the residence time in the digester, V/F, in the form of solution diagram [4] (dependence of the chosen norm of the solution on a chosen parameter).

The solution diagram of steady state solutions of the Model (S) is shown in Fig. 3. It is evident from the Figure, that in the range of parameters $V/F \in (2.32 ; 3.76)$ three steady states exist for $K_i = 5$, two of them are stable. The choice of initial conditions will determine which one of the stable steady state will be realized in the digester. The operating state with high conversion is a desirable one, the state with low conversion characterizes nearly completely inactive digester. Fluctuations in the flow-rate can cause jump from the one steady state to the other one. The regions of multiplicity can be relatively wide, as it can be inferred from the bifurcation diagram [4] in Fig. 4, where the curves of the limit points in dependence on the inhibition

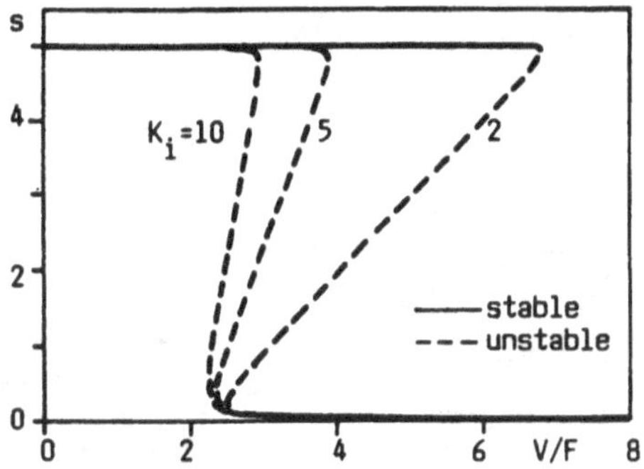

Fig. 3. Solution diagram, Model (S).
$\hat{\mu} = 0.5$, $x_o = 0.005$, $s_o = 5$,
$Y = 0.5$, $K_s = 0.03$.

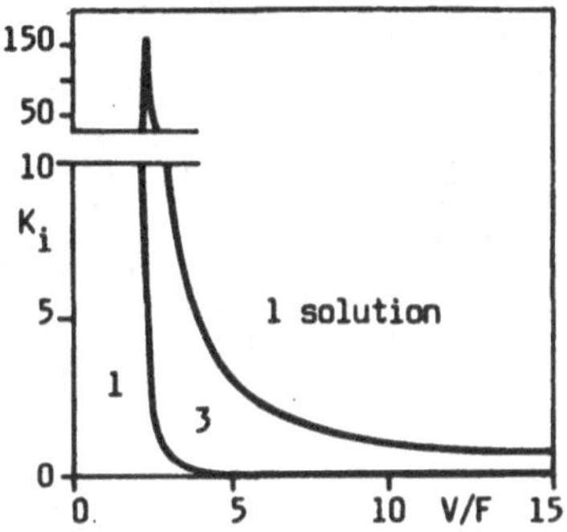

Fig. 4. Bifurcation diagram, Model (S).
$\hat{\mu} = 0.5$, $x_o = 0.005$, $s_o = 5$,
$Y = 0.5$, $K_s = 0.03$.

constant K_i are given. Three steady states exist for values of V/F and K_i located inside the depicted region.

Multiplicity of steady state solutions of the Model (A) are shown in the solution diagram presented in Fig. 5. The situation is similar to Model (S), fluctuations in F can again cause transition from the state of high conversion into the state with low effectivity of the substrate degradation.

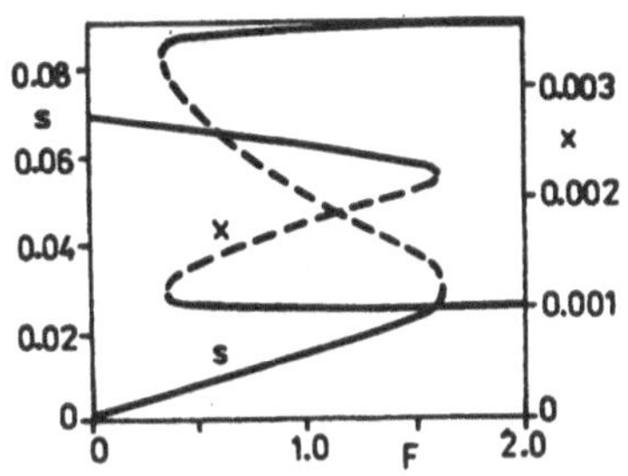

Fig. 5. Solution diagram, Model (A). Values of the parameters given in the text, ——— stable, ----- unstable.

3.2 Effects of periodic feeding

The digester is often operated in a fed-batch manner or with the periodic feeding. It results from the approximately periodical production of waste materials, for example, reflecting circadial cycles or shift work. Hence it is necessary to study response of the system on periodic changes of flow-rate or substrate concentration.

The response of the Model (S) to rectangular periodic changes of the flow-rate F is depicted in Fig. 6a. The outlet substrate concentrations are also periodic with the period equal to the inlet (forcing) period. The response to the periodic variation of the inlet substrate concentration s_o with a 1 day period for Model (A) is given in Fig. 6b. The resulting outlet concentration is again periodic with the period equal to the forcing period. The results of the dynamical simulation of the Model (M) with periodically varied inlet substrate concentration are presented in Fig. 6c. The response outlet substrate concentration is again periodic with the period equal to the inlet period, i.e., 40 days. In the simulation we used the following values of the operation parameters : mean residence time V/F = 10 days , volume of the liquid phase $V = 1\ m^3$ and volume of the gas phase $V_G = 2.24\ m^3$. The simulation,performed in double accuracy,

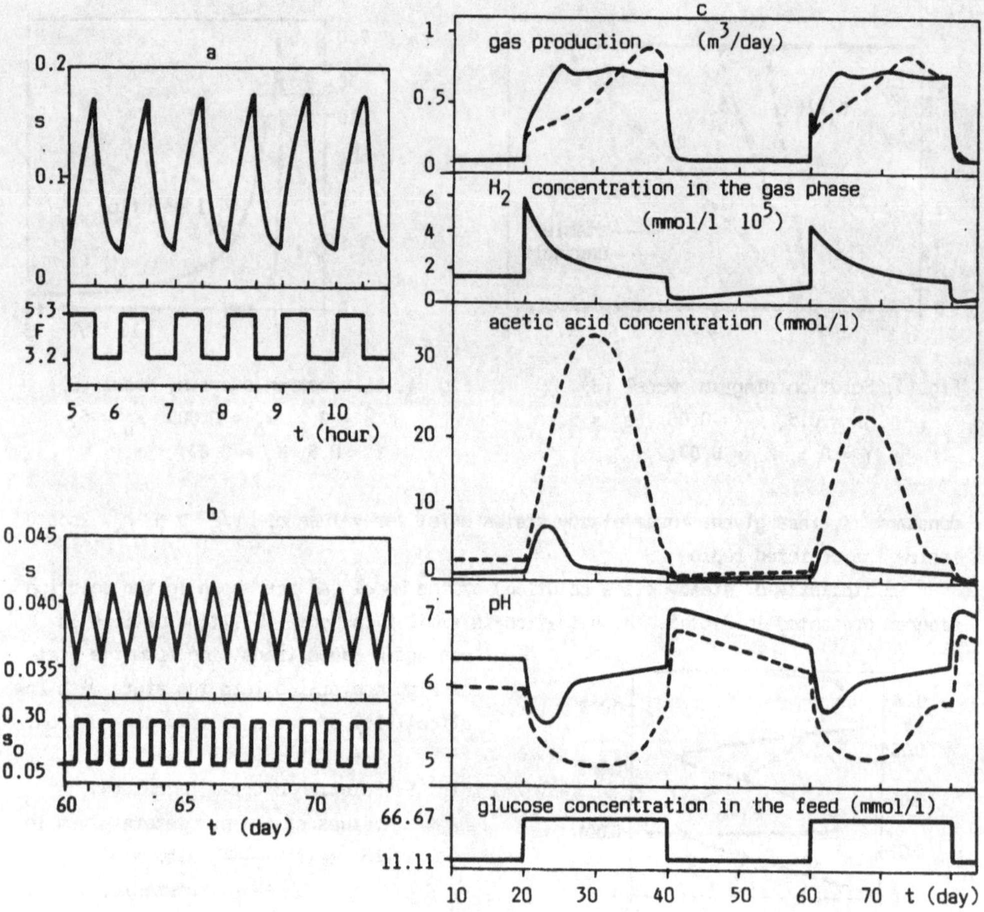

Fig. 6. Effect of periodic feeding.

a) Model (S). $\hat{\mu} = 0.5$, $s_o = 5$, $x_o = 0.005$, $V = 12$, $Y = 0.5$, $K_i = 5$, $K_s = 0.03$. Period of F is 1 day .

b) Model (A). Parameter values see text. Inlet concentration of the substrate, s_o , varies with the 1day period. $K_i = 3\ 10^{-5}$, $K_a = 1\ 10^{-5}$

c) Model (M). Period of feeding is 40 days. [1)]

is a time consuming task, because of both the sharp variations of variables resulting from the periodic feeding of substrate and the stiffness of the system. We can infer from the Fig. 6c that stationary operation is reached already after second period. We can observe sharp changes both in the values of acetic acid concentration, methane and hydrogen production and pH .

The variation of the parameters can be used to reach qualitative agreement with the responses of the actual digester in question.

---

1) In Fig. 6c the full lines correspond to the basic case, the dashed lines to the situation where the value of the rate constant was halved.

3.3 Effects of the time-delay in the growth function

As we have mentioned it in the Introduction, the growth rates of the methanogenic anaerobes are relatively slow. It means that the reactions (adaptation) of the culture to changes in the substrate concentration can be also slow and thus can substantially affect the behaviour of the digester. To test these effects, we have used two simple forms of description of the time delay t_L in the growth rate function of the Model (S)

$$\mu = \frac{\hat{\mu}}{1 + K_s/s(t - t_L) + s(t - t_L)/K_i} \qquad (12a)$$

and

$$\mu = \frac{\hat{\mu}}{(1 + \frac{K_s}{s(t)})\ (1 + \frac{s(t - t_L)}{K_i})} \qquad (12b)$$

The time delay in the inhibition effects of the substrate only is considered in relation (12b). Similarly we can introduce the effects of the time delay in the Model (A) in the relation for μ , Eq. (9), taking $c_{HS}(t - t_L)$ instead of $c_{HS}(t)$.

The time delay t_L can destabilize the original stationary state and this destabilization then usually leads to oscillatory behaviour of the digester. Let us denote t_L^* the critical time delay such that for $t_L < t_L^*$ the stationary solution is stable and for $t_L > t_L^*$ is unstable and periodic behaviour results. The value of the critical time delay t_L^* can be for the stationary state in question evaluated iteratively in such a way that the pair of roots of the characteristic quasipolynomial lies at the imaginary axis. Special decomposition technique can be used for this purpose [9] . When the dependence of t_L^* on some additional parameter is computed, we obtain a bifurcation diagram [4] . Such a bifurcation diagram is for the Model (S) depicted in Fig. 7 for steady state solution with high conversion of substrate [9] .

The periodic solution of the Model (S) resulting when $t_L > t_L^*$ is shown in Fig. 8. Undamped oscillations exist in the outlet values of substrate concentration. Qualitatively similar results can be obtained also for the Model (A) [3] , [8] .

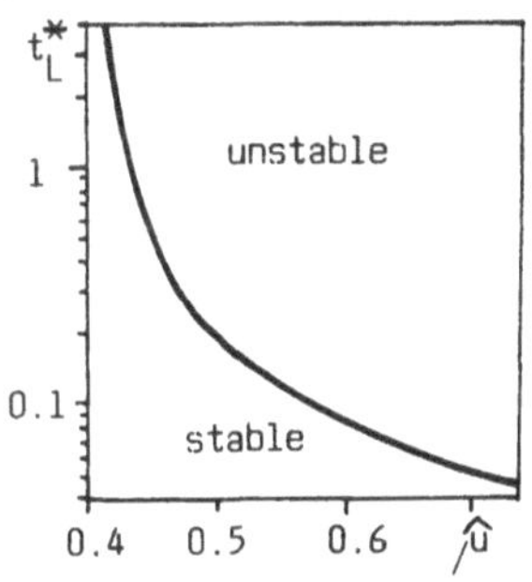

Fig. 7. Bifurcation diagram for the Model (S), Eq.(12a) - dependence of critical time delay on $\hat{\mu}$. $V/F = 3$, $Y = 0.5$, $K_s = 0.03$, $K_i = 2$, $s_o = 5$, $x_o = 0.005$.

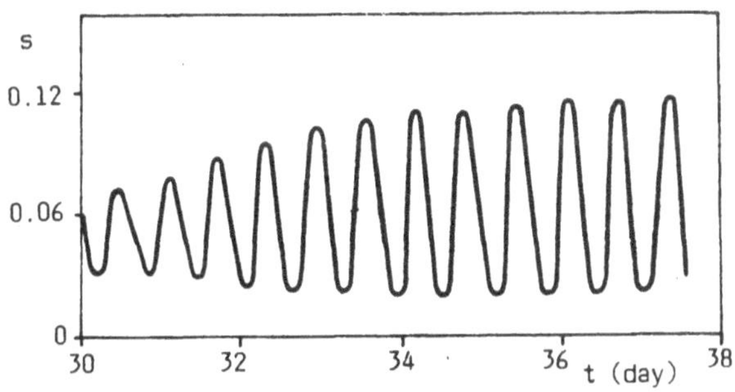

Fig. 8. Periodic solution of the Model (S) for $t_L > t_L^*$, Eq.(12a). $V/F = 3$, $Y = 0.5$, $K_s = 0.03$, $K_i = 2$, $s_o = 5$, $x_o = 0.005$, $\hat{\mu} = 0.51$, $t_L^* = 0.185$, $t_L = 0.19$.

3.4 Fed-batch operation

Fed-batch or periodic operation of bioreactors (digesters) may be economically advantageous in cases, where by using a proper schema of mixing of the feed components with the products of reaction in the reactor we reach higher reaction rates, improvement of the selectivity or better controllability of the reactor.

Let us consider a continuous stirred tank reactor (CSTR). We shall denote the intensive state variables in the reactor and outlet as $x = (x_1, \dots, x_n)$. The proper mass balances can be then written in the vector form

$$\frac{dx}{dt} = \frac{F}{V}(x_o - x) + f(x, \alpha) \quad , \tag{13}$$

where α denotes kinetic and other parameters. Let us describe now a complete cycle of the fed-batch operation, consisting of four steps. In the first step the reactor is operated batch-wise for a time t_s ; the time course of the outlet variables is described by equations

$$\frac{dx}{dt} = f(x, \alpha) \tag{14}$$

with initial conditions

$$x(0) = x^o \tag{15}$$

The values of the considered variables are at the end of first step equal to the values

$$y = x(t_s) \tag{16}$$

In the second step a volume V_s of the reaction mixture is drained from the reactor; its state is characterized by the values y (outlet concentrations). In the third step the reactor is filled up to the original level with the feed characterized by the quantities x_o. The reaction mixture is mixed in the last step and the initial conditions for the new cycle are described by the relations

$$x^o = \omega x_o + (1 - \omega)\, y \quad , \tag{17}$$

where $\omega = V_s/V$, $\omega \in (0\,;\,1]$. We shall consider that the time for feeding, draining and mixing operations is negligible compared to the length of the batch operation. Similar formulation

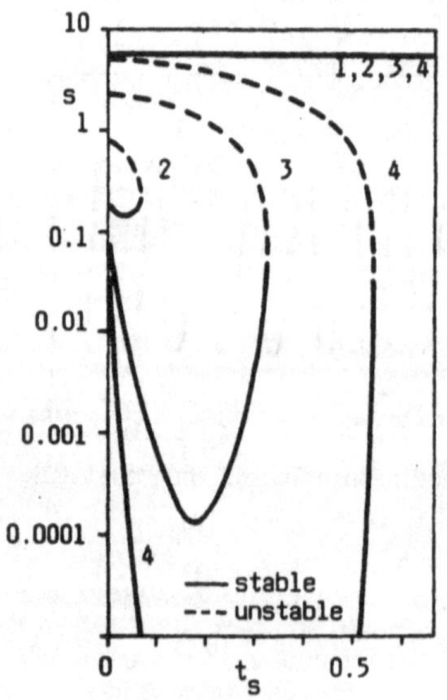

Fig. 9. Dependence of the outlet substrate concentration $s(t_s)$ on the parameter t_s . Model (S). One-periodic solution. $V/F = 3$, $x_o = 0.005$, $s_o = 5$, $K_s = 0.03$, $K_i = 5$, $Y = 0.5$. 1 : $\hat{\mu} = 0.35$, 2 : $\hat{\mu} = 0.4$, 3 : $\hat{\mu} = 0.5$, 4 : $\hat{\mu} = 0.65$.

of the fed-batch problem can be found in [1] . For the sake of possible comparison between the fed-batch operation and CSTR we shall consider equal mean residence time $\overline{V}/F$ in both systems. Then also

$$F\,t_s = V_s \qquad \text{or} \qquad t_s = \omega V/F \tag{18}$$

The value of t_s can be thus chosen from the interval $(0\,;\,V/F)$, hence $\omega \in (0\,;\,1]$. The value $\omega \to 0$ corresponds to CSTR and $\omega = 1$ denotes repeated batch-wise operation.

If we obtain solutions, where y after an m-th cycle equals to x_0 , we call it an m-periodic solution. Computational algorithm for such a solution (based on the shooting technique) and a method of determining its stability is described in, [5] . An algorithm for continuation of such a solution in dependence on t_s (or on other chosen parameter of the problem), leading to the effective way of the construction of the solution diagram is also discussed in [5] . An example of such a solution diagram is shown for the Model (S) in Fig. 9. The steady state with a low substrate conversion ($s \sim 5$) does not depend on t_s . The steady state with high substrate conversion ($s \sim 0$) is, on the contrary, strongly dependent on t_s , i.e., on the way of operation. We can see from the Figure that intermediate values of t_s give highest values of conversion, hence the fed-batch operation will be more effective, when compared with the flow-through reactor. Similar conclusions can be reached also for the Model (A), cf. [3] , [8] .

4. Conclusions

The anaerobic digestion is an example of the process, where relatively complex mixed culture of microorganisms adapts itself to the composition of the particular substrate and thus careful control strategies are necessary to assure safe operation of digesters. Studies of mathematical models including typical form of growth functions have shown that number of essentially nonlinear phenomena, which may cause instabilities of the digester, may occur. Developed computational techniques enable effective construction of the dependence of the digester characteristics on relevant parameters and we believe that they can be useful also in studies of other complex ecological models.

5. Literature

1 Ausikaitis, J., Engel, A.J. : Steady State Multiplicity and Stability in an Adiabatic Controlled Cycled Stirred Tank Reactor. AIChE J. 20 (1974), 256-263.

2 Graef, S.P., Andrews, J.F. : Stability and Control of Anaerobic Digestion. J. Water Pol. Control. Fed. 46 (1974), 666-683.

3 Kubíček, M., Holodniok, M., Marek, M., Lutcha, J. : Anaerobic digester - steady states, transients and control. In :"Digital Computer Applics to Process Control", Pergamon Press, Oxford 1980, pp. 419-426.

4 Kubíček, M., Marek, M. : Computational Methods in Bifurcation Theory and Dissipative Structures. Springer, New York, 1983.

5 Kubíčková, Z., Kubíček, M., Marek, M. : Fed-batch operation of stirred reactors. Chem. Engng. Sci., in press.

6 Mosey, F.E. : Mathematical modelling of the anaerobic digestion process. Wat. Sci. Tech. 15 (1983), 209-232.

7 Mosey, F.E. : Anaerobic processes. Chapter 5 in : "Used-Water Treatment", Vol. 2, Academic Press, London 1983.

8 Svobodová, Z., Kubíček, M., Holodniok, M., Marek, M. : Sci. Papers of the Prague Inst. of Chem. Technol. K18 (1983), 227-246.

9 Svobodová, Z., Kubíček, M., Marek, M. : Paper on CHISA Conference, 1983 (in Czech).

EXAMPLES OF COMPUTER-AIDED EXPLORATION AND DESIGN OF DYNAMICAL SYSTEMS IN NEUROSCIENCES:DESIGN OF OPTIMAL NETS.

E.Labos and E.Nogradi [1)]

1.INTRODUCTION.

The relation of neurosciences to the practice and theory of computation is manifold and not restricted to the examples mentioned here.

The subject of this presentation is to illustrate how conjectures,theorems and computer simulations as tools of rapid mathematical experience interplay and result in definite existential statements.The main conclusion is that the practical tractability of certain problems is possible only by such an "interactive" way.

A special kind of neural network optimization problem will be formulated and partially solved.The practical importance of the content seems to be double:(1) Since their first formulation (McCulloch and Pitts,1943) the formal neural nets(FNNs) were not eliminated from the theoretical neurobiology,because the tool-kits available for similar purposes are either non-tractable complicated(e.g. HH-modules in networks) or often non-effective(e.g.verbal). (2) The second reason is the possible technical significance.Computers whose circuits simulate the behaviour of nerve cells while keeping other technical advantages represent an old desire.It was even explicitly enclosed into von Neumann's "First Draft of a Report on the EDVAC" (Section 4.2).Later,the principle was applied in parametron computers(see Muroga,1971) of which a renaissance seems to be appearing(Loe and Goto,1986).

2.NETWORK PERFORMANCES WITH THE FEWEST POSSIBLE NEURONS.

Definition:An FNN is an $F \in B^n \rightarrow B^n$ function:

$$F(b)=u(bN-T)$$

where $b \in B^n$;$N=(a_{ij})$ is a square matrix;$T=(t_i)$ is the threshold vector;the function u is applied by coordinates and $u(r)=1$ if $r>0$ or $u(r)=0$ otherwise.

If F is iterated,then a sequence of vectors b_1 ,b_2 ,...,b_k appears.Since B^n is finite,thus such sequences are ended either by fixed points (equilibria) or cycles of states.

The problem is the following:Given n,the dimension of the state space B^n .How to choose (N,T) so that the cycle length in the state transition graph would exceed 2^{n-1}.By other words,how to synthetize FNNs with the smallest number of neurons?

E.g.in B^9 the number of states is 512 with 9 neurons(i.e. when N is a 9X9 matrix) it may happen that the state-transition graph consists of numerous short cycles into which various transients engulf.However,if N and T are such that they generate a flow with cycle length exceeding 256,then this network is optimal because 8 or fewer neurons are incapable of doing this.Moreover,in certain cycles of which the length is compatible with the size of neuronal pool,the internal structure may be incompatible with the size.Such sequences are similar to those examined by DeBruijn(1946).

The formal neurons are special truth functions.Other synonims are as follows:threshold gates,linearly separable truth functions,majority gates (see Muroga,1971).To make a diagnosis about a truth function whether it is a neuron or not,it is necessary to solve a system of linear inequalities.This severe condition is a source of difficulties emerging in Threshold Logic.E.g.a direct "statistical" description of cycles,transients according to length as it is done in other cases(see Gelfand,1981,Sonntag and coworkers,1981) is blocked by the simple fact that the number of formal neurons is not yet known.Estimations permit to say only that the order of the number of neurons with n inputs or networks of n neurons is a 2nd or 3rd degree polynomial of n power of two respectively.

[1)] Semmelweis Medical School,1st Dept.of Anatomy,1450 Budapest,Tuzolto-u.58,Hungary

3.REDUCTION OF THE SIZE OF SEARCHING SPACE

The number of all mappings of $B^n \to B^n$ is 2^n .In order to find absolute optimum cycle lengths(i.e.longest cycles),it is necessary to deal with transient-free,i.e.invertible nets.Their number is still high.For example at n=6 it is necessary to hunt among 64! or 63! permutations.However,it is not guaranteed that even a suitable permutation might be obtained by an (N,T),i.e.as a formal neuronal network.On the contrary,as n grows,the ratio of FNN realizable $B^n \longrightarrow B^n$ functions becomes negligible compared to the number of all $B^n \to B^n$ maps.

With the help of a theorem(Labos,1984) the problem becomes much more tractable:only self-dual permutations are permitted in the invertible functions generated by formal neural nets.Since the number of self-dual components in $B^n \longrightarrow B^n$ is equal to the possible components in $B^{n-1} \longrightarrow B^{n-1}$, thus the profit is very significant.Compare the value of $(2^n-1)!$ - estimated by its Stirling-approximation - to the number of FNNs of this lower dimensional space using Muroga's estimation(1971).

Another observation is equally helpful.Whenever we have a transient-free FNN, a great amount of transformations keep this property invariant,however the length of cycles usually change.These transformations are the 2^n partial negations of the components,and independently the n! component permutations.These last ones correspond to permutations of columns in the matrix of network together with a corresponding change in the ordrer of threshold components.Thus for example if at n=7 we have one suitable "start matrix" or (N,T) pair,then immediately we have 5040x128=645120 good ones.Among them - as it was first innocently believed - a great amount of absolute optimal nets may be found.Such a set of nets belonging to a particular start item,(N,T) was called PN block since it was obtained by permutations and negations.However,as computer experiences shows,certain PN-blocks contains many of the "good" nets and certain ones are without maximal cycle length.At n=4,5 and 6 this important phenomenon has been demonstrated.Consequently,the crucial step is to find non-trivial invertible set of selfdual FNNs.Short remarks of construction has been published (Labos,1986).The principles utilized at particular n-s,are generally applicable:the transient-freeness is guaranteed,while the availability of the maximum length is not.

4.START MATRICES AT n=4.

Instead of 16! or 15! only about 90000 or 10000 nets can be designed as FNNs.For the remaining permutations no (N,T) pairs exist.No more than 22 matrices are sufficient to derive the more than 200 PN-blocks.This number of critical matrices was obtained by unifying the cases derived by relabelling the neurons(i.e.the variables).These 22 matrices (and PN-blocks) together with the numbers of L(0) cycle lengths are demonstrated in TABLE 1.Columns of the matrices are coded and listed here as rows:x=(0,0,0,1);y=(0,1,0,0,);z=(0,0,1,0);w=(0,0,0,1);a=(-1,0,1,1);b=(0,-1,1,1);c=0,1,-1,1);d=(0,1,1,-1).The thresholds are zeros.Further:t=(0,1,1,1) or q=(1,1,0,1) with thresholds of 1.The remaining columns were coded as described elsewhere (Labos,1986):a column with code number s in the i-th position is decoded as follows: (1) Carry out binary to decimal conversion; (2) Choose for the diagonal entry of the matrix,i.e. instead of the i-th digit n-2 (here n-2=2);(3) Substitute 0s and 1s of the binary expression by -1s or 1s resp.The same principle is applicable at higher dimensions too.E.g. at n=6 and code 48 (or 110000) the corresponding columns are (1,1,4,-1,-1,-1) or (1,1,-1,4,-1,-1) in the 3rd or 4th columns resp.Example:the results of decoding of matrix D,S and P are as follows:

$$D=\begin{array}{rrrr} 0 & -1 & 1 & 1 \\ 1 & 0 & 1 & -1 \\ 1 & 1 & 0 & 1 \\ -1 & 1 & 1 & 0 \end{array} \quad S=\begin{array}{rrrr} 2 & -1 & 1 & 1 \\ 1 & 2 & -1 & 1 \\ 1 & 1 & 2 & -1 \\ -1 & 1 & 1 & 2 \end{array} \quad P=\begin{array}{rrrr} 2 & -1 & -1 & 1 \\ 1 & 2 & -1 & -1 \\ 1 & -1 & 2 & -1 \\ 1 & 1 & 1 & 2 \end{array}$$

The thresholds vectors belonging to these networks are as follows:

$$\begin{array}{rrrr} 0 & 0 & 1 & 0 \end{array} \quad \begin{array}{rrrr} 1 & -1 & 1 & 1 \end{array} \quad \begin{array}{rrrr} -1 & 0 & 0 & 0 \end{array}$$

In the TABLE 1 the last column shows how many renumbering of the variables lead to different PN-blocks.These numbers are divisors of 4!=24.By the TABLE 1 233 blocks,9984 L(0)=16 generators or 89472 invertible FNNs may be constructed.

TABLE 1 PRIMITIVE NETWORKS GENERATInG ALL THE INVERTIBLE FNNs AT n=4.FREQUENCY DISTRIBUTION OF L(0) LENGTHS OF CYCLES WHICH GOING THROUGH THE "RESTING" STATE 0.

Code	Columns	L=1	2	3	4	5	6	7	8	9	10	11	12	13	14	15	16	PN
A	x y z w	24	96	24	120	-	72	-	48	-	-	-	-	-	-	-	-	1
B	x d c b	24	36	30	72	18	66	-	96	-	18	-	24	-	-	-	-	4
C	x t d b	42	76	22	56	14	50		112	-	14	-	16	-	-	-	-	12
D	d a q b	24	84	24	32		118	-	72	-	-	-	30	-	-	-	-	8
X	0 0 0 0	24	48	24	120	-	24	-	96	-	-	-	48	-	-	-	-	1
Y	3 3 12 12	24	96	24	72	-	40	-	96	-	-	-	32	-	-	-	-	3
Z	1 1 1 0*	24	84	24	132	-	36	-	48	-	-	-	36	-	-	-	-	4
P	6* 1 1 8	24	64	22	36	30	34	44	24	-	30	-	32	-	44	-	-	12
T	5 0 13 0	the same as P..............................																12
N	0 8 8 8	24	24	30	12	18	54	-	72	-	18	-	36	-	-	-	96	4
Q	0 0 0 14	the same as N..............................																4
R	0 9 9 0	24	72	24	56	0	24	0	88	-	-	-	-	-	-	-	96	6
G	3 3 1 0	24	36	30	24	36	54	36	24	-	36	-	48	-	36	-	-	12
K	7 0 4 4	the same as G..............................																12
S	6 3* 9 12	24	96	24	48	-	24	-	72	-	-	-	-	-	-	-	96	6
L	7 2 8 2	24	84	24	100	-	24	-	32	-	-	-	-	-	-	-	96	12
F	6 9 4 6	24	76	22	32	14	42	-	40	-	14	-	24	-	-	-	96	12
J	4 11 9 4	the same as F..............................																12
E	7 3 1 0	24	56	24	40	20	42	-	20	-	20	-	42	-	-	-	96	24
M	6 1 5 8	the same as E..............................																24
I	1 11 9 4	24	76	22	32	27	40	44	16	-	27	-	32	-	44	-	-	24
H	3 9 4 6	the same as I..............................																24

Certain abnormal columns are marked by * :at 0*,6*,3* the thresholds are -3,-1,+1 resp.,while otherwise threshold=the amount of symbol 1 in the columns minus 1. The number of L=16 length generators is computable by taking into account multiplicity of matrices of a given class.E,g.matrix E yields 96x24=2304 networks,of this kind.

TABLE 2 ILLUSTRATION OF THE 3 CRUCIAL STEPS IN THE DESIGN OF NETWORKS GENERATING MAXIMUM CYCLE LENGTH.AN 8 DIMENSIONAL CASE.

Column codes of the START MATRIX are :(127,63,31,15,7,3,1,0).
PARTIAL NEGATION code:6,that is 00000110.It is a command to negate the 6th and 7th columns of the start matrix.This,means changes of signs,but the negated of t threshold is here -(t+1).
C-PERMUTATION OPERATOR:(42856137).It is a command to permute the columns of the matrix and entries of the threshold after having partially negated.
The metamorphose of the start matrix is as follows:

```
6-1-1-1-1-1-1-1     6-1-1-1-1 1 1-1     -1-1-1-1 1 6-1 1
1 6-1-1-1-1-1-1     1 6-1-1-1 1 1-1     -1 6-1-1 1 1-1 1
1 1 6-1-1-1-1-1     1 1 6-1-1 1 1-1     -1 1-1-1 1 1 6 1
1 1 1 6-1-1-1-1     1 1 1 6-1 1 1-1      6 1-1-1 1 1 1 1
1 1 1 1 6-1-1-1     1 1 1 1 6 1 1-1      1 1-1 6 1 1 1 1
1 1 1 1 1 6-1-1     1 1 1 1 1-6 1-1      1 1-1 1-6 1 1 1
1 1 1 1 1 1 6-1     1 1 1 1 1-1-6-1      1 1-1 1-1 1 1-6
1 1 1 1 1 1 1 6     1 1 1 1 1-1-1 6      1 1 6 1-1 1 1-1
```

The thresholds:

```
6 5 4 3 2 1 0-1     6 5 4 3 2-2-1-1      3 5-1 2-2 6 4-1
```

At the start the length of cycles are L=16 or L=1 (240 such fixed points) resp.At the end,the L=256 maximum is reached.The behaviour of this particular network is illustrated by two different methods in Fig.1A.At the present status of art this and similar networks cannot be effectively searched without systematic or random scanning carried out with computer.Despite of the drastic reduction of the searching space a PN-block at n=8 consists of more than 10 millions of networks.

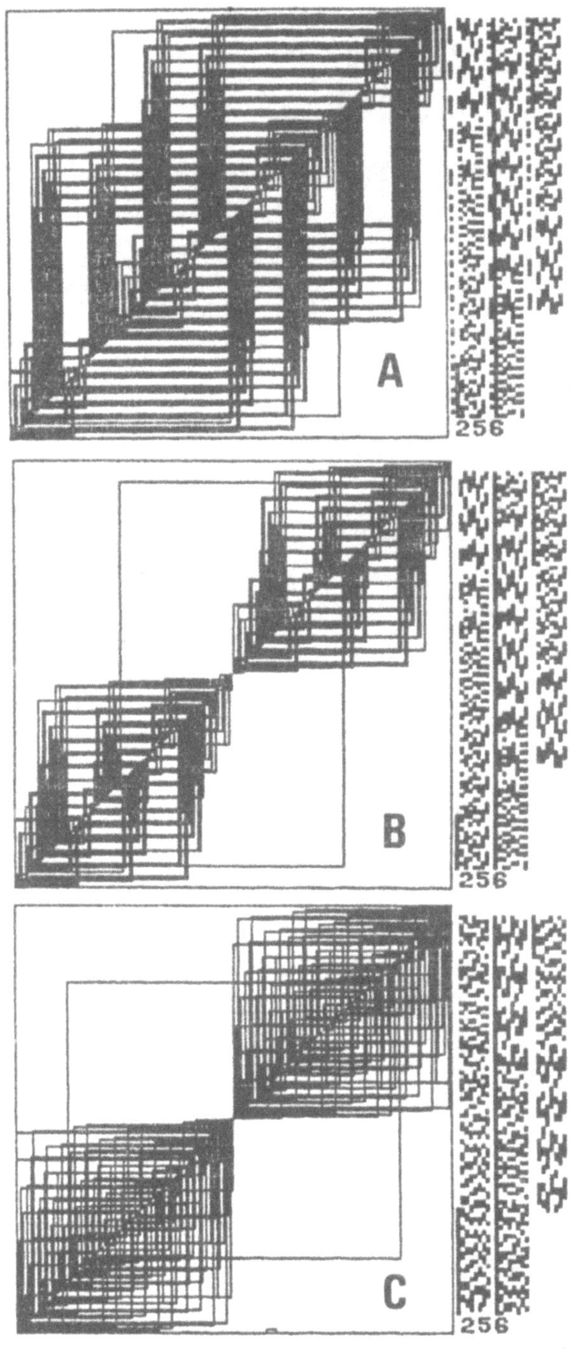

FIGURE 1 BEHAVIOUR OF OPTIMAL FORMAL NEURONAL NETWORKS.
Three 8 neuronal nets were obtained after thousands of trials.
CASE 1A - Start matrix columnar codes:M=(127,63,31,15,7,3,1,0). Code of partial negation: N=6.Columnar permutation: P=(42856137).The result is seen in TABLE 2.The behaviour of the net is depicted in 2 various ways:(1)Punch tape of activity shows the state flow generated by the iteration of the network operator.
(2) Code trajectory: code(b_{k+1})is plotted against code(b_k).The code of state is simply a binary to decimal conversion of state vectors regarded as "numbers".
CASE 1B - M =(63,191,31,15,7,3,1,0),N=6,P=(14856237).Compare to Case 1A It was obtained by exchanging the 2nd and 1st components.This relabeling permutation equals to the 21345678 permutation of rows and columns.
CASE 1C - A new case which cannot be derived in simple way from the previous ones:M = (31,159,223,15,7,3,1,0), N=20,P=(1867325).
The central symmetry of the code trajectory diagrams is a consequence of the selfduality of components.After 128 iterations the negated of the bit-sequences are repeated by each neuron.Observe in the punch tapes that the recursive order in the component flows is very high.At the 1st component flow of C it is 128 while the order of the vector flow is only one!

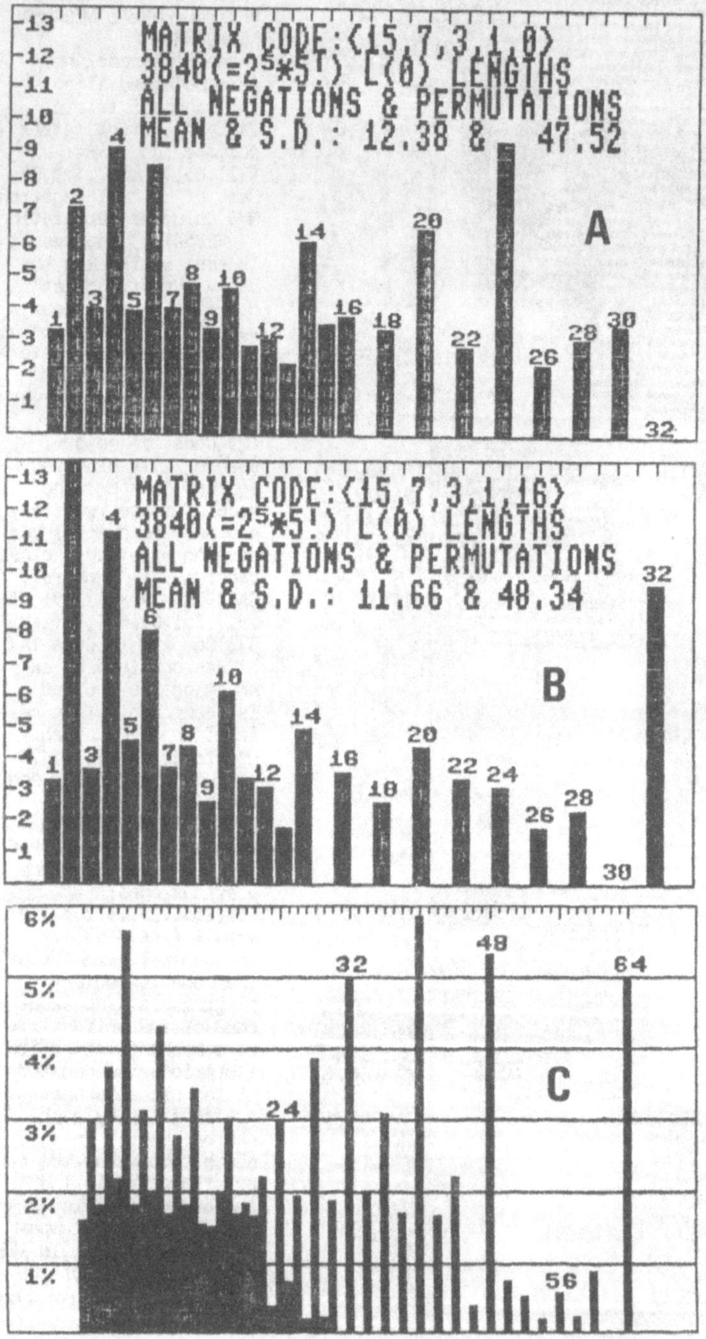

FIGURE 2
PN-BLOCKS.OUTPUTS OF A COMPLETE ANALYSIS.
SPECTRUM 2A: 3840 networks of 5 neurons were tested.The lengths enclosing state 0 were counted and classified according to their magnitude. It is remarkable that this PN--block is "empty" i.e. it does not contain L=32 cycle generator at all.
SPECTRUM 2B: Again 3840 nets were investigated. The frequency of the L=32 length is significant:about 10 per cent.The start matrices in cases A or B are different.
SPECTRUM 2C: 64 720=46080 six-dimensional nets were analysed. They represent negated and C-permuted derivatives of start matrix (31,15,7,3,1,0). Altogether 2304 absolute optimal cases were counted. This means that 5% of this block belongs to the class of "wanted" nets. Observe the lengths above half-maximum, i.e.above 32 or 64 resp.The odd lengths are prohibited because of selfduality.A similar trend is seen also below 32 and 64.

5.RESULTS IN HIGHER DIMENSIONS.

The maximum cycle lengths were reached at n=1,.,9 and a part of the results was published(Labos,1984,1986).Certain cases turned out to be easy,while other dimensions quite difficult.This is due to the phenomenon observable already at n=4:certain start matrices at certain n-s do contain,but other ones do not include nets of the desired property at all.E.g. at any n, the $c_i=2^{n-i}-1$ is suitable code of the i-th columns.Nevertheless,experience shows that at odd dimensions the PN-blocks of these matrices are not good for absolute maximum.

In Fig.1 at n=8 2 essentially different cases are depicted.In Fig 2. the frequencies of different cycle lengths (traversing the state 0) are displayed for three different PN-blocks:two at n=5 and one at n=6.It is very instructing to see that the search in an adequate block leads to numerous results,but the same hunting in a block without the "important" nets is obviously ineffective.This observation we cannot still formulate as a theorem.Since at n=7 the phenomenon inhibited the finding of L(0)=128 we list the cases actually available and which were obtained only after a "good" choice of start matrix.

TABLE 3 LIST OF OPTIMUM (L=128) NETS AT n=7.

1. MATRIX:(63,31,15,7,3,1,64).Permutations and negations:1475326-4,59/1746325-13,50/1642753-25,38/1473625-21,42/2467351-12,26/3167425-55,58/3512746-51/3657142-27/4571326-27/4762351-45,59/4176325-12,13,24,42,47,58,63/4715326-49/4761325-24/5173624-38,63/5762314-30,55/5417326-2,17,18,41,57,58,62/5741326-10/6147325-50/6471325-2,6,23,42,47,58,63/6547123-63/7642351-7,57
2. MATRIX:(31,64,96,48,56,60,62).Permutations and negations:1642753-25,38/4517623-46/4751623-12,59/514762338,63/5471623-9,24/5627431-1/7541623-16,39,54/7415623-5,20,43,58
3. MATRIX:(62,30,14,6,3,1,124).Permutations and negations:7365124-1/5417326-11,18,31,32,52,57
4. MATRIX:(62,30,15,7,3,1,112).Permutations and negations:6471325-6,18/5417326-2,17,18,21,57,58,62
5. MATRIX:(60,30,15,7,67,97,112).Permutations and negations:5417326-17,21,58,62.

The number of listed cases(above 80) may duplicated,since the completely negated nets of L=128 generators also yield this length:(e.g. 4761325 permutations with the partial negations 24 and 101).

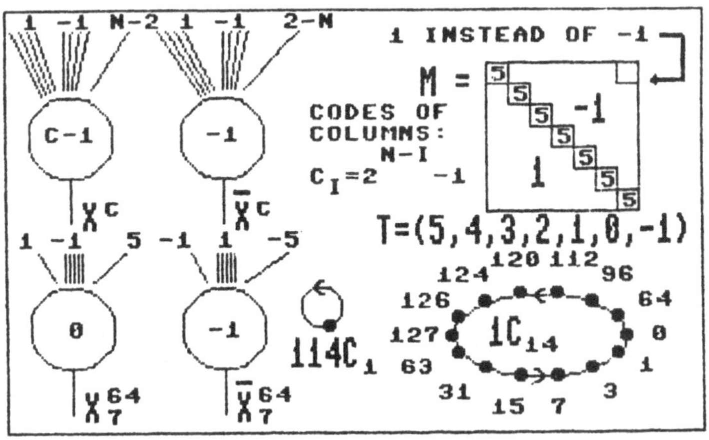

FIGURE 3.DESIGN OF EFFECTIVE START MATRIX.
The matrix depicted at 0 negation and at 1234567 permutation yields L=14 and L=1 lengths. Input lines to the component neurons and their negated are illustrated. The matrix becomes effective if the 7th component exchanged by the column of code 64.See the lines to the corresponding unit.

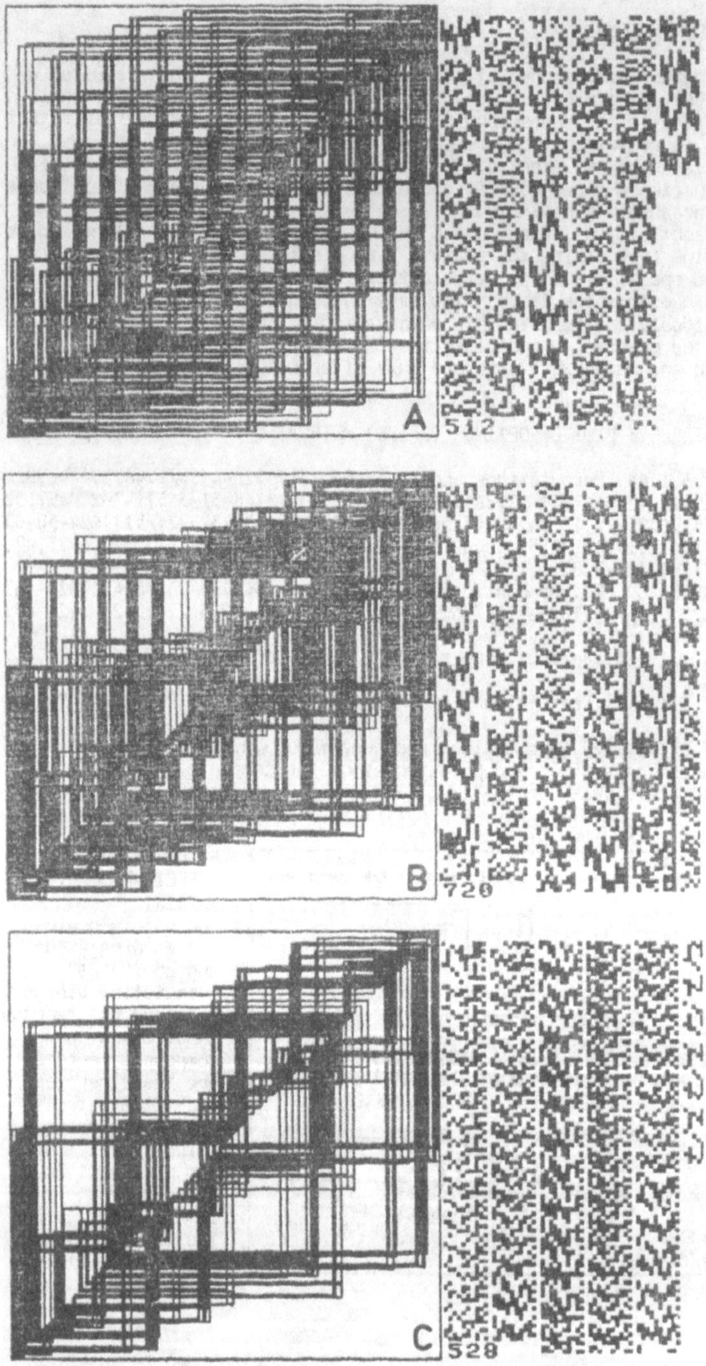

FIGURE 4
OPTIMAL CYCLES
IN NETWORKS OF 9
AND 10 DIMENSIONS.

CASE 4A
After about 6000 trials the first net of 9 variables which yielded maximal length:L=512. Start matrix:
M=(255,127,63,31,15, 7,3,1,256)
Negation code:40
Column-permutation: (579243168)
A PN-block at n=9 includes about 186 millions of networks. The practical efficiency of this method is decreased near to 0.2% illustrating intractability of the problem at the present knowledge.

CASE 4B
After a similar amount of trials,the longest cycle reached at n=10 was L=720. Since n!x2 n is the size of searching space,therefore a step from n-1 to n makes the computation slower at least by 2xn.
M=(511,...1,512)
Negation:92
Permutation:(8,5,1,3, 9,2,4,7,6,10)

CASE 4C
A net of 10 neurons with just optimal length of cycle(L=528) Start matrix as in CASE 4B.Negation: 0000011101(code 29) Permutation of columns as in 4B.

6. SEQUENCES OF 1 AND 0 EMITTED BY THE COMPONENT NEURONS.

Networks of n neurons emit n simultaneous finite words $W_1,\ldots W_n$, each consists of $L=L(W)$ letters (0 or 1) and each is characterized by the $r_1,\ldots r_n$ integer numbers called here orders of the corresponding words. These words may have transients or not.

The order of a word is the smallest integer by which the k-th symbol in the $W=h_1h_2\ldots h_L$ word can be expressed as a recursive function of the preceding symbols:

$$h_k=H(h_{k-1},\ldots,h_{k-r})=H(h) \qquad h_i\in B^1 \ ,h\in B^r$$

Both L and r are defined as minimum values. In the actual context, the 3rd characteristical value is n, the dimension of the state space in which these words are generated by a network.

The following relations are consequences of the definitions:

$$0\leq r\leq L-1 \ ,L\leq 2^n \ ,L\leq 2^r \ ,n<2^n, r<2^r, \ r\lesseqgtr n, n\geq 1$$

Analyzing these elementary relations it comes out that among the 5 numbers, $n,L,r,2^n,2^r$ only 12 relations are permitted. E.g. $2^r = 2^n > L > r=n$ is a legal arrengement.

The order number may be regarded as one of the possible measures of "complexity" which may be attributed to the "temporal" sequences of networks.

If n is given, then the range of length L is between 1 and 2^n. It was difficult to reach the case $L=2^n$.

A different extremal problem arises with respect to the so called DeBruijn sequences, where at fixed order r, the $L=2^r$ is reached. The difficulties arising when such sequences are wanted to synthetize in FNNs, was discussed by Labos(1980). If r is given, the range of L length is between 1 and r+1.

6.1 ENUMERATION THEOREMS FOR 0-1 WORDS.

This subject is closely related to the problems of word-emission by certain formal neuronal nets, the so called "permutation nets"(Labos,1984).

Beside L,n,r,s the p period of a W word has to be defined: it is the smallest natu ral number so that $h_i=h_{i+p}$ holds for every i, where i+p is taken by modulo L. It is obvious that p is a divisor of the L length. If p(W)=L(W), then the W word is called aperiodic.

The shift-equivalence of the words (or tapes) is a concept unifying those words into a single class which are derivable from a $W_1=h_1h_2\ldots h_L$ by a "shift": $W_i=h_ih_{i+1}\ldots h_{i+L-1}$. The consequence of this concept is a $g(p)=h(p)/p$ number of words with period p which is already independent of the L length. To compute g(p), the Moebius inversion-theorem is necessary to apply (Labos,1984):

$$g(p)=\frac{1}{p}\sum_{d|p}\mu(d).2^{p/d}$$

For example the "period decomposition" of the 4096 words of L=12 into 352 classes is as follows: $2P_1 1P_2 2P_3 3P_4 9P_6 335P_{12}$ (i.e. $g(p)P_p$)

The analogous "order-decomposition" or "weight-decomposition" problems include non-trivial enumerations as well. These are related to the determination of aperiodic word-distributions in (r,s) classes. E.g. (again at L=12) the weigth-decomposition is more difficult to understand: $1S_0 1S_1 6S_2 19S_3 43S_4 66S_5 80S_6 66S_7 43S_8 19S_9 6S_{10} 1S_{11} 1S_{12}$ (352 classes as above).

A formula may also be derived for words of length L, period p and weight s as follows:

$$f(L,s,p)=\sum_{a|(p,q)}\mu(a)\binom{p/a}{q/a}$$

where $\mu(a)$ is the Moebius-function, (p.q) is the g.c.d. of q and p.

At last in TABLE 4 the sizes of (p,r,s) classes of the words of L=10 are summarized.

TABLE 4 DISTRIBUTION OF THE 1024 WORDS OF LENGTH 10 ACCORDING TO PERIOD,ORDER AND WEIGHT.

Weight	Period	Order	Words or classes
0 or 10	1	0	1+ 1 (1+1)
1 or 9	10	9	10+10 (1+1)
2 or 8	5	4	5+ 5 (1+1)
2 or 8	10	6	10+10 (1+1)
2 or 8	10	7	10+10 (1+1)
2 or 8	10	8	20+20 (2+1)
3 or 7	10	5	50+50 (5+5)
3 or 7	10	6	30+30 (3+3)
3 or 7	10	7	20+20 (2+2)
3 or 7	10	8	10+10 (1+1)
3 or 7	10	9	10+10 (1+1)
4 or 6	5	3	5+ 5 (1+1)
4 or 6	5	4	5+ 5 (1+1)
4 or 6	10	4	50+50 (5+5,
4 or 6	10	5	90+90 (9+9)
4 or 6	10	6	30+30 (3+3)
4 or 6	10	7	10+10 (1+1)
4 or 6	10	8	20+20 (2+2)
5	2	1	2 (1)
5	10	4	70 (7)
5	10	5	100 (10)
5	10	6	40 (4)
5	10	7	40 (4)

weight(s):	0	1	2	3	4	5	6	7	8	9	10
binom.c.:	1	10	45	120	210	252	210	120	45	10	1
s-class:	1	1	5	12	22	26	22	12	5	1	1
period:		2	2			30					990
p-class:		2	1			6					99
order:	2	2	0	10	190	380	180	120	100	40	0
r-class:	1	1		2	21	38	18	12	10	4	0

6.2 WORDS APPEARING AT SELF-DUAL NETWORKS.

The self-duality means that either pairs of cycles with odd and even lengths or glued cycles of even length appear.The structure of glued sequence is simple:the 2nd half of the word is the negated of the 1st half:e.g 00001111 or 10110100 etc.

An example illustrates how high the recursive order of sequences may be in a maximum length generator networks.The case is 6-dimensional.

Start-matrix:M=(1,33,49,57,61,0).C-permutation:P=(413265).Negation code:38

The six component sequences (self-dual ones) are as follows:

W_1 = 0100100100101101101101111111110110110110110100100100100000000010
W_2 = 0010010010110110110110111111111011011011010010010010010000000001
W_3 = 0000000000000000000000000011111111111111111111111111111111000000
W_4 = 0110110110100100100100000000000010010010010110110110111111111111
W_5 = 0011001110011001100110011001100111001100011001100110011001100110
W_6 = 0001100011001100110011001100110011100111001100110011001100110011

The orders of these six words are as follows:

$$r_1 = 14 \ , \ r_2 = 14 \ , \ r_3 = 32 \ , \ r_4 = 13 \ , \ r_5 = 29 \ , \ r_6 = 27 \ .$$

A one-dimensional record like a component-output may have very high recursive order and this phenomenon is due to the high dimensional and moreover extremal network in the "background".

REFERENCES

1. Amari,S.:Characteristics of Randomly Connected Threshold-Element Networks and Nervous Systems.Proc.IEEE, 59 (1971) 35-47.
2. DeBruijn,N.G.:A Combinatorial Problem.Nederl.Akad.Wetensch.,Proc.49 (1946) 758-764.(see in Hall,M.:Combinatorial Theory.Blaisdell,Waltham,1967.Chapter 9:De Bruijn Sequences).
3. Dertouzos,M.L.:Threshold Logic:A Synthesis Approach. MIT Press, Cambridge,.Mass.,1965.
4. Garey,M.R. and Johnson,D.S.:Computers and Intractability.Freeman & Co.,San Francisco,1979.
5. Hu-Sze-Tseu:Threshold Logic.University of California Press,Berkeley,1965.
6. Gelfand,A.E.:A behavioural summary for completely random nets. Bull.Math.Biol. 44 (1982) 3 309-320.
7. Griffith,J.S.:Mathematical Biology.Chapter 3.,Academic Press,New York,1971.
8. Harrison,M.H.:A census of finite automata.Canad.J.Math. 17 (1965) 100-113.
9. Kleene,S.C.:Representation of events in nerve nets and finite automata.In:Automata Studies.(Shannon,C.E. & McCarthy,J. eds.) 34-41.
10. Labos,E.:Periodic and non-periodic motions in different classes of formal neuronal networks and chaotic spike generators.In Cybernetics and System Research 2(R.Trappl.,ed.) 237-243.Elsevier,Amsterdam,1984.
11. Labos,E.:Selfdual and other Structures of Optimal Cycles in Neural Network Behaviour.In Cybernetics and Systems'86.(r.Trappl,ed.),Reidel,Dordrecht,1986.
12. Labos,E.:Optimal Design of Neuronal Networks.In:Neural Communication and Control.(Szekely,Gy.& al.,eds.). Adv.Physiol.Sci.30 (1980) 127-153.
13. Labos,E.:Effective Extraction of Informatiopon Included in Network Descriptions and Neural Spike Records. In:Biomathematics in 1980.(Ricciardi,L.M. & A.C.Scott,eds.).North-Holland,Amsterdam.
14. Labos,E.:Spike Generating Dynamical Systems and Networks.Proc.of a Symposium on The Mathematics of Dynamic Processes held in Sopron (Hungary),1985.In Press.
15. Levine,D.S.:Neural Population Modeling and Psychology.A Review. Math.Biosci. 66 (1983) 1-86.
16. Lewis,P.E.II. and Coates,C.L.:Threshold Logic.Wiley,New York,1967.
17. Loe,K.F. and Goto,E.:DC Flux Parametron. World Scientific Ser. in Computer Science Vol.6.,Singapore,1986.
18. McCulloch,W.S. and Pitts,W.H.:A Logical Calculus of the Ideas Immanent in NERvous Activity.Bull.Math.Biophys. 5 (1943) 115-133.
19. Muroga,S.:Threshold Logics and Its Application.Wiley-Interscience,New York,1971.
20. Nijenhuis,a.& Wilf,H.S.:Combinatorial Algorithms.Academic Press,New York 1975.
21. Randell,B.(ed.):The Origins of Digital Computers.Selected Papers.Springer-Verlag,Berlin 1973.
22. Rapoport,A.:Ignition phenomena in random nets.Bull.Math.Biophys. 14 (1952) 35-44.
23. Sheng,C.L.:Threshold Logic.Academic Press,New York.,1969.
24. Sonntag,I,Feistel,R. and Ebeling,W.:Random Networks of Catalytic Biochemical Reactions.Biom.J. 23 (1981) 5:501-515.

A DYNAMICAL MODEL OF SYNAPTIC TRANSMISSION

*S. Markov and T. Kostova-Vasilevska**

Abstract

The transmission of nerve impulses in a cholinergic synapse is a biochemical process. Basic steps in this process are: the variation of the presynaptic Ca^{++} concentration as a consequence of the depolarization of the presynaptic membrane; the Ca^{++} - promoted liberation of a neurotransmitter mediator (acetylcholine) in the synaptic cleft; the hydrolysis of this mediator by an enzyme (acetylcholineesterase); and the reaction of the mediator with the receptors on the postsynaptic membrane. This process is modelled by means of a set of ordinary differential equations. The system is studied with respect to stability; numerous computer experiments are performed. The application of the above model is briefly discussed.

1. Introduction

It is believed that the formation of special patterns of neuronal activity constitutes the gist of the mechanism of short-term memory and that some permanent changes in the biochemical mechanism of synaptic transmission underlie the phenomenon of long-term memory. The synapse is a plastic structure which may gradually change its properties depending on the frequency of transmission. This can be interpreted as the appearance of short-term memory. A set of neurons with interconnecting plastic synapses by certain conditions could be specified as a memory unit. The first step in studying the behavior of such memory unit is to model a plastic synapse. To model the synapse as a plastic structure many components affecting plasticity are involved. We construct a model describing the evolution of postsynaptic membrane depolarisation as a consequence of presynaptic exitation by means of a number of successive steps.

The process of synaptic transmission can be briefly described as follows. The presynaptic potential (pre $V = V$) causes the opening of special presynaptic membrane channels and consequently the generation of inside calcium ions (Ca^{++}) current (I_{Ca}). The calcium current changes the inside calcium concentration $[Ca_i]$. The latter is basically regulated also by three other phenomena: the active outlet of Ca^{++} out of the presynaptic area; the buffering of Ca^{++} by certain buffering systems (the most significant ones are the mytochondria) - the bounded thereby Ca is further denoted by CaM; the binding of Ca to certain substances resulting in a special type of complex (denoted below as CaX).

* Bulgarian Academy of Sciences, Research Group for Mathematical Modelling in Biology, "Acad. G. Bonchev" st., Block 25, 1113 Sofia, Bulgaria.

The role of the complex CaX is to promote the liberation of certain neurotransmitter chemical, say acetylcholine (ACh), into the synaptic cleft. The transmitter partakes in two substrate-enzyme reactions: the hydrolysis of ACh by acetylcholineesterase (E) and the reaction of ACh with the receptors (R) on the postsynaptic membrane. The conductivity $g(t)$ of the postsynaptic membrane is altered as a consequence of the formation of the complex ACh-R. The postsynaptic potential (post $V = V_1$) directly depends on the conductivity $g(t)$.

The process of synaptic transmission described above consists of a series of successive steps and can be schematically represented as follows.

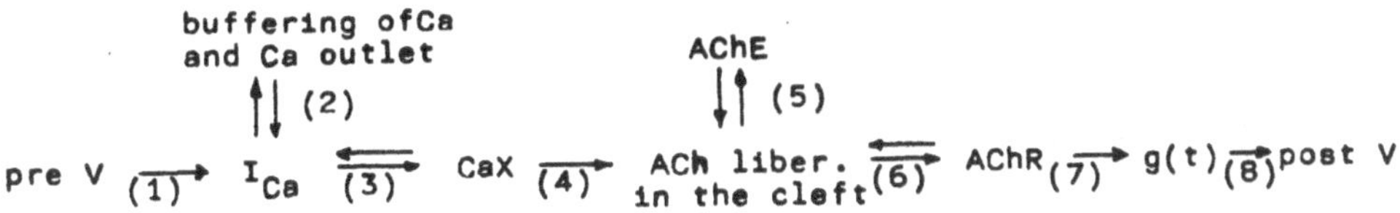

2. The mathematical model

Our modelling of step 1 is based on the work [1] with a small modification allowing for the possibility of saturation of I_{Ca} with respect to the inside and outside calcium concentration ($[Ca_i]$, resp. $[Ca_0]$). We have

$$I_{Ca} = (P[G_C](t)/([Ca_0][Ca_i](t)))([Ca_0] - [Ca_i](t)e^{-2\varepsilon V/RT}) \quad ,$$

where $P, Ca_0, \varepsilon, R, T$ are positive constants and $[G_C](t)$ is the concentration of the open Ca^{++} channels at time t. According to [1].

$$[G_C](t) = [S]^5 \quad ,$$

$$d[S]/dt = -(Ke^{C_sV} + \bar{K}e^{\bar{C}_sV})[S] + Ke^{C_sV},$$

where $K, \bar{K}, C_s, \bar{C}_s$ are positive constants and the meaning of the variable S is explained in [1].

Steps (2) and (3) are modelled by the kinetic equations

$$d(Ca_i]/dt = I_{Ca}(t) - K_M[Ca_i](C_M - [CaM]) - K_X[Ca_i](C_X - [CaX])$$

$$+ \bar{K}_M[CaM] + \bar{K}_X[CaX] - K_0[Ca_i] \quad ,$$

$$d[CaM]/dt = K_M[Ca_i](C_M - [CaM]) - \bar{K}_M[CaM] \quad ,$$

$$d[CaX]/dt = K_X[Ca_i](C_X - [CaX]) - \bar{K}_X[CaX] \quad ,$$

where C_M, C_X are the saturation concentrations of CaM, resp. CaX, and $K_M, \bar{K}_M, K_X, \bar{K}_X, K_0$ are positive constants.

Denoting the concentration of the transmitter in the presynaptic area by $[A_V]$, step (4) is modelled by

$$d[A_V]/dt = C_P - C_A[A_V][Ca_X], \; C_P, C_A = \text{const} > 0$$

Steps (5) and (6) represent two substrate-enzyme reactions: the hydrolysis of ACh by acetylcholineesterase (E) in the cleft and its reaction with the receptors on the postsynaptic membrane. This reactions can be modelled [5] by

$$d[ACh]/dt = C_A[A_V][CaX] - K_R[ACh](C_R - [ACh - R]) - K_E[ACh](C_E - [ACh - E]) + K_{-R}[ACh - R] + K_{-E}[ACh - E] ,$$

$$d[ACh - R]/dt = K_R[ACh](C_R - [ACh - R]) - K_{-R}[ACh - R] - \bar{K}_R[ACh - R] ,$$

$$d[ACh - E]/dt = K_E[ACh](C_E - [ACh - E]) - K_{-E}[ACh - E] - \bar{K}_E[ACh - E] ,$$

where $K_R, K_{-R}, \bar{K}_R, K_E, K_{-E}, \bar{K}_E, C_R, C_E$ are positive constants.

Step (7) is modelled by

$$g(t) = K_g[ACh - R] , \quad k_g = \text{const} > 0 ,$$

and step (8) can be modelled according to [4] by

$$dV_1/dt = (1/C)[\tilde{V}g(t) - (K + g(t))V_1]. \quad C, \tilde{V}, K = \text{const} > 0$$

3. Stability of the model

Renaming the variables by $X_0 = S, X_1 = Ca_i, \quad X_2 = CaM, X_3 = CaX, X_4 = ACh, X_5 = ACh - R, X_6 = ACh - E, X_7 = V_1, x_g = A_V$, we obtain the following system

$$\begin{cases} \dot{x}_0 = -(kl^{C_sV} + \bar{K}l^{\bar{C}_sV})x_0 + kl^{C_sV} \\ \dot{x}_1 = (Px_0^5/(C_o x_1))(C_0 - x_1 e^{-C_1V} , \\ \quad - K_M x_1(C_M - x_2) - K_x x_1(C_x - x_3) \\ \quad + \bar{K}_M x_2 + \bar{K}_x x_3 - K_0 x_1 , \quad C_0 = [Ca_0], C_1 = -2\varepsilon/RT , \\ \dot{x}_2 = K_M x_1(C_M - x_2) - \bar{K}_M x_2 , \\ \dot{x}_3 = K_x x_1(C_x - x_3) - \bar{K}_x x_3 , \\ \dot{x}_8 = C_p - C_A x_8 x_3 , \\ \dot{x}_4 = C_A x_8 x_3 - K_R x_4(C_R - x_5) - K_E x_4(C_E - x_6) + K_{-R}x_5 + K_{-E}x_6 , \\ \dot{x}_5 = K_R x_4(C_R - x_5) - K_{-R}x_5 - K_R x_5 , \\ \dot{x}_6 = K_E x_4(C_E - x_6) - K_{-E}x_6 - \bar{K}_E x_6 , \\ \dot{x}_7 = (1/C)(\tilde{V}K_g x_5 - (K + K_g x_5)x_7) . \end{cases}$$

It is interesting to study the above system with respect to how the output potential reacts to the input potential. The following can be proved [3]:

Theorem. System (1) has the following three properties:

(i) For any $V = V_0$ there exists a unique stationary solution $(\bar{x}_0, \bar{x}_1, \ldots, \bar{x}_8)$ belonging to the set

$$Z = \{(x_0, \ldots, x_g): x_i > 0,\ i = 0,1,4,7,8,\ 0 < x_2 < C_M,\ 0 < x_3 < C_x,\ 0 < x_5 < C_R,\ 0 < x_6 < C_E\} .$$

(ii) If $V(t)$ is bounded, then Z is an invariant set of (1).

(iii) Let $V(t) = V_0$ for $t = t^*$. Let $x(t) = (x_0,...,x_8)$ be a solution of (1) with $x(t^*) \in Z$ and $V(t) = V_0, t \geq t^*$. Then $\lim_{t \to \infty} x_i(t) = \bar{x}_i, i = 0,...,8$.

Remarks. The restrictions on the domain Z for the variables in part (i) of the theorem are natural from the point of view of the biological interpretation of x_i. Part (ii) states that the variables x_i do not leave the domain Z, whenever V is a bounded function, which is also a natural requirement. In order to explain the meaning of part (iii) consider the following.

Suppose that the system is perturbed for a short time after which $V(t)$ retains a constant value V_0 (this situation corresponds to the experimental voltage clamp study of the synaptic transmission), i.e.,

$$V = \{V(t), t \in [t_0, t^*]; V_0 = \text{const}, t \geq t^*\}$$

Important for the synaptic transmission is that post $V(t)$ should tend to its rest value when $t \geq t^*$. Thus it is necessary that the solution

$$\{\dot{y}_i = f_i(y, V_0), y_i(t^*) = x_i(t^*)\}$$

satisfy the conditions $y_i(t) \to \bar{x}_i, t \to \infty$.

A great number of computer experiments with various sets of parameters of (1) are performed [3]. It is shown [3] that different types of plasticity according to [2] can be demonstrated.

REFERENCES

[1] Llinas, R., Steinberg, I. and K. Walton. Presynaptic calcium currents and their relation to synaptic transmission: Voltage clamp study in squid giant synapse and theoretical model for the calcium gate. Proc. Nat. Acad. Sci. USA, V. 73, 8, 1976, 2918-2922.

[2] Kandel, E.R. Calcium and the control of synaptic strength by learning. *Nature*, V. 293, 697-700, 1981.

[3] Kostova, T. Asymptotic stability and other problems generated by a model of synaptic transmission. Ph.D. Thesis Sofia University, 1984.

[4] Martin, A. The effect of membrane capacitance on non-linear summation of synaptic potentials. J. Theor. Biol., V. 59, 179-189, 1976.

[5] Venkov, L. and S. Markov. A model of synaptic transmission. Cell. Mol. Biol. V. 26, 541-546, 1980

GRADIENTS FOR REPLICATOR SYSTEMS

*Karl Sigmund**

Abstract: In many models of evolutionary biology, the state follows an equation given by a gradient, and some fitness-function increases according to a maximum principle.

1. Introduction

Many biological models are described by "ecological equations" of the form

$$\dot{x}_i = x_i f_i(\mathbf{x}) \tag{1}$$

on $\mathbb{R}^n_+$, or by "replicator equations"

$$\dot{x}_i = x_i (f_i(\mathbf{x}) - \bar{f}) \tag{2}$$

on S_n, where S_n is the unit simplex

$$S_n = \{\mathbf{x} \in \mathbb{R}^n_+ : \sum x_i = 1\}$$

and

$$\bar{f} = \sum_{j=1}^{n} x_j f_j(\mathbf{x}) . \tag{3}$$

For a survey of such systems, we refer to Schuster and Sigmund (1983) and Sigmund (1985b). The x_i are the densities, or frequencies, of replicating units like individuals, genes, phenotypes or polynucleotides.

A pervasive image in theoretical biology is that of a fitness landscape: the state of the replicating population evolves in certain cases in such a way that the corresponding average fitness increases steadily. In order to view fitness as a potential of this evolutionary process, however, and the corresponding equations as gradients, one has to introduce an appropriate non Euclidean metric in the state space $\mathbb{R}^n_+$ or S_n.

Let us explain this in a general context. Assume that V is a real valued continuously differentiable function on an open subset U of R^n. The increase of V at a point $\mathbf{x} \in U$ in the direction given by a vector $\mathbf{y}$ of unit length is given by

$$\lim_{h \to 0} \frac{V(\mathbf{x} + h\mathbf{y}) - V(\mathbf{x})}{h} = \sum_{i=1}^{n} \frac{\partial V}{\partial x_i} y_i = \text{grad } V(\mathbf{x}) \cdot \mathbf{y} \tag{4}$$

---

* International Institute for Applied Systems Analysis and Institute for Mathematics, University of Vienna

where $\underline{x} \cdot \underline{y}$ denotes the Euclidean inner product of $\underline{x}, \underline{y} \in \mathbf{R}^n$. By the Cauchy Schwarz inequality, this quantity is bounded by $\|\text{grad } V(\underline{x})\|$ and this bound is attained precisely if $\underline{y}$ points into the same direction as grad $V(\underline{x})$. Thus the vector field

$$\underline{x} \to \text{grad } V(\underline{x}) = \left(\frac{\partial V}{\partial x_1}(\underline{x}), \ldots, \frac{\partial V}{\partial x_n}(\underline{x}) \right) \tag{5}$$

points into the direction of greatest increase of V, and the orbits of $\dot{\underline{x}} = \text{grad } V(\underline{x})$ follow the lines of steepest ascent of the "landscape" given by V.

The potential V is a Ljapunov function $\dot{V}(\underline{x})) \geq 0$, with equality at the rest points. The orbits cross the constant level sets orthogonally, and the stable rest points are the local maxima of V.

It turns out that equations of type (1) or (2) are not gradients in this sense, except in the most trivial cases. The reason, roughly speaking, is that the boundary faces $x_i = 0$ of the state space are invariant. What is needed is a non Euclidean metric which reflects this feature of the state space.

Such a metric was first proposed by Shahshahani (1979) and thoroughly investigated by Akin (1979). It turned out that the classical selection equation of Fisher-Haldane-Wright is a Shahshahani gradient, and hence that Kimura's maximum principle is valid. We shall describe this in Section 2, and discuss some characterizations of such gradients among replicator equations. In Section 3, we discuss Hofbauer's results on the mutation selection equation, and in Section 4 recent aspects of the game dynamics of mixed strategies in terms of gradient systems. In Sections 5, 6 and 7 we give a survey of biomathematical applications of related non-Euclidean metrics in genetics, chemical kinetics and game dynamics.

2. The Shahshahani metric

Let $\underline{x}$ be a point in *int* S_n. The tangent space $T_{\underline{x}} S_n$ is isomorphic to the space

$$\mathbf{R}_0^n = \{\underline{\xi} \in \mathbf{R}^n : \sum_{i=1}^{n} \xi_i = 0\}.$$

If $\underline{\xi}$ and $\underline{\eta}$ are two vectors in $T_{\underline{x}} S_n$, their Shahshahani product is defined by

$$\langle \underline{\xi}, \underline{\eta} \rangle_{\underline{x}} = \sum_{i=1}^{n} \frac{1}{x_i} \xi_i \eta_i \tag{6}$$

which differs from the Euclidean product

$$\underline{\xi} \cdot \underline{\eta} = \sum_{i=1}^{n} \xi_i \eta_i \tag{7}$$

by the weights $\frac{1}{x_i}$ which are large near the boundary faces $x_i = 0$ of S_n.

Let V be a smooth map from a neighborhood U of S_n into $\mathbf{R}$. The derivative $D_{\underline{x}} V$ of V at $\underline{x}$ is a linear map from $T_{\underline{x}} S_n$ into $\mathbf{R}$. By the Riesz representation theorem, there exists a unique vector $\underline{f} \in T_{\underline{x}} S_n$ such that

$$D_{\underline{x}} V(\underline{\xi}) = \underline{f} \cdot \underline{x} \tag{8}$$

holds for all $\underline{\xi} = T_{\underline{x}} S_n$. Similarly, there exists a unique $\underline{f}_{\underline{x}} \in T_{\underline{x}} S_n$ such that

$$D_x V(\xi) = \langle l_x, \xi \rangle_x . \tag{9}$$

The vector l is just the Euclidean gradient grad$V(x)$. The vector l_x is called the Shahshahani gradient and denoted by grad$_x$ $V(x)$. (There are, of course, many other inner products and hence gradients: we shall meet a few of them in later sections).

We shall denote the right hand-side of (2) by $\hat{f}$. This is a vector field with components

$$\hat{f}_i(x) = x_i(f_i(x) - \bar{f}) . \tag{10}$$

There are several useful criteria for $\hat{f}(x)$ to be a Shahshahani gradient.

(a) In Sigmund (1985a) it is shown that (2) is a Shahshahani gradient iff there exist functions V and Ψ from some neighborhood U of S_n into $\mathbf{R}$ such that

$$f_i(x) = \frac{\partial V}{\partial x_i}(x) + \Psi(x) \tag{11}$$

holds for $i = 1, \ldots, n$ and all $x \in int\ S_n$. Thus if $\dot{x} = f(x)$ is an Euclidean gradient with potential V, i.e., if $f(x) = \text{grad}\, V(x)$, then the corresponding replicator equation $\dot{x} = \hat{f}(x)$ is a Shahshahani gradient with the same potential, i.e., $\hat{f}(x) = \text{grad}_x V(x)$.

For example, if f_i depends only on x_i, for $i = 1, \ldots, n$, then (2) is a Shahshahani gradient with potential

$$V(x) = \sum_{i=1}^{n} \int f_i(x_i) dx_i . \tag{12}$$

This greatly simplifies the global analysis of systems of self-replicating macromolecules with replicating rates f_i which are constant (see, e.g., Jones et al. (1976) or Thompson and McBride (1974)) or which are decreasing functions of x_i (see, e.g. Epstein (1979) who studied functions f_i of the form $a - bx_i$ or $ax_i(b + cx_i)^{-1}$).

(b) In Akin (1979) it was shown that (2) is a Shahshahani gradient iff the bilinear function $H_x \hat{f}$ given by

$$H_x \hat{f}(\xi, \eta) = \langle \xi, D_x \hat{f}(\eta) \rangle_x \tag{13}$$

is symmetric for $\xi, \eta \in T_x S_n$.

(c) Probably the most useful characterization is due to Hofbauer (1985b): the replicator equation (2) is a Shahshahani gradient iff

$$\frac{\partial f_i}{\partial x_j} + \frac{\partial f_j}{\partial x_k} + \frac{\partial f_k}{\partial x_i} = \frac{\partial f_i}{\partial x_k} + \frac{\partial f_k}{\partial x_j} + \frac{\partial f_j}{\partial x_i} \tag{14}$$

holds on $int\ S_n$ for all i,j,k. This "triangular integrability condition" (which has obviously only to be checked for pairwise different indices i,j and k) corresponds to the classical integrability condition $\frac{\partial f_i}{\partial x_j} = \frac{\partial f_j}{\partial x_i}$ for Euclidean gradients on simply connected sets. It was first derived in Sigmund (1985b) for replicator equations with linear functions f_i:

$$f_i(x) = \sum_{j=1}^{n} a_{ij} x_j = (Ax)_i . \tag{15}$$

In this case, (14) becomes

$$a_{ij} + a_{jk} + a_{ki} = a_{ik} + a_{kj} + a_{ji} . \tag{16}$$

This is obviously valid for every symmetric matrix A. This shows in particular that the classical selection equation of population genetics

$$\dot{x}_i = x_i \left[\sum_{j=1}^{n} a_{ij} x_j - \bar{f} \right] \tag{17}$$

with $a_{ij} = a_{ji}$ is a Shahshahani gradient (cf. Shahshahani (1979) and Akin (1979)) and hence that Kimura's Maximum Principle is valid: the average fitness $\bar{f} = \mathbf{x} \cdot A\mathbf{x}$ increases at the maximal rate, i.e.: the evolution of (17) proceeds along the lines of steepest ascent in the fitness landscape.

More generally, (16) is valid iff there exists a symmetric matrix B (with $b_{ij} = b_{ji}$) and numbers c_j such that $a_{ij} = b_{ij} + c_j$, or, equivalently, iff there exist constants c_i such that $a_{ij} - a_{ji} = c_i - c_j$ holds for all i and j.

3. The Selection Mutation Equation

Two equations have been proposed to study the joint effects of selection and mutation upon a single gene locus

$$\dot{x}_i = (\mathbf{x} \cdot W\mathbf{x})^{-1} \sum_{j,k} \varepsilon_{ij} x_j w_{jk} x_k - x_i \tag{18}$$

and

$$\dot{x}_i = x_i \left[\sum_j m_{ij} x_j - \mathbf{x} \cdot M\mathbf{x} \right] + \sum_j (\varepsilon_{ij} x_j - \varepsilon_{ji} x_i) \tag{19}$$

the first one by Crow and Kimura (1970) and the second one by Hadeler (1981). The x_i, here, denote the frequencies of alleles A_i in the gene pool, the ε_{ij} are the mutation rates from A_j to A_i, the w_{ij} the survival probabilities of the genotype $A_i A_j$ from zygote to adult and the m_{ij} the differences between birth and death rates for $A_i A_j$. As Hofbauer (1985b) has shown, (19) is a limit case of (18).

If there is no selection, i.e. if all w_{ij} or m_{ij} are equal, then both (18) and (19) reduce to the mutation equation

$$\dot{x}_i = \sum \varepsilon_{ij} x_j - x_i . \tag{20}$$

One can write this simple linear equation as a replicator equation (2), with

$$f_i(\mathbf{x}) = \sum_{j=1}^{n} \varepsilon_{ij} \frac{x_j}{x_i} . \tag{21}$$

For $i \neq j$, one has $\frac{\partial f_i}{\partial x_j} = \frac{\varepsilon_{ij}}{x_i}$. It is easy to see that the triangular integrability condition is satisfied iff ε_{ij} is independet of j, i.e. iff the mutation rate depends only on the "target gene" (see Hofbauer (1985b)). In this case, we shall write $\varepsilon_i = \varepsilon_{ij}$ (for $j \neq i$) and $\varepsilon = \varepsilon_1 + \cdots + \varepsilon_n$. The mutation selection equations (18) and (19) can be written as replicator equations (2) with

$$f_i(\underline{x}) = (1-\varepsilon)(\underline{x}\cdot W\underline{x})^{-1}\left[\sum_j w_{ij}x_j\right] + \frac{\varepsilon_i}{x_i} \tag{22}$$

resp.

$$f_i(\underline{x}) = \sum_{j=1}^{n} w_{ij}x_j + \frac{\varepsilon_i}{x_i}. \tag{23}$$

Both equations are Shahshahani gradients: the potential of (22) is given by

$$V(\underline{x}) = \frac{1-\varepsilon}{2}\log(\underline{x}\cdot W\underline{x}) + \sum_i \varepsilon_i \log x_i \tag{24}$$

and that of (23) by

$$V(\underline{x}) = \underline{x}\cdot M\underline{x} + \sum_i \varepsilon_i \log x_i$$

(Hofbauer (1985b)). In particular, this implies that if the model without mutation ($\varepsilon_{ij} = 0$ for all $i \neq j$) admits a stable interior equilibrium, then for every choice of mutation rates $\varepsilon_{ij} = \varepsilon_i$ depending only on i such that $\varepsilon<1$, (18) and (19) have a globally stable equilibrum in S_n. This result of Hofbauer (1985b) generalizes a theorem by Hadeler (1981).

If the mutation rates ε_{ij} depend on j, a general result of Akin (1979) implies that there exist selection coefficients m_{ij} or w_{ij} such that periodic orbits can occur in (18) resp. (19). Hofbauer (1985b) constructed examples which even display stable limit cycles.

4. Mixed Strategies and the Sex Ratio

Maynard Smith has developed a game theoretic approach to evolutionary biolgy which has found many interesting aplications (see, e.g. Maynard Smith (1982)). This approach is based on the idea that successful strategies will spread. An adequate explicit "game dynamics" was introduced by Taylor and Jonker (1978): if R_1 to R_N denote the different strategies, y_1 to y_N their frequencies and $f_i(\underline{y})$ the payoff for strategy R_i against the mixed strategy $\underline{y}$ which consists in playing R_j with probability y_j, then one assumes that the rate of increase $\frac{\dot{y}_i}{y_i}$ of R_i is given by the difference between the payoff $f_i(\underline{y})$ of R_i and the average payoff $\bar{f}(\underline{y}) = \underline{y}\cdot f(\underline{y})$. This leads to the replicator equation

$$\dot{y}_i = y_i(f_i(\underline{y})-\bar{f}(\underline{y})) \tag{25}$$

on S_N.

Now let us assume that there are n phenotypes E_1 to E_n in the population, each corresponding to some (pure or mixed) strategy: E_i plays strategy R_j with probability p_j^i and can hence be characterized by a vector $\underline{p}^i \in S_N$. We shall denote the frequency of E_i in the population by x_i and by

$$m_k = \sum_{i=1}^{n} x_i p_k^i \tag{26}$$

the frequency of strategy R_k. Thus $\underline{x} \in S_n$ and $\underline{m} \in S_N$ denote the distributions of phenotypes and of strategies, respectively.

In many situations the payoff for strategy R_j against a population in state $\underline{x}$ depends only on the distribution $\underline{m}$ of the strategies, and hence is given by $f_j(\underline{m})$. (Such games, which Thomas (1985) calls degenerating, are indeed quite common: we shall presently encounter two simple examples of them). In this case, the payoff for phenotype E_i is given by

$$\underline{p}^i \cdot \underline{f}(\underline{m}) = \sum_{j=1}^{N} p_j^i f_j(\underline{m}) \tag{27}$$

while the average payoff in the population is

$$\sum_{i=1}^{n} x_i(\underline{p}^i \cdot \underline{f}(\underline{m})) = \underline{m} \cdot \underline{f}(\underline{m}) . \tag{28}$$

If the payoff corresponds to biological success, it is natural to use the game dynamical "Ansatz": the rate of increase $\frac{\dot{x}}{x_i}$ of phenotype E_i corresponds to the difference between its payoff and the average payoff. This leads to the replicator dynamics

$$\dot{x}_i = x_i[(\underline{p}^i - \underline{m}) \cdot \underline{f}(\underline{m})] . \tag{29}$$

In Sigmund (1986) it is shown that if the "auxiliary dynamics" (25) is a Shashahani gradient, then so is the "mixed strategist dynamics" (29). Both equations have the same potential. In particular, if there are only two pure strategies R_1 and R_2, then (14) is trivially satisfied for (25), since there are no three pairwise different indices i, j, k. Thus the corresponding mixed strategist games are always gradients and satisfy some maximum principle.

As an example, let us consider the "sex ratio game" where the two pure strategies R_1 and R_2 are "produce only males" and "produce only females". The phenotypes E_i correspond to different sex ratios p_i. The number of children is supposed to be independent of the sex ratio: as Fisher (1930) has pointed out, the number of grandchildren is then a function of the sex ratio and can be viewed as payoff for the sex ratio.

Let us denote the average sex ratio $\sum x_i p_i$ in the population by m, and assume random mating. The probability that a given male (resp. female) is the parent of a given child is proportional to $\frac{1}{m}$ (resp. $\frac{1}{1-m}$). The average number of grandchildren produced by a son (resp. a daughter) is therefore also proportional to $\frac{1}{m}$($resp.$ $\frac{1}{1-m}$), so that

$$\frac{p_i}{m} + \frac{1-p_i}{1-m} \tag{30}$$

is a measure for the reproductive success of the sex ratio p_i. Equation (29), then, turns into

$$\dot{x}_i = x_i(p_i - m)\left(\frac{1}{m} - \frac{1}{1-m}\right) . \tag{31}$$

It is easy to check, now, that

$$V(m) = \log m(1-m) \tag{32}$$

is a Shahshahani potential for (31). Thus the product $m(1-m)$ increases to its maximum at $m = \frac{1}{2}$ (this principle was first formulated by Shaw and Mohler (1953)) and it increases at a maximal rate.

Another example is given by the Hawk-Dove-game (cf. Maynard Smith (1982)). If the different phenotypes correspond to different probabilities of escalating a fight, then their evolution is described by a gradient system (see Sigmund (1986)).

5. More Gradients in Selection

Hofbauer (1985a) has shown that gradients occur in certain fertility equations.

If there are two alleles A_1 and A_2 and hence three genotypes $G_1 = A_1A_1$, $G_2 = A_1A_2$ and $G_3 = A_2A_2$, with frequencies x, y, and z, and if one assumes that F_{ij} is the fertility of the mating between a G_i-male and a G_2-female, then one obtains the fertility equation on S_3

$$\dot{x} - x\bar{F} = f_{11}x^2 + f_{12}xy + \frac{1}{4}f_{22}y^2 \tag{33}$$

$$\dot{y} - y\bar{F} = \frac{1}{2}f_{22}y^2 + f_{12}xy + f_{23}yz + 2f_{13}xz$$

$$\dot{z} - z\bar{F} = f_{33}z^2 + f_{32}yz + \frac{1}{4}f_{22}y^2$$

where $f_{ij} = (F_{ij} + F_{ji})$ and $\bar{F}$ is the sum of the right hand sides. With $u = \frac{x}{y}$ and $v = \frac{z}{y}$, this yields, up to a change in velocity, the equations

$$\dot{u} = a(u) - vb(u) \tag{34}$$

$$\dot{v} = c(v) - ub(v)$$

with

$$a(u) = (f_{11} - f_{12})u^2 + (f_{12} - \frac{1}{2}f_{22})u \tag{35}$$

$$b(u) = f_{23}u + 2f_{13}u^2 > 0$$

and similar expressions for $c(v)$ and $d(v) > 0$. One may define an inner product in the tangent space to $int\ \mathbf{R}^2_+$ at $(u,v) \in int\ \mathbf{R}^2_+$ by

$$<(\xi_1,\eta_1),(\xi_2,\eta_2)>_{(u,v)} = \frac{1}{b(u)}\xi_1\xi_2 + \frac{1}{d(v)}\eta_1\eta_2 . \tag{36}$$

Then (34) is a gradient with respect to this non-Euclidean metric. The potential is given by

$$V(u,v) = \int \frac{a(u)}{b(u)}du - uv + \int \frac{c(v)}{d(v)}dv . \tag{37}$$

Hofbauer (1985a) has also investigated fertility equations with n alleles. If the fertility coefficient can be written either (a) as the sum of a male and a female contribution, or (b) as the product of the (sex-independent) contributions of the parental genotypes, then the differential equation for the gene frequencies reduces to the classical selection equation (17) and hence is a gradient.

6. Chemical and Ecological Gradients

In Schuster and Sigmund (1985) several molecular mechanisms of autocatalytic selfreproduction in a continuously stirred tank reactor are analyzed by means of gradients. An example is given by the reactions

$$A + I_i \underset{f'_i}{\overset{f_i}{\rightleftarrows}} 2I_i \tag{38}$$

$(i = 1, \ldots, n)$, where I_i denotes some type of polynucleotide and A is an energy rich compound. The constants f_i and f_i' denote the rates of formation and decay. One assumes that A is fed into the reactor at a constant concentration a_0, and that the rate of the flow through the reactor is given by r. The concentrations x_k of I_k then obey the equations

$$\dot{x}_i = x_i[f_i(a_0 - \sum x_i) - f_i' x_i - r]. \tag{39}$$

If one introduces as inner product for the tangent space to $int\ \mathbf{R}_+^n$ at $\mathbf{x} \in int\ \mathbf{R}_+^n$

$$(\underline{\xi}, \underline{\eta})_{\mathbf{x}} = \sum \frac{1}{f_i x_i} \xi_i \eta_i \tag{40}$$

then

$$V(\mathbf{x}) = \sum_{i=1}^{n} \left[a_0 - \frac{r_i}{f_i} - \frac{f_i' x_i}{2f_i} - \frac{1}{2} \sum_{j=1}^{n} x_j \right] x_i \tag{41}$$

is a potential for (39). This system leads to a globally stable equilibrium.

Similarly, the three-molecular autocatalytic mechanisms

$$A + 2I_i \underset{f'_i}{\overset{f_i}{\rightleftarrows}} 3I_i \tag{42}$$

$(i = 1, \ldots, n)$ is described by

$$\dot{x}_i = x_i[f_i(a_0 - \sum x_j)x_i - f_i' x_i^2 - r]. \tag{43}$$

An appropriate inner product, this time, is given by

$$[\underline{\xi}, \underline{\eta}]_{\mathbf{x}} = \sum_{i=1}^{n} \frac{1}{f_i x_i^2} \xi_i \eta_i . \tag{44}$$

Equation (43), then, is a gradient with potential

$$V(\mathbf{x}) = \sum_{i=1}^{n} \left[(a_0 - \frac{f_i' x_i}{2f_i} - \frac{1}{2} \sum x_j) x_i - \frac{r \log x_i}{f_i} \right]. \tag{45}$$

This system may lead to multiple stable equilibria.

Similarly, the ecological equation (1) is a gradient with respect to the Shahshahani metric given by (6) (but this time in $int\ \mathbf{R}_+^n$) iff $\dot{x}_i = f_i(\mathbf{x})$ is an Euclidean gradient. In particular, the Lotka Volterra equation

$$\dot{x}_i = x_i(r_i + \sum a_{ij} x_j) \tag{46}$$

is a gradient with respect to this metric iff $a_{ij} = a_{ji}$. If there exist constants $f_i > 0$ such that

$$a_{ij}f_j = a_{ji}f_i \tag{47}$$

for all i and j, then the appropriate metric is given by (40). Thus every Lotka Volterra equation modelling competition or symbiosis between two species is a gradient.

7. Gradients and Partnership Games

In classical game theory, the two players have usually different roles. Similarly, one frequently encounters in evolutionary game theory conflicts between two different populations (e.g., between males and females, or predators and preys, or owners and intruders of a territory). If one denotes by E_1 to E_n the phenotypes of one population (with frequencies x_1 to x_n), and by F_1 to F_m the phenotypes of the other one (with frequencies y_1 to y_m) then the state of the populations is given by $\underline{x} \in S_n$ and $\underline{y} \in S_m$. The payoff is described by the matrices A and B : if an E_i-strategist meets an F_j-opponent, then a_{ij} will be the payoff for the former and b_{ji} for the latter. The usual game dynamical "Ansatz" now yields

$$\dot{x}_i = x_i((A\underline{y})_i - \underline{x}\cdot A\underline{y}) \tag{48}$$

$$\dot{y}_j = y_j((B\underline{x})_j - \underline{y}\cdot B\underline{x})$$

for $i = 1,\ldots,n$ and $j = 1,\ldots,m$. These games are investigated thoroughly in Hofbauer and Sigmund (1984) and Hofbauer (1986).

A bimatrix game is said to be a rescaled partnership game if there exists an $n \times m$ matrix C, vectors $\underline{c} \in \mathbb{R}^m$ and $\underline{d} \in \mathbb{R}^n$, and positive constants α and β such that for all i and j

$$a_{ij} = \alpha c_{ij} + c_j \qquad b_{ji} = \beta c_{ji} + d_i\,. \tag{49}$$

It is easy to see that in this special case, the interests of the two players essentially coincide. It is not astonishing, therefore, that there is a maximum principle for the payoff. Indeed, Hofbauer (1986) has shown that for rescaled partnership games, (48) is a gradient with respect to $\underline{x}\cdot C\underline{y}$ (essentially the payoff for both populations). As an inner product in the tangent space at $(\underline{x},\underline{y}) \in int\ S_n \times S_m$, one uses

$$\langle(\underline{\xi},\underline{\eta}),(\underline{\xi}',\underline{\eta}')\rangle_{(\underline{x},\underline{y})} = \sum_i \frac{\xi_i \xi_i'}{\alpha x_i} + \sum_j \frac{\eta_j \eta_j'}{\beta y_j}\,. \tag{50}$$

A bimatrix game described by A and B is a rescaled partnership game iff there exists a constant $\gamma > 0$ such that

$$a_{ij} - a_{il} - a_{kj} + a_{kl} = \gamma(b_{ij} - b_{il} - b_{kj} + b_{kl}) \tag{51}$$

for all $i,k \in \{1,\ldots,n\}$ and $j,l \in \{1,\ldots,m\}$. This corresponds to the "triangular condition" (16).

Acknowledgement: Research for this paper was partly supported by the Austrian "Forschungsförderungsfonds" Project P5995.

8. References

Akin, E. (1979) The geometry of population genetics. Lecture Notes in Biomathematics 31. Springer-Verlag, Berlin-Heidelberg-New York.

Crow, J.F., and M. Kimura (1970) An introduction to population genetics theory. New York: Harper & Row.

Epstein, I. (1979) Competitive coexistence of selfreproducing macromolecules. J. Theor. Biology 78, 271-298.

Fisher, R.A. (1930) The genetical theory of natural selection. Oxford: Clarendon Press.

Hadeler, K.P. (1981) Stable polymorphisms in a selection model with mutation. SIAM J. Appl Math. 41, 1-7.

Hofbauer, J. (1985a) Gradients versus cycling in genetic selection models, in "Dynamics of Macrosystems", eds. J.P. Aubin, D. Saari and K. Sigmund, Lecture Notes in Economics and Mathematical Systems 257, Springer-Verlag, Berlin-Heidelberg-New York.

Hofbauer, J. (1985b) The selection mutation equation. J. Math. Biol. 23, 41-53.

Hofbauer, J. (1986) Dynamics of asymmetric conflicts, to appear.

Hofbauer, J. and K. Sigmund (1984) Evolutionstheorie und dynamische Systeme. Verlag P. Parey, Hamburg-Berlin.

Jones, B., R. Enns, and S. Ragnekar (1976) On the theory of selection of coupled macromolecular systems. Bull. Math. Biol. 38, 15-28.

Maynard Smith, J. (1982) Evolutionary game theory. Cambridge University Press, Cambridge, U.K.

Schuster, P. and K. Sigmund (1983) Replicator Dynamics. J. Theor. Biology 100, 533-538.

Schuster, P. and K. Sigmund (1985) Dynamics of evolutionary optimization. Ber. der Bunsen Gesellschaft für phys. Chemie, 89, 668-682.

Shahshahani, S. (1979) A new mathematical framework for the study of linkage and selection. Memoirs AMS 211.

Shaw, R.F. and J.D. Mohler (1953) The selective significance of the sex ratio. Amer. Nat. 87, 337-342.

Sigmund, K. (1985a) The maximum principle for replicator equations, in "Lotka-Volterra approach to dynamical systems", ed. M. Peschel, Akademie-Verlag, Berlin.

Sigmund, K. (1985b) A survey on replicator equations. In "Complexity, Language and Life", Ed. J. Casti and A. Karlquist, Biomathematics 16, Springer- Verlag, Berlin-Heidelberg, New York.

Sigmund, K. (1986) Game dynamics, mixed strategies and gradient systems, to appear.

Taylor, P. and Jonker, L. (1978) Evolutionarily stable strategies and game dynamics. Math. Biosciences 40, 145-156.

Thomas, B. (1985) Evolutionarily stable sets in mixed strategist models. Theor. Pop. Biol. 28, 332-345.

Thompson, C. And J. McBride (1974) On Eigen's theory of selforganization of matter and evolution of biological macromolecules. Math. Biosciences 21, 127-145.

THE LIMITING FACTOR CONCEPT IN RELATION TO STABILITY, SENSITIVITY AND BIFURCATION IN PLANKTON MODELS

Tom Aldenberg*

INTRODUCTION

In biological water quality modelling, much attention is paid to the dynamics of the planktonic components of the aquatic system (e.g. Canale, 1976, Scavia & Robertson, 1979). Although deterministic dynamic models are considered to be the most successful, they may fail badly in certain circumstances. One reason may be the inadequacy of experimental methods in describing process mechanics (DePinto, 1979). Another, possibly related to the first, may be the fact that the models lack portability: a calibration based on one situation may not hold for another. In The Netherlands, we are facing the question, as to whether state-of-the-art models developed elsewhere, for example those for the US Great Lakes, can be applied to Dutch shallow lakes, which on occasion may have chlorophyll levels as much as 25 to 50 times higher. Without re-calibration certainly not. But not only the values of the coefficients are at stake. We don't know to what extent their mathematical structure can be retained.

To be able to deal with these kind of questions, it is imperative to have a clear picture of the possible dynamic behaviour of plankton models for large variations of coefficient values. In this paper, I will investigate two simple constant coefficient plankton models, one for the lab and one for the field. The limiting factor concept will be generalized with the aid of sensitivity analysis and related to the stability and branching behaviour (bifurcation) of steady state solutions.

THE CHEMOSTAT

The single-nutrient-single-species chemostat or continuous culture (Herbert et al., 1956) yields a convenient play-model to illustrate some of the concepts and tools I want to use. The equations are:

$$\begin{cases} dx/dt = G.x - D.x, \\ G = g(s(t)), \\ s(t) = s_0 - Q.x(t). \end{cases} \quad (1)$$

Here, x(t) is the concentration of algal biomass in a culture vessel as a function of time. Coefficients G and D (both time^{-1}) are growth-rate and dilution rate (i.e. flow rate divided by the volume of the vessel), respectively. The growth-rate is a function g of s(t), the inorganic ambient nutrient concentration in the vessel, which

* RIVM (National Institute of Public Health and Environmental Hygiene)
P.O.Box 1, 3720 BA Bilthoven, The Netherlands.

is equal to the reservoir nutrient concentration, s_0, minus the concentration of nutrient tied to the biomass: Q.x. (Q is also called cell quota.) The latter algebraic identity follows from a mass balance on the total nutrient concentration (cf. Waltman, 1983, p.8 and 9). So the system is basically one-dimensional.
Let us assume g to be monotonically increasing and g(0)=0. The dilution rate acts as an overall death-rate for the biomass. We will now consider variations of D and s_0.

STABILITY OF STEADY STATES AND BIFURCATION

There are two possible steady state points satisfying:

$$f(x) = (g(s_0 - Q.x) - D).x = 0. \quad (2)$$

These are:

E_0: $x_0 = 0$, $s = s_0$;

E_1: $x_1 \neq 0$ satisfying $g(s_1) = D$, $s_1 = s_0 - Q.x_1$.

In Fig. 1, the situation with $D < g(s_0)$ is displayed. In this case, the dilution rate is smaller than the highest growth-rate possible, given s_0. For steady state x_e (e=0,1) to be stable, neighbourhood stability analysis (May, 1973) requires the community "matrix" (here scalar):

$$a_e = (\partial f/\partial x)_e = g(s_e) - D - g'(s_e).Q.x_e < 0, \quad (3)$$

where the prime denotes differentiation.
Hence, E_0 is stable iff:

$$a_0 = g(s_0) - D < 0,$$

and E_1 is stable iff:

$$a_1 = -g'(s_1).Q.x_1 < 0.$$

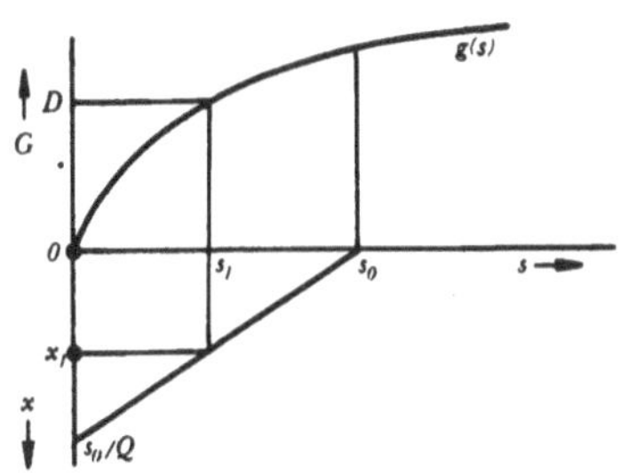

Fig. 1. Steady state in a chemostat.

Hence, for $D < g(s_0)$ as in Fig. 1, E_0 is unstable. In this case, also

$$0 < x_1 \leq s_0/Q,$$

that is, x_1 is feasible: not negative and not greater than the maximum possible biomass given s_0. (The maximum biomass is obtained in a batch culture: $D = 0$.) Since $x_1 > 0$ and $g' > 0$, $a_1 < 0$ and therefore E_1 is stable.

For $D > g(s_0)$, E_0 is stable (wash-out state). There is no feasible E_1 type of steady state now, but we could introduce a virtual steady state, as in optics, for infeasible points satisfying $f(x) = 0$.

Consequently, if $g(s_0) < D < g(\infty)$, E_1 is a virtual steady state with $x_1 < 0$ and $s_1 > s_0$. In this case, $a_1 > 0$. Hence, a virtual E_1 is unstable.

The advantage of considering a virtual steady state becomes apparent when the transition between the two cases considered is interpreted as a bifurcation problem. In Fig. 2, both steady states x_0 and x_1 are plotted as a function of D. Two branches arise, labelled E_0 and E_1, which intersect at $D = g(s_0)$, a branch point or bifurcation point (Diekmann, 1976). When D passes $g(s_0)$, the stability of E_1 is transferred to E_0. However, the unstable E_1 branch continues as a virtual steady state. So the branches really cross. In bifurcation theory parlance, a solution of the implicit equation:

$$f(x_e,D) = (g(s_0-Q.x_e) - D).x_e = 0$$

has index

$$\text{sign}(\partial f/\partial x)_e = \text{sign}(a_e).$$

Focussing on E_0 only, the index switches from +1 for $D < g(s_0)$ to −1 for $D > g(s_0)$. A bifurcation theorem says (Diekmann, 1976) that if the index changes sign in a point, that point must be a bifurcation point, i.e. another solution is crossing. This can only be E_1. The sum of the indices of all solutions to the implicit equation is a topological invariant called the degree of a map (here f). The degree does not depend on the value of the bifurcation parameter (D). In this case, the degree is $-1 + 1 = 0$. When the degree is used as a check on the number of solutions found, frequently virtual steady states have to be considered to complete the number of solutions: they are there, mathematically.

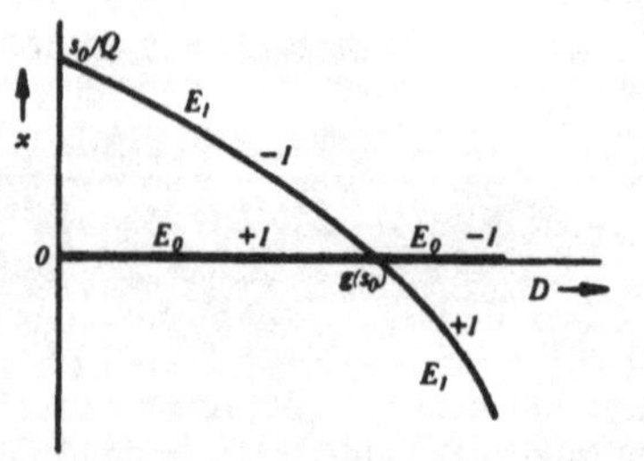

Fig. 2. Bifurcation diagram with D varying in the chemostat.

SENSITIVITY ANALYSIS AND THE LIMITING FACTOR CONCEPT

The slope of the E_0 and E_1 branch in Fig. 2 is recognized as the sensitivity coefficient, i.e. partial derivative:

$$\partial x_0/\partial D,\ \partial x_1/\partial D,$$

respectively. The analytic determination of these slopes is a matter of implicit differentiation of (2):

$$\partial x/\partial D = -(\partial f/\partial D)/(\partial f/\partial x) = x/a.$$

Hence,

$$\partial x_0/\partial D = 0,$$

$$\partial x_1/\partial D = x_1/a_1 = -1/(g'(s_1).Q) < 0.$$

By increasing the flow rate, the experimentalist can reduce the biomass until wash-out. These sensitivity coefficients can also be calculated in non-steady state (transient state) conditions. Suppose:

$$dx/dt = f(x,p) \tag{4}$$

is the dynamical system and p a coefficient (parameter). The sensitivity coefficient:

$$z(t) = \partial x/\partial p$$

satisfies the linear differential equation, or variational equation:

$$dz/dt = \partial f/\partial p + \partial f/\partial x.z \tag{5}$$

$$= \partial f/\partial p + a.z$$

(Himmelblau and Bischoff, 1968, p.270, Hale, 1969, p.21), with "a" the community matrix (3). The initial condition is $z(0) = 0$. One may either solve this equation simultaneously during simulation, or afterwards, x and possibly other time functions acting as forcing functions. The steady state sensitivity coefficient is:

$$z_e = -(\partial f/\partial p)_e/(\partial f/\partial x)_e$$

$$= -(\partial f/\partial p)_e/a_e, \tag{6}$$

(provided a_e is not zero), which conforms to the implicit differentiation of $f(x_e,p) = 0$.

Let us apply this method to the calculation of the sensitivity of the concentration of biomass to an increase in the input nutrient concentration in the chemostat:

$$z = \partial x/\partial s_0.$$

Now (5) reads:

$$dz/dt = g'(s).x + a.z,$$

with:

$$a = g(s) - D - g'(s).Q.x.$$

Since $g'.x > 0$ for $x(0) > 0$ and because $z(0) = 0$, z first gets positive. But, if $D < g(s_0)$, $g'.x$ will ever prevent z from becoming zero. This means that the biomass will always be positively sensitive to an increase in nutrient input, also during transients. Great care must be taken to correctly interpret transient sensitivity coefficients. They describe how much the biomass would have been different at a certain moment of time, if a small change had occurred in the value of the parameter <u>all the way from the beginning</u> of the simulation or the experiment. In steady state, we have:

$$z_e = -g'(s_e).x_e/a_e \qquad (e = 0,1), \text{ i.e.}$$

$$z_0 = \partial x_0/\partial s_0 = 0,$$

$$z_1 = \partial x_1/\partial s_0 = 1/Q.$$

These slopes can be visualized in a branching diagram (Fig. 3) similar to Fig. 2. D is supposed to be fixed and

$$s_1 = g^{-1}(D),$$

with g^{-1} the inverse function of g, is the bifurcation point. For $s_0 < s_1$ only the zero-state E_0 is stable. No life is possible below threshold input concentration s_1: x_1 is a virtual steady state. Beyond s_1 the stable steady state biomass is linearly related to s_0.

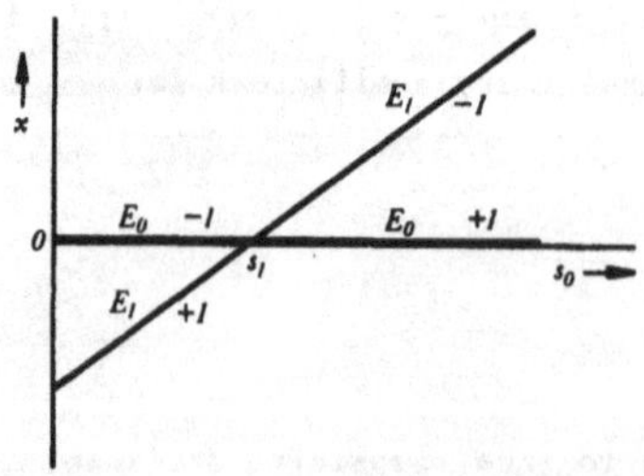

Fig. 3. Bifurcation diagram with s_0 varying.

In a batch culture ($D = 0$), we have $s_1 = 0$, i.e. no free nutrient remains. Fig. 3, with s_1 in the origin, then visualizes the sensitivity of the resulting biomass to the initial total nutrient concentration, which is the basic idea behind nutrient enrichment tests.
We observe that the above treatment yields an interesting connection between z, the sensitivity of the biomass to the nutrient input concentration, g', the sensitivity of the growth-rate to the ambient nutrient concentration level and the experimental procedure to enforce a nutrient limitation in a culture. The approach generally followed experimentally is to take the input concentration of one nutrient relatively low with respect to the concentrations of other nutrients and light intensity and monitor the result through the biomass level. This is essentially what z expresses: the response of the biomass to variations in the input medium concentration. Then, given the composition of the reservoir medium, the dilution rate is used to scan the growth-rate g in relation to the ambient nutrient level in het culture vessel. This kinetic information is generally taken to be the most transportable, e.g. with regard to estimates of growth-rates in a lake. Either $g/g(\infty)$ or g' is an indicator of how much the nutrient is limiting the growth-rate. We just found out in the simple chemostat equations that the sensitivity of the biomass with respect to the input concentration is strongly related to the slope of the growth-rate. So, good old sensitivity analysis is the missing link between nutrient limitation on the basis of growth-rates and nutrient limitation on the basis of biomass. I propose to regard

$$\partial x / \partial s_0$$

as a measure of the limitation of the biomass by a nutrient in all kinds of different systems receiving s_0 as input concentration. In simple test systems like chemostats and batch cultures (enrichment tests), this conforms to the experimental approach and the intuitive idea behind the limiting factor and relates mathematically to the process-oriented point of view. In more complex systems like more advanced nutrient uptake models, models of microcosms and lakes, this relationship is more intricate and subtle and therefore of great interest. These models are built on a process-oriented basis, so growth-rates, half-saturation constants, inhibition levels, etc. are all there. But then, we are interested in the overall response of the biomasses -- not only those of the algae -- to input concentration levels, certain stresses, treatment design in lake restoration and so on, and these are readily calculated as sensitivity quotients, which were to be calculated already for systems analysis or statistical reasons anyway.
If a biomass level is sensitive to an input nutrient concentration, we call that species or trophic level limited by the nutrient with regard to the particular system under study. The latter stipulation is essential. The growth-rate based limitation often has a physiological basis, i.e. is organism oriented, while the biomass based limitation is (eco-)system oriented. Identical growth-rates may lead to different sensitivities in different systems, and therefore to different limitations on the basis of biomass.

SENSITIVITY AND STABILITY

The variational equation (5) can be linearized around steady state sensitivity (6) to yield:

$$d(z-z_e)/dt = a_e \cdot (z-z_e),$$

with a_e the steady state community matrix again. Thus the steady state sensitivity coefficient is (un)stable iff the steady state itself is (un)stable.

A transient initial state always has sensitivity coefficient zero, however near to a steady state, since an initial state can be chosen independently of parameter values. It turns out that an unstable steady state may have a positive sensitivity coefficient, while in the direct vicinity, transient sensitivity coefficients may be negative. An easy example is given by the system:

$$dx/dt = x - p$$

having solution:

$$x(t) = p + (x_i - p).\exp(t),$$

where x_i is the initial condition. Steady state $x_e = p$ is unstable. The sensitivity of the steady state to the parameter is positive:

$$z_e = \partial x_e / \partial p = 1$$

and also unstable. <u>All</u> transient sensitivities are:

$$z(t) = \partial x / \partial p = 1 - \exp(t),$$

i.e. $z(0) = 0$ and $z(t) < 0$ for $t > 0$. Clearly, $z_e = 1$, the unstable steady state sensitivity coefficient, is reached for time going to <u>minus</u> infinity. Hence, an unstable steady state may be positively sensitive, while nearby transient states may be negatively sensitive.

A less artificial example is given by a one-humped g(s) in the chemostat equations (1), Fig. 4.

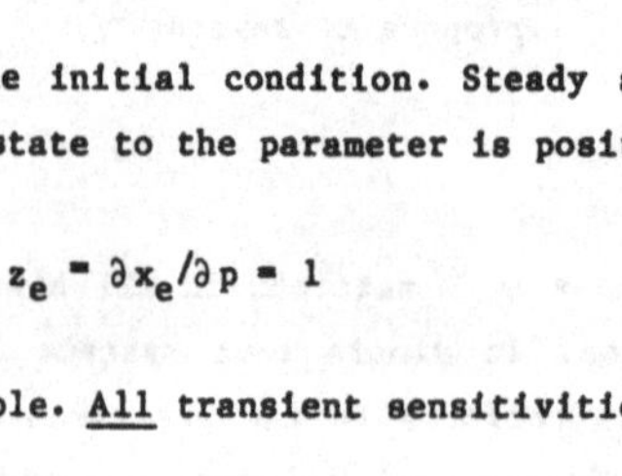

Fig. 4. Steady states in the chemostat with a one-humped growth-rate.

Now there may be three steady states: E_0, E_1 and E_2. Graphically, one can easily see that both x_1 and x_2 are positively sensitive to s_0, i.e. are nutrient limited, which is illustrated in Fig. 4 by the parallel lines. However, transient states near x_2 experience negative sensitivity due to the inhibition of the growth-rate. The leading term in (5) is $g'(s).x < 0$ for x near x_2 and this prevents z(t) from ever becoming positive as long as s is to the right side of the hump. In $x = x_2$, the unstable steady state, the sensitivity coefficient is

$$z_2 = 1/Q > 0.$$

(A bad dream: if a decreasing growth-rate results from a toxicant, while a nutrient stabilizes the steady state, the biomass may become toxicant-limited.)

A SIMPLIFIED WATER QUALITY MODEL

The next two-dimensional system:

$$\begin{cases} dx/dt = g(s).x - c.y.x, \\ dy/dt = h(x).y - m.y, \\ s(t) = s_0 - Q.x(t) - R.y(t) \end{cases} \qquad (7)$$

contains some of the essential functionals of a nutrient (s), phytoplankton (x), zooplankton (y) water quality model (Di Toro et al., 1971). In this simplified model, growth-rate limitation due to light is not considered, as is phytoplankton natural mortality. Light limitation, both in a culture and in the field, can be treated in the same way as we have treated nutrient, but this will not be presented here. Both g and h, the specific rates of primary and secondary production respectively, are increasing functions of their arguments. Furthermore, g(0)=0 and h(0)=0. The term c.y.x is the grazing rate and m.y is the zooplankton mortality rate. Nutrient tied to the biomass, i.e. Q.x + R.y, is not available for growth. There is no detritus.
There are three possible steady states:

$$E_0\colon x_0 = y_0 = 0,\ s = s_0\ ;$$

$$E_1\colon x_1 = s_0/Q,\ y_1 = 0,\ s_1 = 0\ ;$$

$$E_2\colon x_2 = h^{-1}(m),\ y_2 = g(s_2)/c,\ s_2 = s_0 - Q.x_2 - R.y_2.$$

The latter is still in implicit form with respect to y_2. But, y_2, or better: $R.y_2$ can be graphically constructed as follows. Having available a plot or table of g(s), a Monod type function, say, then we have to plot

$$R.g(s)/c$$

as a function of s, which has the same form as g itself (Fig. 5). This establishes a steady state relationship between R.y and s. Now, x_2 is found from the inverse graph of h at the value of m (Fig. 6). So, $s_0 - Q.x_2$ is left over to $s_2 + R.y_2$. Thus the second steady state relationship between R.y and s is

$$s + R.y = s_0 - Q.x_2.$$

The intersection yields s_2 and $R.y_2$ (Fig. 5). But this graph can be inserted into a plot of the phase plane: R.y against Q.x (Fig. 7), with the origin of Fig. 5 at the point $(Q.x_2, 0)$ of Fig. 7 and E_2 is readily found. This procedure may be followed for all kinds of different m, i.e. different specific mortality rates of the zooplankton, which yields a curve of E_2 points with s_0 fixed (Fig. 7). Clearly, E_2 and E_1 meet for $Q.x_2 = s_0$. That means that for too high a zooplankton mortality rate coefficient, E_2 becomes virtual (Fig. 8):

$$Q.x_2 > s_0,$$

$$R.y_2 < 0.$$

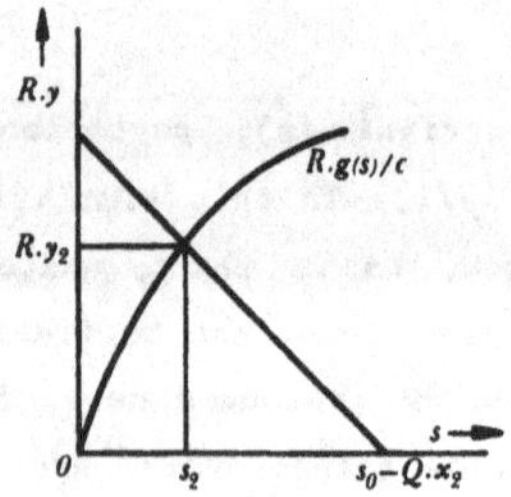

Fig. 5. Graphical construction of $R.y_2$ and s_2.

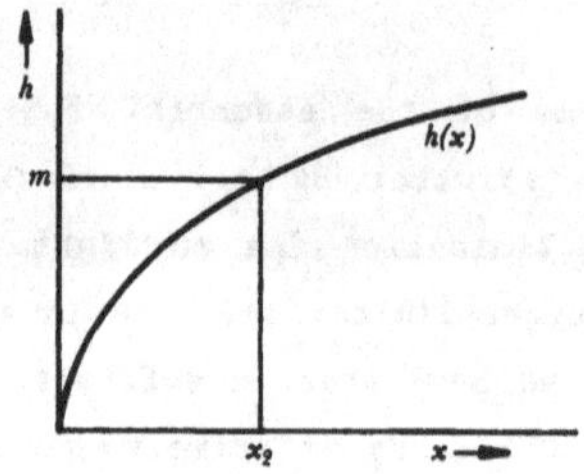

Fig. 6. Phytoplankton determined by the zooplankton parameters.

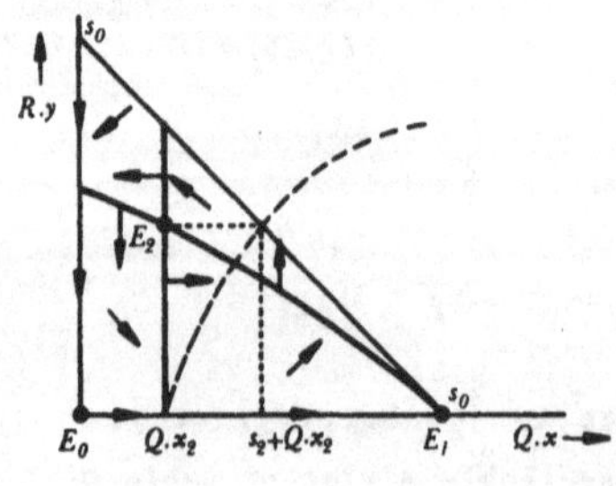

Fig. 7. Phase plane of (7): isoclines and construction of E_2.

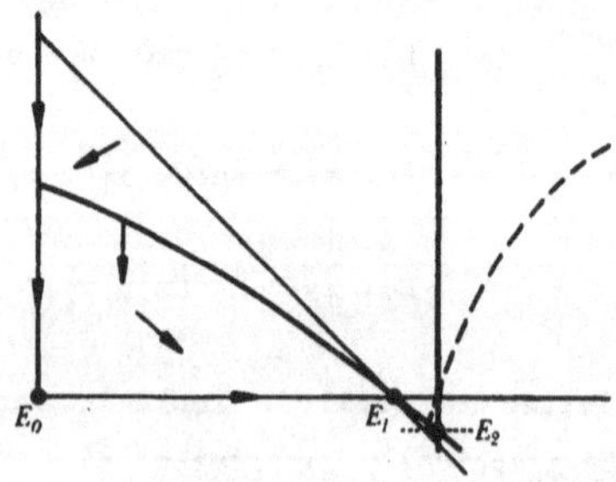

Fig. 8. Phase plane of (7): E_2 as a virtual steady state.

The procedure does not only work for different m's, but of course for different values of x as well. Since

$$R.y = R.g(s_0-Q.x-R.y)/c$$

also defines the phytoplankton isocline dx/dt = 0 for $x \neq 0$, the curve of E_2 points resulting from changing m is exactly the phytoplankton isocline. The zooplankton isocline is a vertical line at

$$Q.x = Q.h^{-1}(m).$$

A neighbourhood stability analysis (May, 1973) reveals that E_0 is unstable, a saddle point in fact, and that, if E_2 is within the triangle, than E_2 is stable and E_1 is unstable (Fig. 7). Here, the fact that both g and h are increasing functions is essential. For example, h increasing means that if $x_2 < x_1$, then $h(x_2) < h(x_1)$ and hence $m < h(x_1)$. That is, the zooplankton has the propensity to grow near E_1 while the phytoplankton wants the same thing: a saddle point. For $x_2 > x_1$, $h(x_1) > m$ and the zooplankton disappears. On the line y = 0, system (7) essentially behaves like the batch culture, system (1), with D = 0. Consequently, if E_2 is outside the triangle, E_1 is stable (Fig. 8). The virtual E_2 is unstable.

Fig. 9 depicts the steady state branching (bifurcation) diagram when m is varied. The zooplankton is at maximum for negligible m. Phytoplankton biomass may be very low then. When m increases, zooplankton diminishes and phytoplankton increases. This continues up to the bifurcation point:

$$m = h(s_0/Q),$$

where zooplankton has disappeared. Increasing m further does not alter the situation. Now E_1 is the stable point and zooplankton, although a potential component of the system, does not play a significant role anymore.

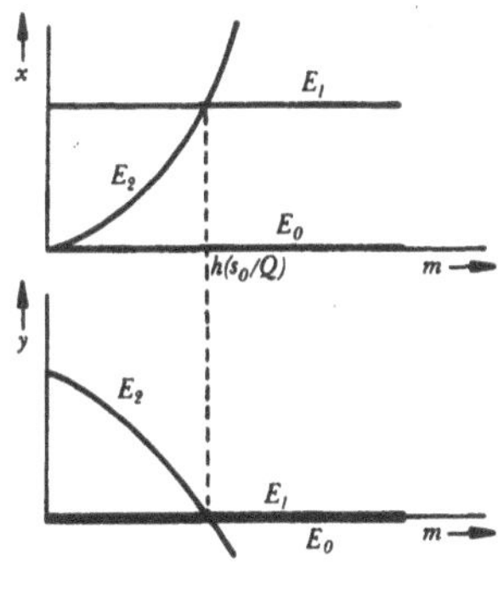

Fig. 9. Bifurcation diagram for m varying.

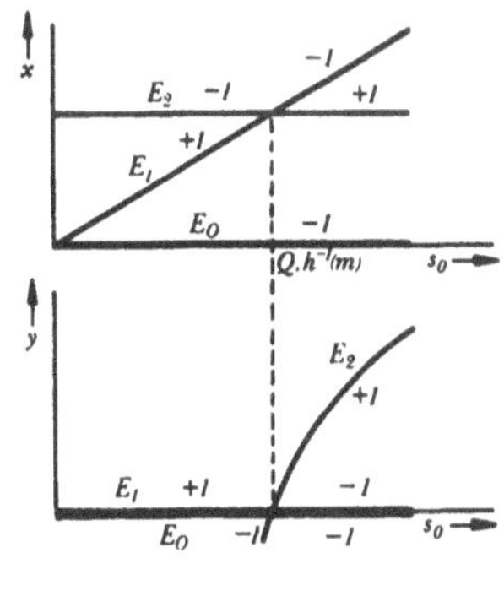

Fig. 10. Bifurcation diagram for s_0 varying.

The bifurcation diagram when varying s_0, the total nutrient concentration, illustrates the usefulness of the sensitivity approach to the limiting factor concept (Fig. 10). Up to the bifurcation point:

$$s_0 = Q.h^{-1}(m),$$

zooplankton does not exist and the system mimics a batch culture. Thus phytoplankton is nutrient limited and

$$\partial x_1/\partial s_0 = 1/Q.$$

Beyond the bifurcation point, there is room for the zooplankton. Now E_2 is stable and the steady state phytoplankton concentration is not limited anymore by the nutrient, i.e.

$$\partial x_2/\partial s_0 = 0.$$

The phytoplankton biomass is lower than it would have been without the zooplankton. In fact, a considerable amount of available nutrient may be left-over, so that the phytoplankton may very well grow at a nearly optimal rate, in contrast to the batch mode, where growth stops completely and the nutrient is depleted.

(The +1 and -1 in Fig. 10 are the indices of the particular branches. In the two-dimensional case, the index is defined as the sign of the determinant of the community matrix, i.e. the sign of the product of the two eigenvalues. A stable point now has index +1. A saddle point has index -1. The degree is -1 + 1 - 1 = -1.)

Beyond the bifurcation point, the zooplankton further increases with increasing s_0:

$$\partial y_2/\partial s_0 > 0 \quad .$$

and since any state variable that is sensitive to the external nutrient input concentration is defined to be limited by that nutrient, now the zooplankton is nutrient limited. Hence, when the internal steady state exists, the system oriented nutrient limitation is in fact carried over from the phytoplankton to the zooplankton.

More explicitly, the variational equation (5) is now a matrix differential equation, with

$$z = \begin{pmatrix} \partial x/\partial s_0 \\ \partial y/\partial s_0 \end{pmatrix} \quad ,$$

$$\partial f/\partial p = \begin{pmatrix} g'.x \\ 0 \end{pmatrix} \, ,$$

and the community matrix:

$$a = \partial f/\partial x = \begin{pmatrix} g-c.y-g'.Q.x & -g'.R.x-c.x \\ h'.y & h-m \end{pmatrix}.$$

The initial conditions are

$$\partial x/\partial s_0(0) = 0,$$

$$\partial y/\partial s_0(0) = 0.$$

Due to the term g'.x, <u>first</u> the nutrient limitation of the phytoplankton gets positive, while the initial slope of the zooplankton nutrient limitation is zero, since the zooplankton differential equation does not depend on s_0. After transients, the system goes to E_2 -- provided E_2 is within the feasibility triangle — and the stable steady state nutrient limitations settle at:

$$\partial x_2/\partial s_0 = 0/(h'.y) = 0,$$

$$\partial y_2/\partial s_0 = -g'.x/(-g'.R.x-c.x)$$

$$= g'/(g'.R+c) > 0.$$

That is, the phytoplankton nutrient limitation has disappeared, while the zooplankton has become nutrient limited.
Hence, the carry-over of the nutrient limitation from the phytoplankton to the zooplankton in fact can be followed dynamically in time.

AVENUE

These effects found in a simplified water quality model must be very common in ecological models and sensitivity analysis applied both in such qualitative analyses like the one here, as well as in numerical simulations, may be very useful in understanding the carry-over of factor or resource limitations to other ecosystem components, e.g. trophic levels. Through sensitivity analysis, the limiting factor concept can be generalized much beyond the seminal ideas of J. von Liebig and others (cf. Rodhe, 1978), at the same time providing a road to a more profound understanding of ecosystem response with respect to enrichment, toxicant stress and impoverishment.

ACKNOWLEDGEMENTS

I would like to thank Dr.Fritz Auert for participating in the very stimulating IIASA workshop and Ingrid Huijgen for typing assistance.

LITERATURE

[1] Canale, R.P. (ed.): Modeling biochemical processes in aquatic ecosystems. Ann Arbor 1976.

[2] DePinto, J.V.: Water column death and decomposition of phytoplankton: an experimental and modeling review. In: Scavia and Robertson, op.cit., 25-52.

[3] Di Toro, D.M., D.J. O'Connor, and R.V. Thomann: A dynamic model of the phytoplankton population in the Sacramento - San Joaquin Delta. Adv.Chem.Ser., 106 (1971), 131-180.
[4] Diekmann, O.: Bifurcation and stability of steady states. Nonlinear analysis, vol.1 (N.M. Temme, ed.). MC Syllabus 26.1, CWI Amsterdam 1976, p.71-103.
[5] Hale, J.K.: Ordinary differential equations. Wiley Interscience 1969.
[6] Herbert, D., R. Elsworth, and R.C. Telling: The continuous culture of bacteria: a theoretical and experimental study. J.Gen.Microbiol., 14 (1956), 601-622.
[7] Himmelblau, D.M., and K.B. Bischoff: Process Analysis and simulation - deterministic systems. Wiley 1968.
[8] May, R.M.: Stability and complexity in model ecosystems. Princeton University Press 1973.
[9] Rodhe, W.: Algae in culture and nature. Mitt.Internat.Verein.Limnol., 21 (1978), 7-20.
[10] Scavia, D., and A. Robertson (eds.): Perspectives on lake ecosystem modeling. Ann Arbor 1979.
[11] Waltman, P.: Competition models in population biology. SIAM, Philadelphia 1983.

EFFECTS OF STRUCTURED ENVIRONMENT TO DYNAMIC BEHAVIOR: SOME ILLUSTRATIONS

*G. Barna and P. Erdi**

Abstract

It is more or less accepted that biological development is driven by the interaction of the "potential of genome" and of "environmental" factors. The first of these was subject of an extended investigation while the environment was considered a disturbing factor.

Here we emphasize the role of environmental effects to autonomous system. Particularly, two examples are given representing quite different approaches. Their motivations come from different hierarchical levels of theoretical neuroscience, but the methods and results might have general heuristic power as well.

Periodic perturbation of the transmitter-recycling system [3]

An illustration is presented for effects of structured environment at synaptic level in connection with interactions of certain neurochemical oscillators. Rhythmic behavior is characteristic for biological systems. The stability of most biological rhythmic phenomena can be understood by the concept of limit cycle. The independence of the amplitudes and frequencies from the initial conditions might be associated to regular temporal patterns. The mutual interaction among different biological oscillators, however, might imply different phenomena like entrainment (phase locking), complex oscillations, quasiperiodicity, or even chaos (see e.g. [5]). The regular periodic operation of synaptic level rhythmic generator requires a *fine-tuned* neurochemical *control system*. Even mild impairment of metabolism leading to acetylcholine synthesis might imply "abnormal" dynamic synaptic activity to be associated to neurological disorders. In general terms, many neutral disorders are characterized by changing the "normal" temporal patterns to "abnormal" behavior. This abnormal dynamics might be interpreted as a result of disease occurring in a physiological control system operating within a range of control parameters.

We restrict ourselves to study the skeleton model of slow oscillation due to the 'integrated synaptic activity'. According to the basic assumptions of our four compartmental model the state of the system is characterized by

- cytoplasmic ACh concentration: X
- ACh concentration at the postsynaptic membrane surface: Z

* Central Research Institute for Physics of the Hungarian Academy of Sciences, P.O. Box 49, H-1525 Budapest, Hungary

- Ch concentration neat postsynaptic cell: W
- Ch concentration near presynaptic cell: Y.

The model is:

$$\dot{x}(t) = -k_1 x(t) + k_4 y(t) + k_5 x^2(t) y(t)$$

$$\dot{z}(t) = k_1 x(t) - k_2 z(t)$$

$$\dot{w}(t) = k_2 z(t) - k_3 w(t)$$

$$\dot{y}(t) = k_3 w(t) - k_4 y(t) - k_5 x^2(t) y(t) + a \cdot \cos(wt + \varphi)$$

To give a rough estimation for the first order rate constants the relations $k_3 \sim k_4 << k_1 < k_2$ were adopted.

(i) **Phase locking** The main entrainment band ($w / w_0 \cong 1$) and its neighborhood was examined (w_0 is the inherent frequency of the undriven system). In a certain range around w_o, its width increasing with a, the perturbing signal ($a / a_o \cong 0.01$), where a_o is the amplitude of the inherent oscillation) the width of the main entrainment band is about 0.02 ($w_o \cong 0.23$). Other entrainment bands have also been studied.

(ii) **Transients: they can be interesting** In principle it is well known that infinite computational accuracy as well as infinite computational length would be necessary to obtain "complicated", asymptotic behavior. The study of attractors does not make superfluous to follow the complicated transient dynamics during intermediate time regimes. In the vicinity of $w / w_o = 2$ for $a = 0.5$ numerical calculations demonstrated the existence of a "phantom attractor" which seems to show 22T periodicity. It can be interpreted as a quasistable temporal structure. Though the "phantom structure" seemed to be stabilized, a very slight spontaneous modification of the trajectory implied "destabilization" leading to "simple" limit cycles.

Transitions induced by external noise: some comments

According to the traditional point of view, fluctuations are averaged out. It was clearly demonstrated that noise can support the transition of the system from one stable state to another stable regime. Since stochastic models might exhibit a qualitatively different behavior, then their deterministic counterparts, external noise can cause transitions to states which are not available (or even do not exist) in a deterministic framework.

The theory of 'noise-induced transition' [4] states that noise might play an active and a constructive role in the organization of ordered structures. Ebeling and his coworkers studied in a series of papers the effect of noise to oscillatory systems (e.g. [2]). Noise superimposed might change the qualitative behavior of the deterministic system by destabilizing temporary, 'metastable' structures. The often used but not always clearly defined concept of 'self-organization' might get a more precise meaning in the light of the results of the theory of noise-induced transition. The concept of environmental noise during growth has successfully been adopted in biological context.

Our particular example deals with formation and plasticity of ocular dominance columns [1].

The visual cortex considered as the best paradigm of modular organization of the neural centers. We are far from understanding the ontogeny of columnar structures in the cortex. Ocular dominance columns or stripes are characteristic examples of ordered neural structures. It is well known, that the term 'ordered' is here in connection with the fact that projections between the eye and brain are arranged retinotopically. This means that neighboring presynaptic cells project through their fibres onto neighboring cells of the postsynaptic sheet. The retinotopic map of the cortex might be considered a point-to-patch rather than a point-to-point representation.

In normal cats and monkeys the afferents from lateral geniculate nucleus (LGN) laminae corresponding to the two eyes innervete common target structure, e.g. layer IV of visual cortex and form partially overlapping bands. This alternating termination of LGN afferents showing about 300-400 μm wide periodicity is thought to be the anatomic substrate for the ocular domains.

The formation of mapping between two one-dimensional presynaptic and one postsynaptic arrays of cells is investigated. The initial state is a homogeneous arrangement. The iteration loop of the algorithm consists of several substeps:

(i) The activity of the presynaptic chains are determined by unstructured stimuli, only one randomly selected element in both presynaptic activity has non-zero value. (The physiological relevance can be improved by taking into consideration correlations among presynaptic activities.)

(ii) The activity of the postsynaptic chain increases due to the transfer of presynaptic information and decreases by a first-order decay process.

(iii) Hebbian rule supplemented with a selective decreasing term: the greater the contribution from the synapse to the output of the postsynaptic cell, the less the weakening of the synapse.

(iv) The normalization procedure prescribes that the sum of the synaptic strength on a cell is constant. This constraint induces competition among synapses of the same cell.

Simulation experiments can be carried out with a deterministic and stochastic "learning rule". In the latter case

(v) superimposed noise phenomena are also taken into consideration.

Deterministic simulation experiments (i.e. $k_6 = 0$) led mainly to locally ordered structures. The *inclusion of external noise* can drastically modify the macroscopic structure: globally ordered near-periodic structure appears during the simulation of ontogeny. The model is also capable of describing plastic properties. A partial postsynaptic lesion leads to a rearrangement of synaptic connections.

According to the model the connections between cells are not preprogrammed, only an algorithm is given to select favorable connections. Our two main points are the following: the formation of ocular dominance columns strongly depends on the activity of the system; *environmental noise may have a positive role in the structure generating mechanism.*

References

[1] Barna, G. and Erdi, P.: Pattern formation in neural systems II. Cybernetics and Systems (Trappl, R. ed.) Reidel 1986, 343-350.

[2] Ebeling. W. et al. ZAMM *66* (1986) 141-146.

[3] Erdi, P. and Barna, G.: Pattern formation in neural systems I. Cybernetics and Systems (Trappl, R. ed.) Reidel 1986, 335-342.

[4] Horsthemke, W. and Lefever, R.: Noise-induced transition. Theory and applications in physics, chemistry and biology. Springer, Berlin-Heidelberg-Tokyo, 1984.

[5] Tomita, K.: Chaotic response of nonlinear oscillators. Phys. Rep. *86* (1982) 113-167.

Modelling and Control of Yeast Growth in Biotechnical Systems

Th. Bley and B. Heinritz [1)]

Microbial growth is one of the fundamental processes of what is now called biotechnology.
The microbial population growing and metabolizing in a technical system - a fermentor - is a special type of an ecosystem. It is special in that we know much more on kinetic laws, on stationary states, and on dynamic behaviour than we do for natural ecosystems. It is special also in an other sense, namely that we can influence the environmental conditions for this ecosystem in a very broad range and so we can learn a lot of ecosystem properties in practice.
Much effort has been made by microbiologists, chemists, physicists, engineers and last not least mathematicians in modelling the ecosystem "microbial population + fermentation system". But we can state after some 40 years of quite intensive modelling that only a small minority of model-builders has been aware that the object described is an ecosystem.
To a first approximation we can say that microbial growth has usually been understood by biotechnologists as a chemical (or biochemical) process of reacting substances. The existence of individual organisms has been neglected. In these models (e. g., the MONOD-model /15/) the microbial population has been a homogeneously distributed, autocatalytically operating substance. Though these models have really been helpful in understanding and running continous fermentation processes they do not have the suprising or exciting dynamic properties which we know from the real processes.
Models of this type are commonly called unstructured models.
With our classification we follow a fundamental paper by FREDRICKSON and co-workers published in 1967 /6/.
Since this theoretical analysis of microbial growth models appeared we have known for sure that describing microbial growth by using unstructured models is not satisfactory in many important respects. The more transient and periodic fermentation processes attract the interest of biotechnologists, the more we need to structure our models if we want describe these processes reasonably.

1) Institute of Biotechnology of the Academy of Sciences of the GDR, 7050 Leipzig, Permoserstraße 15

The frequently applied method consists in linking biochemical entities, such as metabolites, enzymes etc. with the model variables (see e. g. /8/). The result are structured unsegregated models. This way is successful for modelling processes in which cells are in the same physiological state or, if this physiological state does not very much influence the course of the process, as is the case in making some metabolic products.
But, for running many fermentation processes efficiently, it is of great importance that a microbial population should not be a homogeneous broth of reacting enzymes and metabolites but rather a relatively complex ecosystem of individual cells distributed over a broad spectrum of physiological states (cf./13, 19/).
Modelling this quality we must use structured segregated models, that means, we have partial differential equations which describe the motion of the population in a high dimensional state space. Practical use of structured segregated models requires the production of equations which are suitable for mathematical treatment. Most trials have been made on the basis of one-dimensiohal continuous distribution functions describing one feature of a cell which is intended to characterize the cell state. Such a feature is, for instance, the age of a cell, and we have the VON-FOERSTER equation as microbial growth model /22/.
Another evident possibility is to choose the mass of a cell for characterizing the physiological state /21/. But this type of models has not led to the success hoped for in biotechnological practice.

The reason for this is that these models are partial differential equiations which do not lend themselves readily to mathematical treatment - particularly if the birth rate functions cannot be formal ones but must describe real processes, and, what is even more significant, there is no real connection between the age or the mass of a cell and its metabolic activity if we have nonbalanced growth conditions. The description of this connection is indispensable for modelling the dynamics of fermentation processes reasonably.
So we have a gap between practical suitability on the one side and a sufficiently comprehensive picture of the real process in mathematical equations on the other side. This gap could not be bridged by using continuous state spaces.
We tried to find a way out of the impasse by leaving formal thinking and switching to more properly biological considerations. All growth and differentiation processes of living beings are sequences of more or less sharply separated states with different physiological

properties; we may only think of the development stages of an insect or of the different growth characteristics of a cereal plant in the course of a year.
On the basis of theoretical calculations we could show that single cells which grow in distinctly separated phases having different metabolic properties show selection advantages over cells without such phases /5/.
For eukariotic cells such states are unambiguously characterized by the phases of the cell cycle and we could prove experimentally that these phases are connected with different physiological properties such as specific growth rate, carbon consumption rate, and efficiency of producing cell mass /4, 9/.
We refer to yeasts and not to all possible microorganisms not because the modelling framework does not lend itself readily to the description of bacteria, but because

- The demarcation of states is very clear for eukariotic microorganisms; possible demarcations for prokariotic organisms are being discussed.
- Yeasts are organisms which provide a good model for higher developed organisms and are therefore investigated to a large extent by microbiologists.
- Most of our results of new control strategies are given for the efficient running of yeast growth processes.

Lumping the kinetically similar S-, G2-, and M-phases of the eukariotic cell cycle together, we get two-state models as the first and simplest example for this model type. In the case of budding yeasts these states are connected with morphological features called the single and the budding cell. So we can construct our first "state-structure"-model on the basis of the kinetic scheme top left in the figure. This model was investigated previonsly and we could show some new and interesting properties /2/. The stationary solutions were consistent with those described by the MONOD model and observed in continuous fermentation processes; but, in contrast to unstructured modelling we could describe phenomena of fermentation processes such as autonomous oscillations (not generated by feedback inhibition as mostly assumed, but not always confirmed by experimental results) in connection with a synchronization of the population. Furthermore damped oscillations of process caracterizing values after shifting the fermentation conditions and, essential for process control, an entrainment of cell state distribution by periodically feeding inlet substrate are observed.
Together with Dr. B. WEGNER (K.-Weierstraß-Institut of Mathematics of the Ac.of Sci.of the GDR) we carried out a HOPF-bifurcation analysis on

a broader class of two-state growth models and we could show the splitting of periodic solutions for low dilution rates as a function of the substrate consumption kinetics in the different states.
It should be noted that this type of model resembles the VOLTERRA type very closely. A transformation in VOLTERRA equations creates no problems and, remarkably enough, without a loss of obviousness regarding the biological meaning of the variables (cf. /17/).
But we must concede that this type of model is not sufficient in some cases, for instance, if we try to describe the synchronization of a fermentation process by using a special control of carbon substrate supply not only qualitatively but also in the correct timing. This drawback could not be overcome by defining a greater number of states. The reason for this is that we must consider the regulation and control of how the cells pass through these states.
To do this, we use a hypothesis of SMITH and MARTIN (/20/ cf. /12/). This hypothesis assumes that the proliferation cycle of a cell can be divided into a "stochastic" and a "deterministic" state. The deterministic state essentially comprises the S-, G2-, and M-phases of the mitotic cycle, and it is of nearly constant duration. The stochastic state contains the G1-phase, and the transition from this state to the determine the transition probability from x_0 to x_1. It ist high for good conditions and low for unfavourable ones (cf. /18/).
from x_0 to x_1 and from the duration of x_1.
Experimental investigations of microbial growth have shown that even great changes in milieu conditions have only a very small effect on the duration of the deterministic state. But milieu conditions do determine the transition probability from c_0 to x_1. It is high for good conditions and low for unfavourable ones (df. /18/).
The mathematical model which follows is a delay-differential equation with a retardation term $(t - \tau)$, in which τ is the time spent by a cell in the deterministic state. It is shown in the lower part of the figure.
The model can be understood as an age-structured growth model as shown top right with a degenerate division probability function $\Gamma(a)$ /3/.

We should refer to NISBET and GURNEY, who in recent years developed a similar modelling framework for the mathematical description of insect populations and proposed the name "stage-structure" models for them /16/. In their notable publications they show how a mathematical model can be developed which is both graphic and highly flexible as is necessary for the reflection of the properties of biological systems dynamics.
Though the report given until now has only concerned model development, our effort was directed to modelling and simulating new and efficient

strategies of fermentation control. We could show by model simulations and with many experimental results that it is possible to increase, for instance, the specific biomass production per carbon consumption by changing the feed rate of carbon substrate for synchronized cultures according to the physiological properties of the corresponding cell states. /10, 11/. Thus it is possible to avoid an ineffective conversion of carbon substrate to carbon dioxide or e. g. organic acids.

This control concept could also be successfully employed to control the processing of metabolic products when the rate or the efficiency of production was coupled with specific cell states /7/.
Obviously, ecological systems should be controlled periodically, or in a broader sense, dynamically, in many cases because of their inherent properties if we intend a maximal advantage (cf. /1/).
Another approach to such a periodic control of biotechnological processes is given by Prof. KUBICEK in this volume.
Our intention for the further optimization of periodic control of the growth of microbial populations consists in the development of adaptive control algorithms. On the basis of cell-state structured growth models we construct a guiding control strategy for which an actual optimization is obtained on the basis of the on-line processing of suitable measuring values. Using an adaptive filter algorithm such an efficient control of a fermentation process was realized in co-operation with the Technical University of Leipzig /14/.

It should be noted that this control realizes the necessary synchronization of the process by use of system inherent signals. This means that our ecological view on the modelling and control of biotechnical processes does not only provide a new approach to theoretical considerations on microbial growth but also to the efficient running of commercial plants.

Acknowledgement
We thank Prof. K. P. HADELER for a helpful discussion.

<u>References</u>

/1/ BAILEY, J. E.: Chem. Eng. Comm. <u>1</u> (1974) 111

/2/ BLEY, Th. and SCHMIDT, A.: stud. biophys. 78 (1980) 11

/3/ BLEY, Th., HEINRITZ, B. and SCHMIDT, A.: stud. biophys. <u>98</u> (1984) 119

/4/ BLEY, Th. et al.: Z. Allg. Mikrobiol. <u>20</u> (1980) 275

/5/ BOGNER, H. L., BLEY, Th., SCHNEIDER, K. R.: Syst. Anal. Model. Simul. <u>1</u> (1984) 425

/6/ FREDRICKSON, A. G., RAMKRISHNA, D., and TSUCHIYA, H. M.: Math. Biosci. <u>1</u> (1967) 327

/7/ GUTHKE, R. and KNORRE, W. A.: Biotech. Bioeng. <u>24</u> (1982) 2129

/8/ HARDER, A. and ROELS, J. A.: Adv. Biochem. Eng. <u>21</u> (1982) 55

/9/ HEINRITZ, B. et al.: Z. Allg. Mikrobiol. <u>21</u> (1981) 583

/10/ HEINRITZ, B. et al.: Acta Biotech. <u>3</u> (1983) 125

/11/ HEINRITZ, B. et al.: Acta Biotech. <u>4</u> (1984) 275

/12/ LASOTA, A. and MACKEY, M. C.: J. Math. Biol. <u>19</u> (1984) 43

/13/ MALEK, I.: in FEMS Symp. No. 13, Academic Press, London 1982

/14/ MÜCKEL, B., BÖHME, B., and BLEY, Th.: Acta Biotech. to appear

/15/ MONOD, J.: Ann. Rev. Microbiol. <u>3</u> (1949) 371

/16/ NISBET, R. M. and GURNEY, W.S.C. Lect. Notes Biomath. <u>54</u> (1984) 97

/17/ PESCHEL, M. and MENDE, W.:"Leben wir in einer Volterra-Welt" Akademie-Verlag Berlin 1983

/18/ SAMOKHIN, G. P. et. al.: Exp. Cell. Res. 131 (1981) 267

/19/ SCHMIDT, A.: Acta Biotech. <u>3</u> (1983) 171

/20/ SMITH, J. A. and MARTIN, L.: Proc. Natl. Acad. Sci. USA <u>70</u> (1973) 1263

/21/ SUBRAMANIAN, G. et al.: Bull. Math. Biophys. <u>32</u> (1970) 521

/22/ VON FOERSTER, H.: in "The Kinetics of Cellular Proliferation" Grune and Stratton, New York 1959

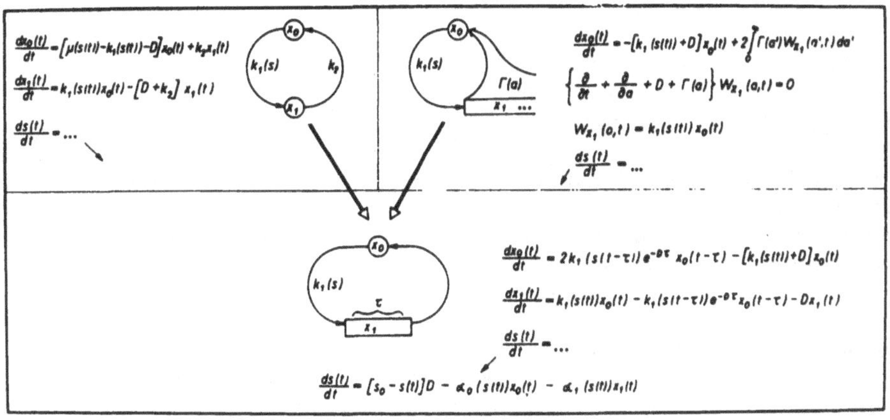

Scheme of Model Development

<u>Symbols</u>

a	- age
α_i (s)	- stoichiometric coefficient of the i-th cell state
D	- dilution rate of the CSTR
Γ(a)	- age dependent division rate
k_i (s), k_i	- transition rate between cell states
s	- (carbon) substrate concentration
s_o	- " in the influent medium
t	- time
τ	- retardation time, i.e., the time which a cell spends in the deterministic state
x_i	- concentration rsp. number of cells in the i-th cell state
W_{x_1} (a,t)	- nonnormalized age density of cells of state x_1

A DYNAMIC SIMULATION MODEL OF TREE DEVELOPMENT UNDER POLLUTION STRESS

Hartmut Bossel*

DESCRIPTION OF THE SIMULATION MODEL

The tree is a living system. The dynamic effects of leaf or root impairment on the life processes of the tree can therefore only be studied by looking at the tree as a dynamic system composed of many interrelated and functionally different components. The present paper describes a dynamic simulation model for a coniferous tree. Further related work is reported in Bossel 1982, Bossel/Metzler/Schäfer 1985, Metzler 1985 and Bossel 1986.

The objective of this study was to model essential processes of tree development, and to study how impairment of essential functions (photosynthesis, feeder root renewal) would affect this development. The basic structure of the model is shown in Fig. 1.

The possible effects of pollutants enter this system in two ways: (1) The decrease in specific photoproductivity of the leaves reduces the available photosynthate, and (2) a higher feeder root turnover rate drains the assimilate pool. The quantification for the SPRUCE model is fully based on empirical data for the different processes (e.g. specific photoproduction as a function of light and temperature, transpiration coefficient, respiration demands, etc.). Since the model can easily be run on a small microcomputer, the model equations are presented here as BASIC statements (in the correct order of computation, and omitting program control statements). The results presented here were computed using the DYSYS procedure for microcomputers (Bossel 1985), but the model statements can obviously also be implemented in DYNAMO or other languages.

The model parameters, their abbreviations, dimensions, numerical values, and respective data sources are listed in Table I. The state variables, their abbreviations, dimensions, initial values, and respective data sources are given in Table II. The table functions and other functions, their abbreviations, dimensions, sources, and tables of values are presented in Table III. The auxiliary and rate equations will be documented and explained in groups. The variables, their abbreviations and dimensions are listed in Table IV.

Leaf aging and leaf efficiency: (see the corresponding BASIC equations below). NLAG, LFAG: The natural leaf aging rate is increased by pollution damage. NLF: The current number of needle age classes is determined by the fact that needles are shed when their efficieny limit SHED has been reached. RPOL: The increase in leaf damage is partially offset by an improvement due to leaf renewal. EFF: From the aging rate, the relative efficiency of each needle age class can be computed; the average foliage efficiency is computed as the weighted average.

---

*Environmental Systems Research Group,
University of Kassel, D 3500 Kassel, F. R. Germany

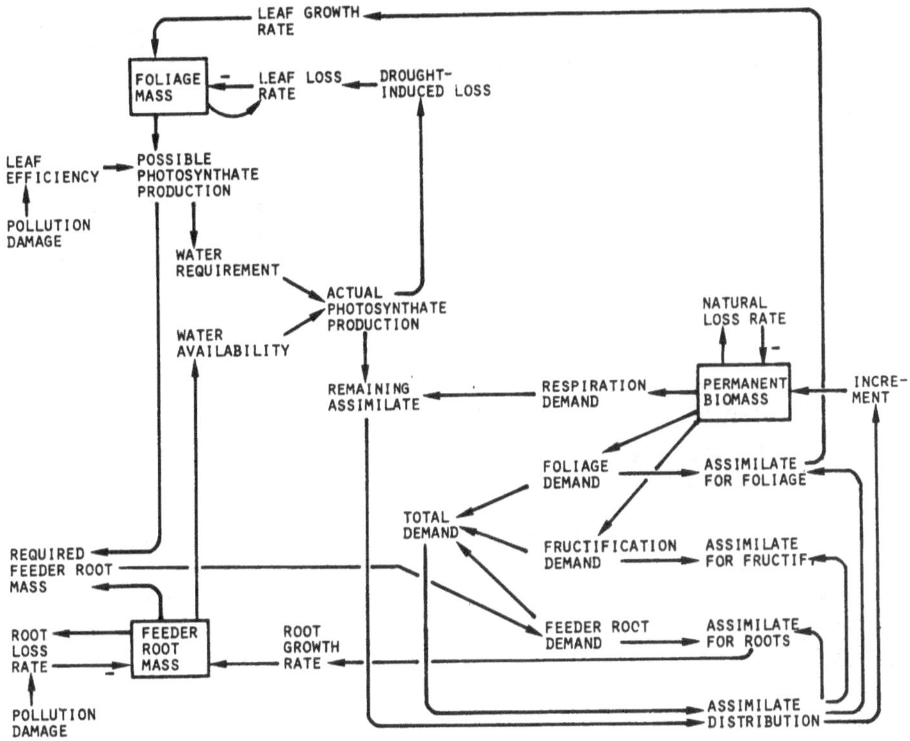

Fig. 1. **Simplified causal loop diagram for the SPRUCE model.** Photosynthate production is determined by foliage mass and transport capacity of the root system. Available assimilate is distributed according to priorities: respiration needs first, renewal needs second. The remainder, if any, is used for increment. If renewal needs are only inadequately satisfied, an accelerating process of deterioration (breakdown) begins.

```
NLAG=(1-SHED)/NDAG
LFAG=NLAG*(1+EPOL)
NLF=INT((1-SHED)/LFAG)
RPOL=LFDM-EPOL/NLF
SUM=0:FOR N=1 TO NLF:
E(N)=1-LFAG*((N-1)+TY):
SUM=SUM+E(N)*L(N): NEXT N
EFF=SUM/LEAF
```

Production of assimilate: MPRD: Assimilate production capacity is a function of leaf mass, net specific photosynthetic production, daylight period, temperature influence on photosynthesis, and relative foliage efficiency. DTRSP: This production capacity determines the required transpiration flow. TRSP: However, the actual flow is a function of feeder root mass, its transport coefficient, and soil moisture. RTRSP, PROD: If less water than required reaches the leaves, then the actual production of photosynthate is correspondingly less. RROOT, RLRT: The required feeder root mass follows from relating transport capacity to need. RTLOS: Natural feeder root decay is aggravated by root damage.

```
MPRD=LEAF*PHN*CLUX*CTEM*EFF
DTRSP=MPRD*TRSPC
TRSP=ROOT*TRANC*DROF
IF TRSP<DTRSP THEN RTRSP=TRSP/DTRSP ELSE RTRSP=1
PROD=RTRSP*MPRD
RROOT=DTRSP/TRANC
RLRT=RROOT/ROOT
RTLOS=NRTO*RTDM*ROOT
```

Respiration: RSLF, RSRT, RSST, RESP: The respiration of leaves, feeder roots, and wood biomass is a function of specific respiration coefficients and temperature. Since daylight respiration of leaves is accounted for in PHN, only dark respiration is considered in this case.

```
RSLF=LEAF*LRSP*(1-CLUX)*CTEM
RSRT=ROOT*RRSP*CTEM
RSST=BIOM*SRSP*CTEM
RESP=RSLF+RSRT+RSST
```

Distribution of assimilate if the supply is not sufficient to meet respiration needs: RAV: The relative availability of assimilate for respiration is computed by comparing the rate at which assimilate will be available to the rate at which it will be used for respiration (RESP). The rate of availability is the current production plus the consumption rate ASSI/CT of stored assimilate. LFLR, RTLR, BILR: The leaf, feeder root, and biomass losses due to insufficient respiration increase with decreasing RAV. The numerical time constants determine the speed of the deterioration due to respiration insufficiency.

```
RAV=(PROD+(ASSI/CT))/RESP
LFLR=LEAF*(1-RAV)*(52/13)
RTLR=ROOT*(1-RAV)*(52/13)
BILR=BIOM*(1-RAV)*(1/3)
RESP=RAV*RESP
```

Distribution of assimilate if the requirement for leaf and feeder root renewal cannot be fully met: NFOL: New foliage is grown only if the seasonal temperature has increased beyond TVEG. The demand for new leaves is determined by the foliage function, the normal number of needle age classes, and the amount of above-ground permanent biomass. DROOT: The assimilate demand for the growth of new feeder roots changes during the year. When the ground is too cold, no feeder root growth takes place. In the spring, before the onset of leaf growth, the demand is larger (by factor RTGF) than would be required for the mere replacement of losses; in the fall only the losses are replaced. During the vegetation period, the demand is also affected by the current relative feeder root requirement RLRT. CASS: If it turns out that the assimilates remaining after deduction of respiration needs are insufficient to meet the full demand for leaf and feeder root renewal (DGROW), then the supply reduction factor CASS<1 is applied to compute the actual amount available for both requirements (SLEAF, SROOT). SLEAF: The assimilate supply rate for leaf renewal has a time constant of 1 week (1/52 year) and is proportional to the remaining 'gap' between the foliage 'goal' (winter foliage plus new leaf demand LOLD + NFOL) and the current foliage LEAF. SINC: Since there is no assimilate surplus, there is no wood increment.

```
IF TEMP < TVEG THEN LOLD=LEAF
IF TEMP > TVEG THEN NFOL=(RFOL/NDAG)*BIOT ELSE NFOL=0
IF TEMP<TROOT THEN DROOT=0
IF TEMP>=TROOT AND TEMP<TVEG AND TY<.5 THEN DROOT=RTGF*RTLOS
IF TEMP>=TROOT AND TEMP<TVEG AND TY>.5 THEN DROOT=RTLOS
IF TEMP>TVEG THEN DROOT=RTGF*RTLOS*RLRT
DGROW=NFOL+DROOT
REST=PROD+(ASSI/CT)-RESP
CASS=REST/DGROW
IF TY<.5 THEN SLEAF=NFOL*CASS*(52/1)*(1-(LEAF/(LOLD+NFOL)))
ELSE SLEAF=0
SROOT=DROOT*CASS
SINC=0
```

Distribution of assimilate if requirements for leaf and feeder root renewal can be met: In this case, there is no reduction in SLEAF and SROOT, and the surplus assimilate is used for wood increment with a rate having a time constant of 2 weeks.

```
IF TY<.5 THEN SLEAF=NFOL*(52/1)*(1-(LEAF/(LOLD+NFOL)))
ELSE SLEAF=0
SROOT=DROOT
SINC=(REST-DGROW)*(2/52)
```

Other rates: LFDR: Leaf losses due to lack of nutrients or water appear if the flow from the feeder roots drops below 50 % of the required value. The loss rate is assumed to have a time constant of 26 weeks. LRST, LSHD: In the fall, after the end of the vegetation period, the oldest needle age class is shed at a rate having a time constant of 8 weeks. BLDRO, NBLR: Small biomass losses occuring even under normal conditions are augmented by losses due to drought conditions.

```
RTGR=SROOT: LFGR=SLEAF
IF RTRSP<.5 THEN LFDR=LEAF*(1-RTRSP)*(52/26) ELSE LFDR=0
IF TEMP>TVEG THEN LRST=LEAF-L(NLF)
IF TY>.5 AND TEMP<=TVEG THEN LSHD=(LEAF-LRST)*(52/8)
ELSE LSHD=0
BLDRO=BIOM*BLDR
NBLR=BIOM*NRBL
```

The **differential equations for the state variables** are written as Euler-Cauchy integrations using BASIC. ROOT: The feeder root mass increases with the assimilates reaching it, and decreases with turnover losses, and additional losses due to insufficient respiration activity. ASSI: The only assimilate gains are from photosynthetic production; losses are due to respiration, leaf and feeder root renewal, and biomass growth. L(i), LEAF: The model uses a variable number of needle age groups, the current number of age classes NLF being computed from the current state of leaf damage. At the beginning of a year, all age classes are shifted to the next higher year. The first age class is initially empty and is filled by new leaf growth during the spring (LFGR). Losses are possible in all age classes due to deficits in nutrient and water supply (LFDR), or due to deficits in assimilate for nighttime respiration (LFLR). In addition, the oldest age class is shed in the fall (LSHD). LEAF: The total foliage is the sum of the leaf mass contained in all age classes. BIOM: The permanent above-ground living biomass has gains from assimilate surpluses and losses from the drying up of wood due to drought conditions, from normal losses, and from possible losses due to a deficiency in the supply of respiration assimilates. EPOL: The pollution damage changes according to the rate computed earlier (RPOL). These Euler-Cauchy integrations converge for timesteps DT=1/52 (1 week) or smaller.

```
ROOT=ROOT+DT*(RTGR-RTLOS-RTLR)
ASSI=ASSI+DT*(PROD-RESP-SLEAF-SROOT-SINC)
L(1)=L(1)+DT*(LFGR-(LFDR+LFLR)*L(1)/LEAF)
L(NLF)=L(NLF)-DT*(LSHD+(LFDR+LFLR)*L(NLF)/LEAF)
FOR N=2 TO (NLF-1): L(N)=L(N)-DT*((LFDR+LFLR)*L(N)/LEAF): NEXT N
LEAF=0: FOR N=1 TO NLF: LEAF=LEAF+L(N): NEXT N
BIOM=BIOM+DT*(SINC-BLDRO-NBLR-BILR)
EPOL=EPOL+DT*RPOL
```

SIMULATION RESULTS

The simulation model produces three behavioral modes: (1) normal growth under subcritical pollution stress, (2) stagnation for somewhat higher (subcritical) stress, and (3) eventual breakdown of the tree under supercritical conditions. The reasons for these distinct behavioral modes can be found by identifying the dominant system loops in each case (Fig. 2).

Due to the complexity of the SPRUCE model (especially the table functions and other nonlinearities), analytic expressions for the stagnation and breakdown limits cannot be easily formulated. However, the numerical computation of these limits is straightforward. Fig. 3 shows the results from 250 simulations for young spruce (initial permanent biomass BIOMO = 50 t ODM / ha). The diagram indicates whether a given combination of leaf and feeder root damage will (1) allow growth (small dots), or (2) result in stagnation (medium dots), or (3) lead to tree breakdown (large dots).

An important conclusion from these studies is the observation that younger trees have a much better tolerance to pollution damage than older trees. In fact, pollution levels leading to certain breakdown of older trees may still allow some growth in younger trees.

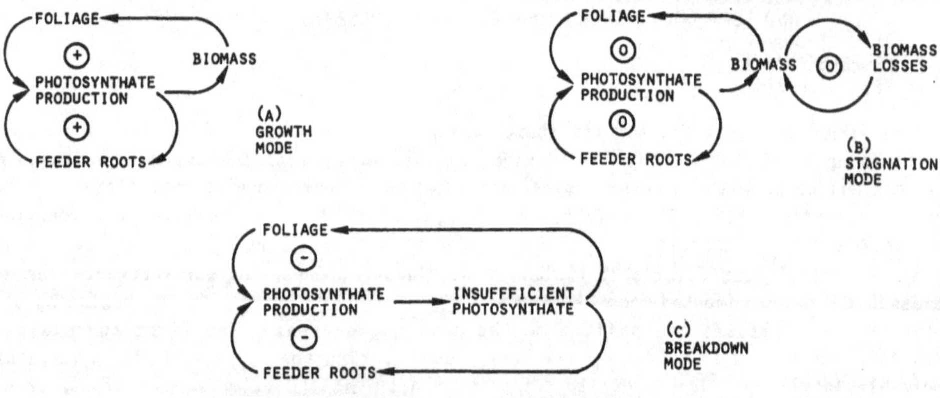

Fig. 2. **Dominant system loops determining the three modes of behavior.** (A) In the growth mode, surplus photosynthate leads to increase in biomass, foliage, and feeder roots, and hence further enhancement of photosynthetic production. (B) In the stagnation mode, photosynthate production just suffices to keep up with natural losses of biomass, leaves, and feeder roots. (C) In the breakdown mode, an insufficient supply of photosynthate causes incomplete replacement of leaves and feeder roots, leading to further accelerated reduction of photosynthetic production, and of leaf and feeder root mass.

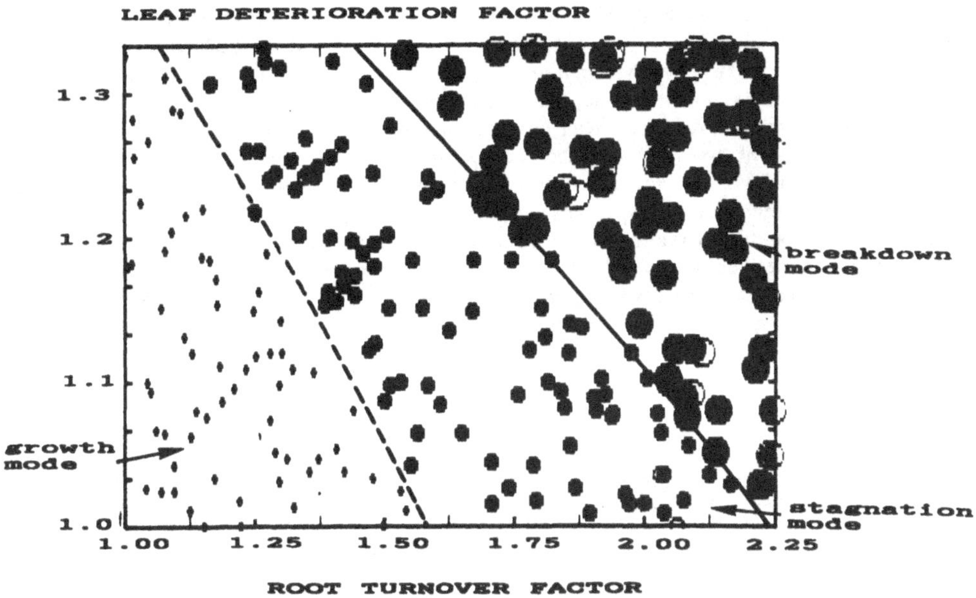

Fig. 3. **Behavioral mode diagram for young spruce**, obtained from 250 simulations using random values for leaf deterioration factor LFDF and root turnover factor RTOF. (1) Growth mode: small dots; (2) Stagnation mode: intermediate dots; (3) Breakdown mode: large dots. Initial conditions: BIOM=50, ASSI=2.5, ROOT=7, LEAF=7, NDAG=7. Simulation over 20 years using the SPRUCE model. The diagram clearly shows the existence of three distinct behavioral regions.

CONCLUSIONS

(1) Systems analysis of the structure and major processes of a tree suggests forest die-back to be a 'natural' reaction of trees to chronic pollution stress: there is probably no need to search for a mysterious 'factor X' which would cause the forest die-back observed in many parts of the world.

(2) The simulations suggest that the breakdown of trees is caused by insufficient assimilate production or excessive assimilate consumption due to pollution stress. Deficient leaf and/or feeder root renewal leads to increasingly rapid deterioration.

(3) The breakdown mode will only occur if the assimilate deficieny reaches a critical level. This may occur as the result of chronic pollution stress over many years.

(4) It therefore makes sense to speak of 'subcritical' stress (no breakdown, but reduced growth) and 'supercritical' stress (eventual breakdown).

(5) In the subcritical regime, the tree still looks 'healthy', even though growth may be absent. Two distinct modes - growth and stagnation - are possible in this regime.

(6) The simulations suggest that there is no difference in principle between the systemic effects of leaf damage and of feeder root damage: both reduce assimilate production and accumulation and act on the same life processes.

(7) Older trees have a much smaller tolerance to leaf and/or feeder root damage.

(8) There is considerable inertia in the response of trees to pollution reduction, especially with conifers with many needle age classes. Also, the dynamics of the breakdown phase are self-sustaining. Hence pollution reduction to subcritical levels will only save those trees which have not yet entered into the breakdown phase.

(9) The results point to the importance of systems analysis and modelling of ecological systems: they may reveal the systemic causes of strange behavior which empirical research of isolated details would not be able to clarify.

(10) While the model presented here cannot be used for precise forecasts of forest development, it can be expected that simulation models improved with respect to structure and data could be used for regional forecasts of forest growth and die-back as the result of a given pollution spectrum.

REFERENCES

Bossel, H., 1985. Umweltdynamik - 30 Programme für kybernetische Umwelterfahrungen auf BASIC-Rechnern. Te-Wi Verlag, Munich.

Bossel, H. et al., 1982. Dynamik von Waldökosystemen - Mathematisches Modell und Computersimulation. Interdisziplinäre Arbeitsgruppe Mathematisierung (IAGM), University of Kassel.

Bossel, H., Metzler, W., Schäfer, H. (Editors), 1985. Dynamik des Waldsterbens - Mathematisches Modell und Computersimulation. Springer Verlag, Berlin/Heidelberg/New York/-Tokyo.

Eidmann, F.E., Schwenke, H.J., 1967. Beiträge zur Stoffproduktion, Transpiration und Wurzelatmung einiger wichtiger Baumarten. Forstwissenschaftliche Forschungen, Heft 23, Parey, Hamburg/Berlin.

Finck, S., 1982. Pflanzenernährung in Stichworten. Hirt, Kiel.

Hunt, R., 1982. Plant growth curves - The functional approach to plant growth analysis. E. Arnold, London.

Larcher, W., 1980. Oekologie der Pflanzen auf physiologischer Grundlage. E. Ulmer, Stuttgart.

Mayer, H., 1977. Waldbau auf soziologisch-ökologischer Grundlage. G. Fischer Verlag, Stuttgart/New York.

Metzler, W., 1985. Dynamical systems modelling and simulation of air polluted forest ecosystems. Proceedings, 5th International Conference on Mathematical Modelling, Berkeley, July 1985.

Mitscherlich, G., 1971. Wald, Wachstum und Umwelt. Vol. 1: Form und Wachstum von Baum und Bestand. Sauerländer, Frankfurt/M.

McClaugherty, C.A., Aber, J.D., 1982. The role of fine roots in the organic matter and nitrogen budgets of two forested ecosystems. Ecology, 63(5), p. 1481-1490.

Rehfuess, K.E., 1981. Waldböden. Parey, Hamburg/Berlin.

Röhrig, E., 1966. Die Wurzelentwicklung der Waldbäume in Abhängigkeit von den ökologischen Verhältnissen. Forstarchiv, 37. Jahrgang, Heft 10, p. 217-229 and Heft 11, p. 237-249.

Schmidt-Vogt, H., 1977. Die Fichte. Parey, Hamburg/Berlin 1977.

Sharpe, P.J.H., 1983. Responses of photosynthesis and dark respiration to temperature. Annals of Botany 52, pp. 325-348.

TABLE I - PARAMETERS FOR THE SPRUCE MODEL

CT = 2/52 (a)
Assimilate consumption time

LFDF >= 1 (-)
Leaf deterioration factor. LFDM = LFDF-1.

LRSP = 1.5 ((kg assimilate/a)/kg leaf ODM)
Specific dark respiration of leaves
(converted from data in Larcher 1980, p. 138)

LUX = 250 (W/m2)
Average insolation during daylight hours
(value not critical; light saturation occurs at 100 W/m2;
Larcher p. 139-141)

NDAG = 8 (a)
Normal needle shedding age
(Schuett 1983, p.34)

NRBL = 0.01 (1/a)
Normal wood biomass loss (relative)
(Larcher 1980, p. 186, modified)

NRTO = 1 (1/a)
Normal feeder root decay rate
(McClaugherty/Aber 1982, p. 1481)

RRSP = 3 ((kg assimilate/a)/kg f.root ODM)
Specific respiration of feeder roots
(after Eidmann/Schwenke 1967, Roehrig 1966, McClaugherty 1982)

RTGF = 2 (-)
Feeder root growth factor
(McClaugherty/Aber 1982, modified)

RTOF >= 1 (-)
Feeder root turnover factor. RTDM = RTOF.

SHED = 0.2 (-)
Relative leaf efficiency for shedding
(Larcher 1980, p. 137)SRSP = 0.05 ((kg assimilate/a)/kg biomass ODM)
Specific respiration of stem and coarse roots
(after Larcher 1980, p. 134, p. 186)

SWP = 0 (bar)
Soil water potential
(Assumption: water saturation)

TRANC = 2000 ((kg water/a)/kg f.root ODM)
Transport coefficient
(recomputed from data in Eidmann/Schwenke 1967)

TROOT = 6 (degrees C)
Average soil temperature marking begin and end of feeder
root growth (from Mitscherlich p. 54, and temperature data
for Kassel)

TRSPC = 241 (kg water/kg assimilate)
Transpiration coefficient
(Eidmann/Schwenke 1967, p. 17)

TVEG = 10 (degrees C)
Average air temperature marking begin and end of vegetation
period (Larcher 1980, p. 87; Mitscherlich 1971, p. 54)

TABLE II - STATE VARIABLES AND INITIAL VALUES FOR THE SPRUCE MODEL

ASSI = 6.25 (t ODM/ha)
Stock of assimilate/photosynthate
(inventory for Jan.1)

BIOM = 250 (t ODM/ha)
Permanent biomass above and below ground

EPOL = 0 (1/a)
Air pollution effect on leaf efficiency
(no damage initially)

L(1)...L(NLF) = 12.5/8 (t ODM/ha)
Leaf mass per needle age class
(corresponding to a foliage mass of LEAF = 12.5 for NLF = 8)

ROOT = 12.5 (t ODM/ha)
Feeder root mass

TABLE III - TABLE FUNCTIONS AND OTHER FUNCTIONS FOR THE SPRUCE MODEL

BLDR (1/a)
Biomass loss due to drought
(after Larcher 1980, p. 343, and Finck 1982, p. 80)

SWP	0	10	15	30
BLDR	0	0	1	1

CLUX = (1/24)*(12+4*sin(6.28*TY-1.57))
Daylight hour fraction
(approximate relationship for the latitude of Kassel)

CTEM (-)
Temperature influence on photosynthesis and dark respiration
(after Larcher 1980, p. 150, p. 152, and Sharpe 1983, p. 337)

TEMP	-5	5	20	30
CTEM	0	.2	1	1

DROF (-)
Drought factor
(after Schmidt/Vogt 1977, p. 413, and Larcher 1980, p. 292)

SWP	0	2	15	20	30
DROF	1	1	.15	.1	0

PHN ((kg assimilate/a)/kg leaf ODM)
Net photosynthesis
(after Larcher 1980, p. 138, p. 141)

LUX	0	5	100	1000
PHN	-1.5	0	24	24

RFOL (kg leaf mass ODM/kg biomass ODM)
Relative foliage function
(after Bossel et al. 1982, p. 106; Mayer 1977, p. 40; Mitscherlich 1971, p. 30; Larcher 1980, p. 168; see also discussion of the Gompertz 1825 function in Hunt 1982)

LBIOT = ln(ln BIOT)	0.0	0.9	2.08
RFOL	0.7	0.18	0.0

where BIOT = BIOM+LEAF+ROOT

TEMP (degrees C)
Air temperature function (monthly average)
(long term weather data for Kassel

TY	.0	.058	.135	.212	.308	.385	.462	.558	.635	.731	.808	.885	.961	1.
TEMP	.7	0.0	0.8	4.6	8.8	13.2	16.4	17.8	17.3	14.2	9.1	4.9	1.5	.7

TABLE IV - AUXILIARY VARIABLES AND RATES FOR THE SPRUCE MODEL

BILR	permanent biomass loss rate to due respiration assimilate deficiency ((kg ODM/a)/ha)
BIOT	total biomass = BIOM+LEAF+ROOT (t ODM/ha)
BLDRO	biomass loss due to drought ((kg ODM/a)/ha)
CASS	assimilate supply reduction factor (-)
DGROW	assimilate demand for feeder roots and leaf growth ((kg/a)/ha)
DROOT	assimilate demand for feeder root formation ((kg/a)/ha)
DTRSP	required transpiration flow ((kg water/a)/ha)
E(n)	relative photosynthetic efficiency of needle age class n (-)
EFF	average relative foliage efficiency (-)
LEAF	total foliage mass (kg ODM/ha)
LFAG	actual leaf aging rate (1/a)
LFDR	leaf shedding rate due to transpiration deficit ((kg ODM/a)/ha)
LFGR	leaf growth rate ((kg ODM/a)/ha)
LFLR	leaf loss rate due to respiration assimilate deficiency ((kg
LOLD	foliage mass before new leaf growth (kg ODM/ha)
LRST	foliage mass remaining after shedding (kg ODM/ha)
LSHD	shedding rate for the oldest needle age class ((kg ODM/a)/ha)
MPRD	photoproduction capacity ((kg assimilate/a)/ha)
NBLR	normal biomass loss rate ((kg ODM/a)/ha)
NFOL	normal new leaf mass (kg ODM/ha)
NLAG	natural leaf aging rate (1/a)
NLF	number of needle age classes (-)
PROD	actual production of photosynthate ((kg assimilate/a)/ha)
RAV	relative assimilate availability (-)
RESP	total respiration ((kg assimilate/a)/ha)
REST	assimilate remaining for leaf, feeder root, and wood growth ((kg/a)/ha)
RLRT	relative required feeder root mass (-)
RPOL	net leaf damage rate (1/a)
RROOT	required feeder root mass (kg f.root ODM/ha)
RSLF	dark respiration of leaves ((kg assimilate/a)/ha)
RSRT	respiration of feeder roots ((kg assimilate/a)/ha)
RSST	respiration of stem and coarse roots ((kg assimilate/a)/ha)
RTGR	feeder root growth rate ((kg ODM/a)/ha)
RTLOS	feeder root decay rate ((kg ODM/a)/ha)
RTLR	feeder root loss rate due to respiration assimilate deficiency ((kg ODM/a)/ha)
RTRSP	relative transpiration flow (-)
SINC	assimilate supply for wood growth ((kg/a)/ha)
SLEAF	assimilate supply for leaf growth ((kg/a)/ha)
SROOT	assimilate supply for feeder root growth ((kg/a)/ha)
T	time (a)
TRSP	actual transpiration flow ((kg water/a)/ha) ODM/a)/ha)
TY	vegetation time (a) (Jan.1 = 0, Dec. 31 = 1)

a	year
ODM	organic dry matter

A BIOTA MODEL TO ASSESS THE INFLUENCE OF HUMAN IMPACTS ON THE GLOBAL CARBON CYCLE

Horst Bröhl-Kerner *

ABSTRACT

A model structure for the terrestrial biota as a subsystem of the global carbon cycle is presented. Growth of vegetation units is governed by internal growth functions, which can be modified to simulate different fertilization effects. The seasonal behaviour of the biota is modelled by introducing simple seasonality functions. Though the model is thereby able to reproduce the main features necessary to assess the consequences of anthropogenic modifications of the vegetation cover, it is simple enough to be introduced into regionalized global carbon cycle models.

INTRODUCTION

The natural global cycle of carbon is significantly influenced by human activities, leading to an increasing atmospheric CO_2 content. Due to the important role of CO_2 in the Earth's climate system, it is essential to understand the causes of this increase, to predict its future course and to find out possible countermeasures to prevent possible negative consequences of a global greenhouse effect.

Today the main source of excess atmospheric CO_2 is the burning of fossil fuels with an annual input of about 5 Gt Carbon, while the oceans have been identified as the most important sink (Bolin, 1981).

The role of the terrestrial biota as the second subsystem exchanging large amounts of carbon with the atmosphere is up to now not very well known. On the one hand, human activities have modified the vegetation cover of the earth to a great extend, releasing carbon originally stored in biomass and soil and reducing carbon storage capacities in significant amounts (Olson, 1982). On the other hand, under certain circumstances, the vegetation may quickly recover from former pertubations, therefore restoring released carbon again. Furthermore, it may exhibit a positive reaction on increased atmospheric CO_2 and increase biomass production and carbon storage (CO_2 fertilization effect, Bacastow and Keeling, 1973), establishing a feedback loop which tends to stabilise the atmospheric CO_2 content.

To assess the role of the terrestrial biota within the global carbon cycle, it is necessary to develop biota models which simulate the time course of biomass production and carbon storage within the different parts of the biota following anthropogenic modifications.

* Institut für Physikalische und Theoretische Chemie, Universität Frankfurt/M., Niederurseler Hang, D-6000 Frankfurt 50, BRD.

MODEL OBJECTIVES

The model presented here was designed to reproduce the main features of an upgrowing vegetation unit in terms of carbon uptake and storage. The compartment structure of the model is shown in Fig.1. The living biota is divided in assimilating material (L), actively growing woody material (W) and structural material including dead plant components (D). The litter layer is divided into herbaceous or leaf (LL) and woody litter (WL), while the soil organic carbon is represented by a single compartment, which may for special purposes exhibit an internal structure. Carbon turnover in assimilating and actively growing woody parts of the system is described by

$$dL/dt = k_{cl}\, L^{\delta} - k_{ll}\, L \tag{1a}$$

$$dW/dt = k_{cw}\, W^{\alpha} - k_{wl}\, W \tag{1b}$$

Integration of these equations yields

$$L(t) = L^{\circ}\,(1 - b_{l}\, e^{-k_{ll} t/n_{l}})^{n_{l}} \tag{2a}$$

$$W(t) = W^{\circ}\,(1 - b_{w}\, e^{-k_{wl} t/n_{w}})^{n_{w}} \tag{2b}$$

with $n_l = 1/(1-\delta)$ and $n_w = 1/(1-\alpha)$, and L° and W° refering to the maximum values of L(t) and W(t), obtained when the system reaches its steady state or climax value. The parameters b are defined by

$$b_{l} = 1 - (L_{o}/L^{\circ})^{1-\delta} \qquad b_{w} = 1 - (W_{o}/W^{\circ})^{1-\alpha}$$

where L_o and W_o refer to an initial state from which growth starts. The function derived in Eq.2a,b are known as Richards-Functions in plant growth analysis (Causton and Venus, 1981).

The input and output terms in Eq.1a,b may be interpreted in different ways. When the system is modelled on the net primary production (NPP) level, then the sum of the output terms is equal to litterfall. When modelling on the gross primary production (GPP) level, equations 1a,b may be written

$$dL/dt = k'_{cl}L^{\delta} - k^{r}_{g}(k'_{cl}L^{\delta}) - k^{r}_{l}L - k'_{ll}L \tag{3a}$$

$$dW/dt = k'_{cw}W^{\alpha} - k^{r}_{g}(k'_{cw}W^{\alpha}) - k^{r}_{w}W - k'_{wl}W \tag{3b}$$

Now the output terms are splitted into litterfall (upper index ') and autotrophic respiration (upper index r, dashed arrows in Fig.1). According to Evans (1975), autotrophic respiration may be modelled as sum of maintenance respiration, which is proportional to the accumulated biomass, and synthesis respiration, which is proportional to gross production. Under these assumptions GPP can be modelled using Eq.1a,b by introducing effective rate constants k.

Growth of different plant components is not independent, but is usually related to meet certain necessities, e.g. structural stability. Under the additional assumption that compartments L and W are related to each other by linear or quasi-linear allometry, a relation often found between growing plant parts (Causton and Venus, 1981), we obtain

$$W(t) = W^{\circ}\,(L(t)/L^{\circ})^{\nu}\ ,\ \nu = (1-\delta)/(1-\alpha)$$

and therefore the sum of the input fluxes

$$S = k_{cl}L^{\delta} + k_{cw}W^{\alpha} = k_{cl}L^{\delta} + k_{cw}(W^{\circ}/L^{\circ\nu})^{\alpha}L^{\alpha\nu}$$

$$= \{\ k_{cl}L^{\delta-1} + k_{cw}(W^{\circ}/L^{\circ\nu})^{\alpha}L^{\alpha\nu-1}\ \}\ L \tag{4}$$

S defines the total sink strength of the vegetation for assimilates and should equal net primary production under optimal growth conditions. On the other hand NPP is usually assumed to be directly proportional to leaf biomass.

$$NPP = k_{p}\ L \tag{5}$$

This condition is obtained when $\delta \ll 1$ and $\alpha = 0.5$, which yields $\alpha\nu = 1$. The relation of sink strength to primary production is of importance when fertilization effects are considered, as will be discussed in a following chapter. To account for this relation an assimilate compartment C is introduced, which is described by

$$dC/dt = NPP - S \tag{6}$$

where S and NPP may be given by Eq.4 and 5, resp..

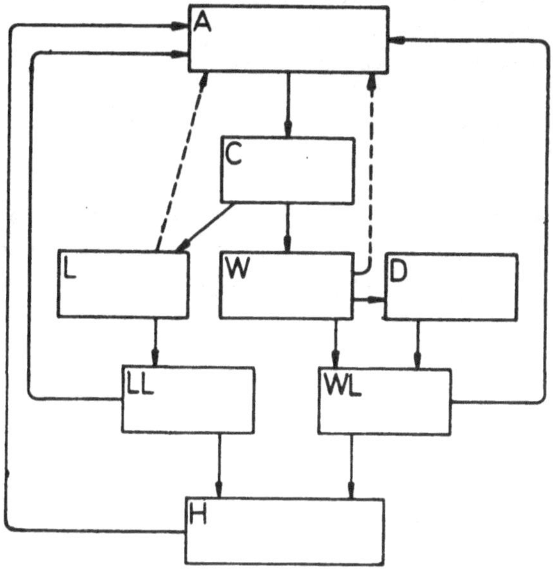

Figure 1 : Compartment structure of the biota model. A: atmosphere, C: assimilate storage, L: leaf biomass, W: actively growing woody biomass, D: structural woody biomass, LL: leaf litter, WL: woody litter, H: soil organic matter.

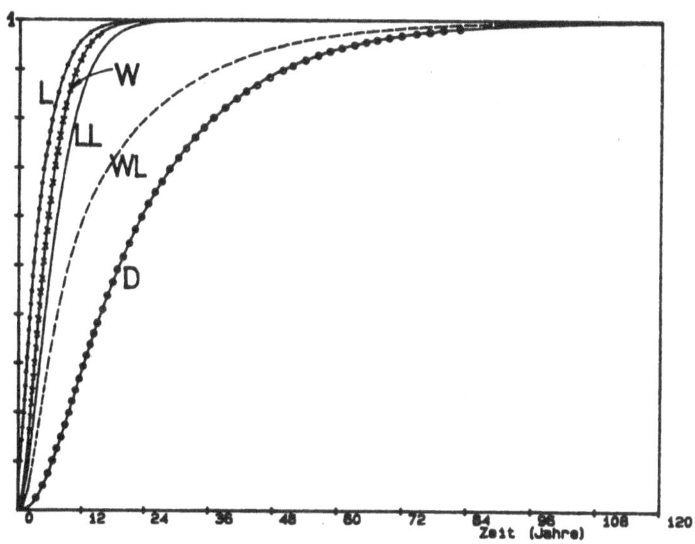

Figure 2 : Growth curves (normalized to unity) for the different components of the biota and litter system for a model forest. Abbreviations as in Fig. 1.

Conversion of woody material to structural, biochemically less active components is assumed here to be a constant part of the output of the woody compartment, such that the equation for compartment D can be written:

$$dD/dt = \lambda k_{wl} W - k_{dl} D \quad , \quad \lambda = \text{constant} \tag{7}$$

The fluxes within the litter and soil system are all assumed to be linearily donor-controlled, with the output of the litter compartments splitted into a respiration flux and a flux of more stable organic compounds (compartment H) in constant proportions, such that

$$dLL/dt = k_{ll} L - k_{la} LL \tag{8a}$$

$$dWL/dt = k_{dl} D + (1-\lambda) k_{wl} W - k_{wa} WL \tag{8b}$$

$$dH/dt = \chi_l k_{la} LL + \chi_w k_{wa} WL - k_{ha} H \tag{8c}$$

completes the model description. Fig.2 shows the growth curves of the different compartments (excluding soil organic matter) for a hypothetical forest biome. The time to reach the climax state is determined by the accumulation of woody material in compartment D.

SEASONALITY MODELLING

One of the basic empirical data sets obtainable for carbon cycle modelling are the time series of atmospheric CO_2 measurements at different points of the globe (Keeling, 1983). A main feature shown by these measurements is the pronounced seasonality of the CO_2 concentration, which to a larger part is caused by the seasonality in carbon uptake and release of the terrestrial biota (Heimann et al., 1985).
The model developed so far yields annual means for pools and fluxes. To obtain a simulation of the seasonal behaviour of the atmosphere, it is necessary to introduce a seasonality function into the fluxes between atmosphere and the biota system. In the following the flux from atmosphere to biota is termed annual ecosystem production, EP, while annual ecosystem respiration, ER, is defined as the sum of all fluxes from biota to atmosphere. The time behaviour of the biota as a whole can then be described by

$$dB/dt = s_{ab} EP - s_{ba} ER \tag{9}$$

where the seasonality functions s_{ab} and s_{ba} are normalized to unity over one year:

$$\int_0^1 s_{ab}\, dt = \int_0^1 s_{ba}\, dt = 1$$

In an assumed steady state, when annual ecosystem production equals annual ecosystem respiration, the annual "amplitude" (i.e. the difference between minimum and maximum value) of the seasonal atmospheric CO_2 signal is given by

$$A = EP \int_0^{t_1} s_{ab}\, dt - ER \int_0^{t_1} s_{ba}\, dt = EP \int_0^{t_1} (s_{ab} - s_{ba})\, dt \tag{10}$$

when the time scale is chosen in such a way that the intervall $(0,t_1)$ covers the whole time when ecosystem production prevails over ecosystem respiration, or $s_{ab} - s_{ba} \geq 0$.

Under non-steady state conditions the first part of Eq.10 defines the "peak-to-trough" amplitude A_{TP}, while the corresponding "trough-to-peak" amplitude is given by

$$A_{TP} = EP \int_{t_1}^{1} s_{ab}\, dt - ER \int_{t_1}^{1} s_{ba}\, dt$$

The annual mean amplitude is given by

$$A_J = A_{PT}\, t_1 + A_{TP}\, (1-t_1)$$

therefore, the model can serve to estimate the part of the measured seasonal amplitude caused by the terrestrial biosphere, when a reasonable regionalization of the modelled biomass is used.

CO_2 FERTILIZATION

A positive effect of increased CO_2 concentrations on plant growth is well established for many agricultural plants, while there is an ongoing debate whether natural ecosystems may respond to increasing atmospheric CO_2, either with accelerating growth or with increasing carbon storage capacity (Lemon, 1983; Strain and Cure, 1985; Kohlmaier 1985).

The model was designed to simulate both possible effects. The primary effect of increased CO_2 concentration is assumed to be an enhancement of net primary production, NPP, which is modelled by introducing a fertilization function into Eq.5.

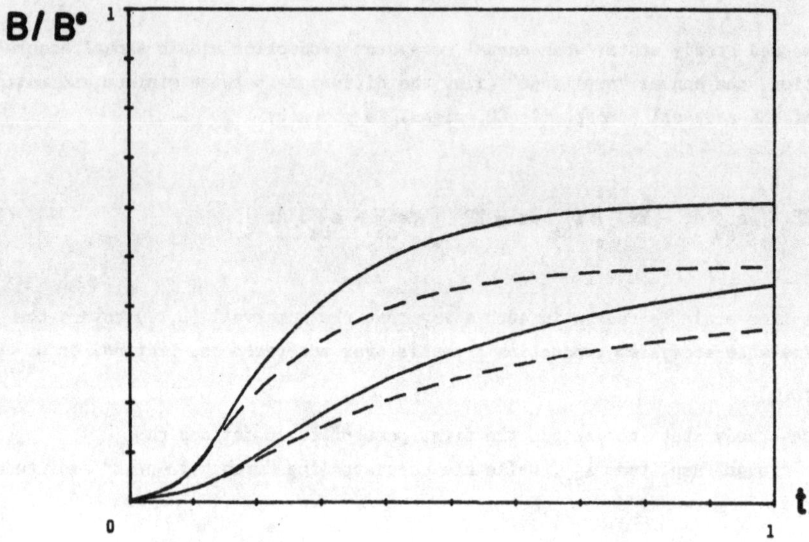

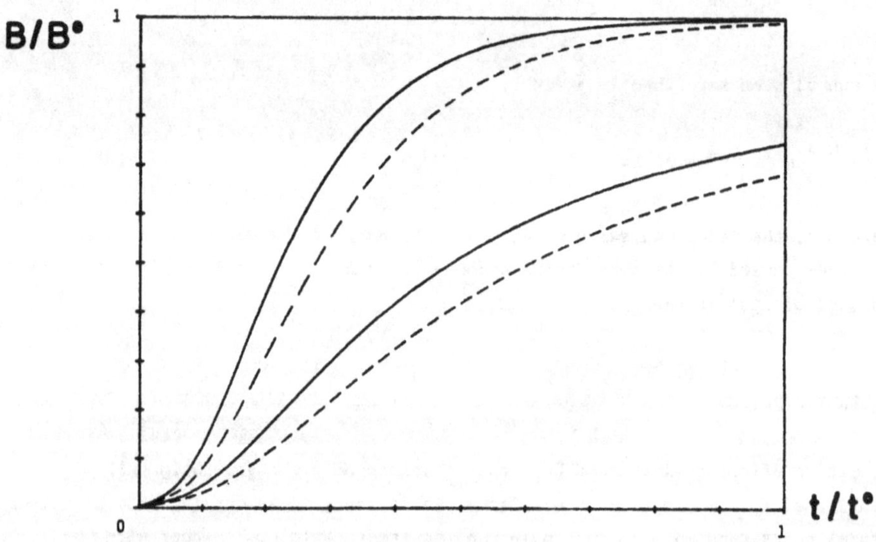

Figure 3 : Total biomass (upper set of curves) and mean biomass over time (lower set).

a) Fertilization leading to increased carbon storage. NPP in the reference case (dashed curves) is half the optimum NPP, while in the fertilization case (solid curves) it is enhanced by 25 %.

b) Fertilization leading to accelerated growth. NPP in the fertilization case is again increased by 25 %.

When a growth limitation due to sub-optimal CO_2 concentration is considered, Eq.1a,b have to be interpreted as describing growth under optimal conditions, i.e. the sink strength (Eq.4) is the maximum incorporable amount of assimilates. Real growth is then limited by NPP, which is lower than sink strength due to suboptimal conditions. In this case the model distributes the assimilates into the different sinks according to their relative sink strengths. Fig.3a shows the growth of total relative biomass (upper curve) and mean biomass over time (lower curve) of a model forest in the reference case (dashed curves) where NPP is assumed to be one half of the optimum, while in the fertilization case (solid curves) it is enhanced by 25%.

When an enhanced NPP should only lead to an acceleration of growth, the internal growth functions determined by Eq.1a,b are modified in such a way, that the sink strength S equals NPP, but the steady state value for each component is maintained (Fig.3b). Though this second type of fertilization effect did not influence the total biomass of the climax state and therefore no change in the potential carbon storage capacity of a hypothetical undisturbed biosphere occurs, there will be an important influence on the actual carbon storage capacity.

For ecosystems periodically disturbed either by natural events like fires or storms or by human impacts like timber harvest, the actual carbon storage capacity is determined by the mean amount of biomass stored over the time between disturbances (Cooper, 1982). As can be seen from Fig. 3, this amount is influenced by both types of fertilization effects.

In Table 1 these effects are compared for forests managed for maximum sustained yield, which means that the forests are harvested when the mean annual increment of biomass (MAI) reaches its maximum and are then allowed to regrowth in the same way as before.

TABLE 1: INFLUENCE OF DIFFERENT CO_2 FERTILIZATION EFFECTS ON FORESTS MANAGED FOR MAXIMUM SUSTAINED YIELD.
Index ° refers to a reference state of optimal growth reached after t°

	NPP / NPP°	Culmination of MAI (t/t°)	B / B°	$\overline{B}$ / B°
ACCELERATED GROWTH	.50	0.68	0.63	0.26
	.60	0.57	0.63	0.26
	.75	0.45	0.63	0.26
	.90	0.38	0.63	0.26
	1.00	0.34	0.64	0.26
INCREASING CLIMAX	.50	0.24	0.27	0.11
	.60	0.26	0.34	0.14
	.75	0.28	0.45	0.18
	.90	0.32	0.57	0.23
	1.00	0.34	0.64	0.26

Table 1 shows that accelerated growth leads to almost constant amounts of biomass, both actual and mean, at time of culmination of MAI, but the time needed to reach this biomass is shortened, leading to shorter rotation periods and therefore reduced demand of area for a given unit of (sustained) yield.
In the case of increasing climax biomass the time to reach culmination of MAI is on the contrary somewhat prolonged, but biomass accumulated up to that time is also considerably enhanced, leading to higher yield per unit area.
Depending on the harvest strategies followed, both effects will therefore lead to an increase in actual carbon storage capacity of these ecosystems. Similar effects may be found in other periodically disturbed ecosystems.

MODELLING ANTHROPOGENIC MODIFICATIONS

Most ecosystems on earth are not in a true climax state, but exhibit a certain distribution of age classes due to more or less periodic disturbances. This is especially pronounced in ecosystems managed or exploited by humans for yield of certain products or damaged due to other purposes. The time course of these human impacts and the response behaviour of the affected ecosystems may exhibit an important influence on atmospheric CO_2 concentration (Emanuel et al., 1985). To assess the consequences of the different impacts in different regions of the world (Kohlmaier et al., 1985), the model has to be used for a variety of biomes and different reservoir changes caused by impacts within the biomes. The number of biomes as well as the number of impacts which has to be considered for a global model, are at least in the order of 5 to 10, while the time frames for the different impacts are tenth to hundreds of years. The model structure presented here is proposed as a compromise between the complexity necessary to assess the main processes correctly, and the simplicity enforced by the limited capacity for numerical calculations. It seems also appropriate with respect to the available empirical data sets for adjustment as well as verification, and for testing assumptions for effects not fully understood up to now, as e.g. the fertilization effects.

REFERENCES

Bacastow, R.B., and C.D. Keeling, 1973. Atmospheric carbon dioxide and radiocarbon in the natural carbon cycle. II. Changes from AD 1700 to 2070 as deduced from a geochemical model. In: Carbon and the Biosphere (G. Woodwell and E. Pecan, eds.), US Atomic Energy Commission, Washington D.C., 86–134.

Bolin, B. (Hrsg.), 1981: Carbon Cycle Modelling. SCOPE 16, John Wiley & Sons.

Causton, D.R. and J.C. Venus, 1981. The biometry of plant growth. London: Edward Arnold (Publ.) Ltd.

Cooper, C.F., 1982. Carbon storage in managed forests. Can. J. For. Res. 13: 155–166.

Emanuel, W.R., I.Y.-S. Fung, G.G. Killough jr., B. Moore III, and T.-H. Peng, 1985. Modelling the global carbon cycle and changes in atmospheric carbon dioxide levels. In: J.R. Trabalka (Ed.): Atmospheric carbon dioxide and the global carbon cycle. U.S. Department of Energy, DOE/ER-0239, Washington D.C.

Evans, L.T., 1975. Beyond photosynthesis - the role of respiration, translocation and growth potential in determining productivity. In: J.P. Cooper (Ed.): Photosynthesis and productivity in different environments. IBP 3, Cambridge University Press.

Heimann, M., C.D. Keeling and C.J. Tucker, 1985: A Three Dimensional Model of Atmospheric CO_2 Transport Based on Observed Winds. 2. Analysis of the Seasonal Cycle of CO_2. Paper presented at a CO_2-Workshop in Kandersteg, Sept. 1985 (submitted to Tellus).

Keeling, C.D., 1983. The Global Carbon Cycle: What we know and could know from atmospheric, biospheric and oceanic observations. In: Proc. Carbon Dioxide Research Conference: Carbon Dioxide, Science and Consensus. US Departm. of Energy, CONF-820970, Washington D.C.

Kohlmaier, G.H., 1985. Exchange of CO_2 between the Atmosphere and the Biosphere: Stimulation of the Terrestrial Vegetation by Fossil Fuel CO_2 in Combination with other Man-Made Changes. In: R.S. Keir and W.H. Berger (Eds.), Carbon Dioxide Transfer in the Atmosphere - Ocean - Terrestrial System. Report on the Conference, Lake Arrowhead, May 5–10, 1985, p.95–98. SIO Reference Series 85–31, Scripps Institution of Oceanography, University of California, San Diego.

Kohlmaier, G.H., H. Bröhl, P. Stock, M. Plöchl, U. Fischbach, A. Janecek, R. Fricke, 1985b. Biogenic CO_2 release and soil carbon erosion connected with changes in land use in the tropical forests of Africa, America and Asia. Mitt. Geol.-Paläont. Inst. Univ. Hamburg, 58, 123 - 136

Lemon, E.R. (Ed.), 1983. CO_2 and Plants. AAAS Selected Symposium 84, Westview Press, Boulder, Colorado.

Olson, J.S., 1982. Earth's vegetation and atmospheric carbon dioxide. In: Carbon dioxide review, W.C. Clark, ed., Oxford Univ. Press, 388–398.

Strain, B.R. and J.D. Cure (Eds.), 1985. Direct Effects of Increasing Carbon Dioxide on Vegetation. U.S. Department of Energy, DOE/ER-0238, Washington D.C.

AN INTEGRATIVE APPROACH TO THE SOLUTION OF ECOLOGICAL PROBLEMS WITH PARTICULAR ATTENTION TO THE REGIONAL CASE

Acad. Leonardas Kairiukstis
Environment Program
IIASA

Abstract

This paper discusses the approach to sustainability of ecological systems. The problems to be solved and the systems in which they interact are analyzed on the local, regional (national), zonal and global levels. In order to mitigate the confrontation between the socio-economic development and the environment, an integrative approach has been suggested. This approach is based on systems analysis of rational use and reproduction of natural resources in combination with a purposeful formation of an optimal environment. The example of the Lithuanian SSR has been used for this purpose.

1. INTRODUCTION

The acceleration of scientific and technological progress and of industrial, metropolitan and agricultural development leads to man's confrontation with nature. The influence of man changes the equilibrium of nature processes which leads to unpredictable consequences. Generally, these are negative, such as desertification, water and atmospheric pollution, increase in soil acidity or salinity, destruction of forests and severe exhaustion of natural resources, impacts on human health and genetic mutations in plants and animals. This primarily occurs at the local and regional levels, sometimes leads to unpredictable negative consequences for the separate regions, and may also lead to negative consequences for the entire biosphere.

The increasing scale and significance of man's role as an agent of global change was forcefully articulated between the two World Wars by a remarkable group of scholars and, more recently, particularly by Professors W. Clark and C.S. Holling (1985). These also included the French theologian and paleontologist, Pierre Teilhard de Chardin, the Austrian-born American biophysicist, Alfred J. Lotka, and above all, the Russian mineralogist, Vladimir Ivanovich Vernadsky. Vernadsky (1926) first formulated the concept of the biosphere as the only terrestrial envelope in which life can exist. In Vernadsky's opinion the most significant aspect of man's development was not his technology per se but rather the sense of global knowledge and communication engendered by that technology. He portrayed this "noosphere" or realm of thought as a new geological phenomenon on our planet. Vernadsky's main concept was strongly developed by Soviet Academician Vladimir Nikolajevich Suckachev (1964) as a complete science of biogeocenology or the science of ecosystems.

Man's role as an agent of global change is associated with the emergence of an increasingly interdependent world economic system (Richards, 1985). Following the second World War, expanding industrial and agricultural development increasingly intensified the global economic interdependence among nations. It also began to introduce issues of ecological and geophysical interdependence between countries. This has led to some significant achievements in monitoring, in defining the issues in scientific and technical terms, in raising public awareness, and in institutional and policy action, on the national and international level. At the same time, research on the biosphere is characterized by an increasing scale and complexity of problems. Examples are the SCOPE program on the major biogeochemical cycles, the ICSU exploration of an international geosphere-biosphere program, NASA's work on global habitability, the WMO's World Climate Programme, UNESCO's Man and the Biosphere Programme, and UNEP's report on the World Environment 1972-82. Furthermore, there is IUCN's World Conservation Strategy, the OECD's work on economic and ecological interdependence, and the WRI's Symposium on The Global Possible.

At the International Institute for Applied Systems Analysis (IIASA) in Laxenburg, Austria, studies are being undertaken in many Projects such as Acid Rain, Ecologically Sustainable Development of the Biosphere, Climate Impacts, Design of Resource and Environmental Policies, Population Aging and Changing Lifestyles. Several methodological themes span across the above-mentioned projects (e.g., "ecologically sustainable/unsustainable regional development in an historical perspective, uncertainty/ environmental risk assessment") and they are deeply concerned with the sustainability of regional development and development of the biosphere as a whole. In collaboration with national research groups, models and software have been developed to analyze the problems above (Moiseev, et al, 1984; Alcamo et al, 1984; Parry and Carter, 1984; Brouwer, Hetling and Hordijk, 1983; Knisel, 1980; Krutko et al, 1982; and Grossmann, 1984 and 1985).

Most of this investigation is characterized by a considerable uncertainty in initial empirical information, a large number of input variables which are collected in different ways (official statistics, experts' estimates, indirect observation, etc.) and have different structures (qualitative, quantitative) and most of which can hardly be forecast. Nevertheless, the preliminary analysis of the investigations on the biosphere shows that:

- the cumulative impact of industrial, agricultural, and social development on the environment has approached a level where it is dangerous for particular regions and can be dangerous for the biosphere as a whole;
- better integrative understanding by international organizations, governments, and the scientific community is now urgently required to plan effective interventions that will bring the above situation under control; and
- the demand for help in creating such integrative understanding is high but is not being met by present isolated research activities around the world.

2. THE HIERARCHY OF PROBLEMS TO BE SOLVED AND THE SCALE OF SYSTEMS IN WHICH THEY INTERACT

The problems to be solved can be divided into groups of environmental development concerns, as suggested by Clark and Holling (1985). Nevertheless, each group of problems interacts on a different space and time scale, and is involved in different systems. Therefore, I suggest that the problems that concern each generation should be analyzed within the systems in which they interact, as follows:

- The **first** group comprises small-scale problems concerning *local environmental systems on the landscape level.* Such problems are: local air or water pollution, soil erosion by water or wind, sustainability of agricultural crops, forest or fish die-off due to pollution, etc. These problems mostly affect the social well being of people and involve only certain branches of the economy and some management bodies. They have proved to be largely controllable. Moreover, they can be analyzed easily with the use of quantitative models. However, local control (e.g., construction of a very tall chimney, cultivation of field protective forest belts, etc.) sometimes contributes to a regional problem.

- The **second** group includes the larger-scale problems of *economic-environmental systems on the regional or national level.* The problems are: regional resources utilization, industrial and agricultural development, assessment of the impact of human activities on the environment, environmentally balanced sustainable regional development strategies. These deeply concern the socio-economic development and living conditions of human populations—ethnic groups in certain regions or nations. Despite the complexity, methods for solving these problems can be identified on the basis of systems analysis by using complete and incomplete models and choosing appropriate regional development strategies. Frequently, however, the conflict between short-term economic benefits and long-term ecological damage is difficult to resolve.

- The **third** group includes large-scale problems of *macro-systems on the transregional, zonal or continental level*. The problems are: long-distance transportation of air pollutants, acid rain, long-distance water transfer (between basins), large international rivers, depletion of tropical forests, loss of genetic resources, desertification, shifts of agricultural and forest productivity onto marginal areas due to climatic changes, depletion of living marine resources, etc. These problems have shown themselves to be more difficult to solve partly because the costs of reversing the trends in some cases are prohibitive. Moreover, they span so many political spheres that the authority and understanding required for concerted action are often lacking.
- The **fourth** group consists of even larger-scale problems directly concerned with the *development of the biosphere on a global scale*. Such problems are: climatic background fluctuations due to sun-earth relations, biogeochemical cycles and evolution leading to a dynamic exchange of chemical constituents among the oceans, the atmosphere and terrestrial biosphere, biogeochemical cycles of soils, global economic-industrial growth, energy consumption and emissions into the atmosphere of significant amounts of active constituents, CO_2, O_3, trace gases, etc., anthropogenic changes in the atmosphere, their climatic and economic consequences, pollution levels in the oceans and their ecological consequences, anthropogenically induced global biospheric change and transition to the noosphere. If these extremely complex problems can be managed at all, it is only with a commitment of resources, and a consistency of purpose that transcend normal cycles and boundaries of scientific research and political action. Nevertheless, there is no other alternative but to try to formulate the approach and find tools (a system of models) with which it would be possible to describe the basic properties of the dynamics of the biosphere which could serve as a point of departure in the efforts made to understand evolutionary trends and biogeocenotic processes in the world and to react to them either by adaptation or confrontation.

3. GENERAL APPROACH TO THE METHODOLOGY

The complexity of the above-mentioned problems and systems in which they commonly interact constitutes the subject of investigation on sustainable/unsustainable development of ecological systems. Such an investigation would require an interdisciplinary multi-scale system approach.

The systems comprising the global biosphere differ greatly in space, time and complexity. They are open, and essentially interact with the ecological and socioeconomic environment. The hierarchical nature of systems is very complicated, with cooperation and conflicts between different subsystems. Possibilities to control the subjects under investigation, on large-scale systems in particular, are very limited. The higher the level of the hierarchy, the fewer the possibilities to experiment with the subsystem. Therefore, it is necessary to use an integrative approach to elaborate a strategy of systems management (control) in a united, hierarchical structure during the process of receiving and collecting data.

The systems under consideration have many facets, each of which can most adequately be addressed with a different set of tools. That is the reason why a "multifaceted" hierarchical approach can be used. This approach has been suggested and widely used by W.D. Grossmann (1984) for small-scale local systems analysis, in contrast to large-scale models which were strongly criticized previously (Lee, 1973). We developed the former approach for the analysis of problems and systems related to the sustainability of the biosphere (Kairiukstis and Umnov, 1985). Complementarily, the main scheme was extended to four levels of problems and factors to be analyzed, methods to be adopted, as well as systems to be covered. Using this approach, we intend to investigate interrelated local, regional, zonal and global systems, according to the strength of the impact felt on each other. The zonal system among them is the turning point. It is expressed in a maximum of outside environmental cross-action and a minimum of management cross-action (Figure 1). For most of the problems discussed above this approach permitted the use of dynamic models in combination with a geographical information system and the generation of a time series of highly precise geographic maps.

Nevertheless, in many cases (particularly in analyzing complex regional socio-economic and environmental systems) it is not realistic to use formal methods and mathematical models. Such models are usually based on the assumption that the models describe these systems exactly and sufficiently. However, it is not always possible to build mathematical models with the required properties and the user must invest much effort in verifying the practical applicability of the solutions obtained by standard schemes. For these cases more practical environmental impact assessments and socio-economic methods, developed by R.E. Munn (1979) as well as incomplete models, can be used (Umnov, 1985).

Using both approaches, an analysis of the main constituents of the global biosphere and verification of the theoretical hypothesis about possible changes will illustrate what the most serious constraints are for the future sustainable development of lower scale, i.e., zonal, regional and local systems. On the other hånd, analysis of local and highly aggregated regional systems will reveal the factors which have negative or positive impacts on the sustainability of the systems of a higher level.

Let us consider an approach to environmentally balanced regional development using the example of the Lithuanian SSR.

THE HIERARCHY OF BIOSPHERE SYSTEMS:
PROBLEMS TO BE SOLVED AND METHODS TO BE USED

INCREASE OF MANAGEMENT CROSSACTION (left, upward) / INCREASE OF MANAGEMENT CROSSACTION (right, downward)

PROBLEMS	THE HIERARCHY OF SYSTEMS	CHARACTERISTICS	DOMINANT FACTORS	METHODS
Understand geosphere–biosphere interactions, environmental fluctuations and biogeochemical cycles. Reveal the factors that seriously constrain sustainability and habitability of all lower systems.	GLOBAL	Great uncertainties in structure. Lack of data.	Fluctuations of the universe.	Theoretical hypotheses
Preserve sustainability. Decide structure of all lower systems. Decrease risks. Explore, recognize and exploit opportunities. Prepare system so that it can cope better with "whatever may happen".	ZONAL (Continental)	Uncertainties in structure and data. High influence of the outside environment. No decision systems.	Climatic.	R&D. Evolution, succession. Biocybernetic approach. Scenarios.
Within the defined structure: Preserve the structure, keep the system going. Problem solving such that interdependencies and feedback reactions are taken into account.	REGIONAL (National)	Uncertainty in data and to a lesser degree in structure. Considerable influence of the outside environment. Many interdependencies. Few decision systems.	Sustainability of human micro population.	Strategic management. Holistic approaches. Considerations must include reactions. Aggregated dynamic models. Preserving of balance.
Prepare counter policies which are best in a specific area and help sustainability of all higher systems. Solve the many routine jobs quickly and precisely.	LOCAL	Preciseness in data and structure. Low influence of the outside environment. Many not interacting decision systems.	Well being of local inhabitants.	Optimization both heuristic and exact. Adaptive control. Real time process control.

Figure 1: The hierarchy of biosphere systems: problems to be solved and methods to be used.

4. THE APPROACH TO SUSTAINABLE REGIONAL DEVELOPMENT IN THE LITHUANIAN SSR

The Lithuanian SSR is one of 15 republics of the Soviet Union with 3.6 million inhabitants living in an area of 6.5 million hectares, i.e., the average population density is 55^2. In the last few decades industry as well as agriculture has been highly developed. The average yearly crop production of cereals for the last five years was between 2.5–3.0 tons/ha., for meat 0.13 tons/ha., and dairy products 3.6 tons/animal.

The Lithuanian Republic is a highly cultivated part of the USSR. Land reclamation by closed drainage was carried out on 60% of the area of collective farms. Extensive road networks were established. Two large (approx. 400–500,000 inhabitants) and 12 small towns (about 100,000 inhabitants) were developed.

In Lithuania, traditionally developed complex silviculture is practiced (the average area of forest preserve is 3,000 ha., forest husbandry - 30,000 ha., with one highly educated specialist each per 1,000 ha.). The forests constitute 1.8 million ha. (27.6%) of the total land surface. This means that approximately 75% of the demand for wood is met. There is a well organized nature conservation service in Lithuania. In addition, Lithuania is well-known for sport hunting and tourism.

Despite great efforts by the State for nature conservation, natural resource utilization and technogenic use of the land have a negative impact on the environment.

4.1 Negative Consequences of Human Activities During Technogenic Landscape Reconstruction

Industrial growth, increased motorization, land reclamation, and intense agricultural activities have destroyed the natural landscape. Extensive land use has caused deterioration of the soil and water. Investigations show that land reclamation and cultivation, as well as the intensive use of pesticides sharply diminish the variety of soil fauna (Atlavinyte, 1978 and Eitmanaviciute, 1982). Hydrophilic species of worms disappear, and other species have diminished fivefold. The variety of ornithofauna also suffers greatly. Insectivorous birds (e.g, thrushes and finches) have been greatly decimated because they use small field bushes for reproduction rather than forests. The disappearance of forests and bushes inside arable lands reveals the negative consequence of surface water runoff. The arable lands in the Aukstaiciu and Zemaiciu uplands especially suffer from erosion. Using the CREAMS* model (which was adapted at IIASA), model calculations of runoff, amount of water evaporation and deep percolation were carried out for the hilly Utena region with regard to soil types, variety of forms of landscape. Model calculations show that in 1972–1980 runoff and deep percolation correspond to about one-third or 204mm precipitation per annum. This determines soil erosion processes which in conformity with the degree of steepness, mechanical particles specification and crops grown there can form 100 tons/ha. of drifts in the nine-year period (Kairiukstis and Golubev, 1982). The aforementioned erosion processes do not happen on slopes with belts of forest cover.

The increase of the plain field area causes wind erosion and depletion of soil fertility. It is a special characteristic of the light soils. For example, in Lithuania, such areas cover about 190,000 ha. The investigations carried out (Pauliukevicius, 1982) show that the annual runoff washes away from every ha. of arable land 50–250 kg of chemicals dissolved in the water. Every year the Nemunas river carries away about 0.5 million tons of drifts into the Kursiu Marios Lagoon. It is partially for this reason, as well as due to a scarcity of sewage treatment plants in large towns and industrial centres, that water pollution has not been sufficiently reduced.

*CREAMS = Chemical, Runoff and Erosion from Agricultural Management Systems (Knisel, 1980).

West winds being predominant, the industrial pollutants from Western and Middle Europe, England, Scandinavia, as well as the GDR and Poland are transported to Lithuania. Acid rain, oxidized deposits, harmful gases, and carcinogenic substances reach the Lithuanian territory in 1–2 days. They cause acidification of soils, water pollution and deterioration of the forest ecosystems. Against the background of the whole atmospheric pollution, the zones of high local pollution around industrial centres like Jonava, Mazeikiai and Keadainiai prove to be dangerous to the forest vegetation, which in its turn is the index of the conditions for human existence. Under the impact of emission processes, litter, soil and water reservoirs accumulate large quantities of dust, fluorine, nitrates, ammonia, chlorite sulphates, phosphates, potassium and other harmful compounds (Vaicys, 1982). Within a radius of 3–5 km from the source of pollution, as shown in the data of the Lithuanian Institute of Forestry, coniferous forests wither. In the places more remote from the pollution sources under the influence of emissions, changes of morphological, physiological, chemical and species composition of vegetation take place. In these areas, the increment of coniferous forests is 30–50% and deciduousness 20–30% less in comparison with non- affected areas. The white fungus disease which affects needles, leaves, branches and trunks is widely prevalent. Cancer of larch and pine-tree sprouts is especially dangerous.

Increased environmental pollution increases mutagenesis. It has become a characteristic of modern civilization. Among some four million pollutants circulating in the biosphere, in the Baltic region alone we found about 300 pollutants bearing mutagenic effects. They expose the genetic stability of species to great danger. Geneticians of the Vilnius State University (Lekevicius, 1980) determined that in the 3–8 km radius of the Mazeikiai oil-refining plant the frequency of point mutations in field vole (Microtus arvalis) increases 10 times. The same phenomena are observed on non-irrigated meadows with a high usage of mineral fertilizers. It is notable that the field vole is an example of chemical pollutants on animal species and possible impacts on man. In many countries there is now sufficient data to prove that human beings who come in close contact with chemicals such as mercury, and with plastics such as polyvinylchloride and who live in a polluted environment, suffer from chromosomal aberrations causing retarded or handicapped offspring.

At the same time, the environment transformation process was and still is an inevitable precondition and consequence of the further development of society. It gives an opportunity to create productive agriculture, to develop industry and transportation networks, to create a good economic basis for higher material, cultural and social population welfare.

The problem faced is how to deal with these conflicts, and in particular:

- How to avoid situations in which certain economic branches and planning organizations disregard local environmental conditions and only solve problems of their own interest?
- How to avoid the creation of bordering territories with contradictory functional purposes, which increase the self-decay of economical branches, prevents the stable economic development of the whole region and has an adverse affect on the environment?
- How to coordinate the interests of the separate economic branches and concerted development of the whole region?

The traditional environment protection agencies are unable to deal with these shortcomings. It is necessary to rationalize natural resources utilization of the region, taking into account redeveloping measures for the sustainability of regional development and the creation of better environmental conditions. While undertaking regional physical planning of the territory, one must not only consider the needs of each branch of the economy, but also the long-term perspectives of socio-economic and environmental development of regional human populations. In other words, sound environmental principles, maintaining the balance between ecological systems and the socioeconomic development in a chain of natural resources utilization and reproduction to ensure long-term economic development.

In order to realize the above, we urgently need a reliable set of tools for economic-environmental situation analysis and scientific experts system for top-level decisionmakers. Only in this case do possibilities really appear to select and implement Sustainable Regional Environmentally Balanced Development Strategies (SREBDS) for each region.

4.2 Environment Formation as a Part of Natural Resources Utilization and Socio-Economic Development of the Region

In the last few years, the universal process of man's intrusion into nature and the spontaneous destruction of ecological systems in Lithuania was met by the rational use and reproduction of natural resources, in combination with the purposeful formation of an optimal environment. To support such an alternative, better interaction between scientific experts and decisionmakers was achieved (Kairiukstis, 1985). The efforts of scientists at the Academy of Sciences, branch institutes and higher schools were united under the program, "Man and the Biosphere". The General Model System (GMS) was created (see Figure 2), making it possible to optimize the development and specialization of the main resource utilizing branches of the national economy on a territorial basis (Kairiukstis, 1982).

Keeping to the tasks of sustainable development of the national economy, efforts have been already made for natural medium optimization (quantity of main landscapes determining its state), to determine the purposeful destination of the landscapes. An evaluation of separate territories was carried out.

It appeared that regions favorable from the point of view of the anthropoclimate such as the continental sea-shore of the Kursiu Marios lagoon, west and middle Suvalkija, the central part of North Lithuania have hardly been used for the development of recreation and health resorts. The Institutes of Construction and Architecture of the Lithuanian SSR together with the Vilnius State University prepared a distribution scheme of recreational regions. While

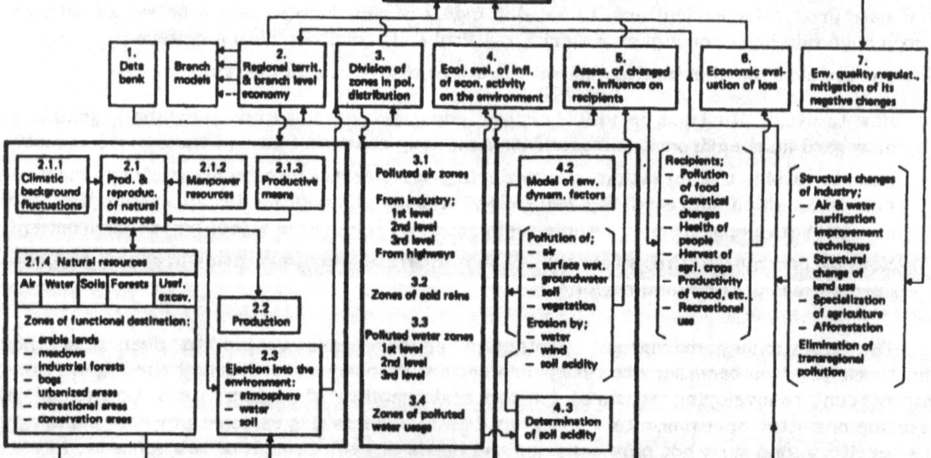

Figure 2: The scheme of a system of models for the analysis of regional-economic-environmental systems.

undertaking such an optimization one has to take into account the local anthropoclimatic peculiarities, air and water pollution, so that the objects destined for recreation are notable for the healthy climate.

The joint efforts of scientific research institutes and scientists of higher schools distinguished those zones being protected for special purposes. They concentrate the natural fund of flora and fauna, landscape variety as well as historical and ethnographical objects. These zones include reserves of grassy and forest vegetation genofund and of separate animal species. Only the concentration of these territories and the legalization of their status can protect them from ever growing industrial penetration and from negative consequences of agriculture and industry. Moreover, part of the protected territories should at least satisfy the utilitary needs of recreation (berry plantations, fields of intensive recreational sport hunting, etc.).

The Department of Geography of the Academy of Sciences of the Lithuanian SSR evaluated the degree of erosion and denudation in Lithuania. Together with the Lithuanian Research Institute of Forestry and the Institute of Botany, those regions were determined which are lacking in forests. Means for afforestation were prepared to protect arable lands and waters (Kairiukstis et al., 1980). The investigations of the Institute of Physics of the Academy of Sciences of the Lithuanian SSR have made it possible to discuss in broad terms the background atmospheric pollution, fall-out of sulphur, fluorine, nitrogen and other compounds, to make model calculations and to reveal the influence of large industrial objects and towns on the neighboring regions. Data are stored on the intensive use of fertilizers and pesticides and their influence on ecosystems links in separate regions. All this is the basis for the monitoring and also the prognosis of inevitable change of forests and arable lands. This also serves as a basis to obtain a sustainable environmentally balanced regional development.

The state of the natural ecosystems depends on climatic background situations which have an impact on all ecosystems. Dendrochronological investigations carried out in the Lithuanian Research Institute of Forestry and in the Institute of Botany reveal the long-term rhythm of the forests' growth. Approximately every 11 and 22 years the maximum and minimum in the growth of trees repeats itself. The long-term (about 50 years) fluctuations of favorable and unfavorable ecological conditions are determined (Figure 3). For example, favourable conditions in 1904, 1915, 1925, 1945, 1957 and 1968, in areas with high humidity the growth of trees was 20–30% above average. Vice versa, in 1908, 1930, 1952, 1963 and 1978 the growth of trees was 20–30% below average. The fluctuations mentioned are in good correlation with crop capacity of agricultural lands. For example, the yield of cereals in that year is higher in which the marsh vegetation increment increases. It is thus possible to forecast forest growth and the growth of other vegetation for at least the next decade. Fluctuations of climatic conditions also bear an influence on the general economic situation in the region. This phenomena can perhaps be explained through the "fluctuation factor of entrepreneurship" developed by W. Krelle (1983). It must be taken into account while modelling the ecological situation and socio-economic development.

The successful use of the ecological situation, forecast for the coming decade, and its adjustment to man's economic activity, make it possible to enlarge the ecosystems suitability and to grow more stable yields. For example, in Lithuania, during the drier and warm period, more productive crops should be sown as in the regions to the south (Ukraine, Byelorussia), giving preference to irrigation and moisture accumulation. On the other hand, for the period of cold and wet weather forecast for the end of this decade, one should be provided with sowing grain produced in the same district or in neighboring northern and western regions, widen the grassy fields and repair and maintain the drainage equipment. An analysis of the cereal yield during the last decades shows that a strategy of adaptive agriculture such as that outlined above might prove valuable.

The accumulated data based on the investigations of various branches of science (Gvishiani, 1977) as well as on data banks using the above mentioned approaches (see section on Analysis of Issues on the Regional Level) enabled the creation of the General Model System (GMS) of the optimal use of natural resources and environment formation based on socio-economic regional development scenarios (Kairiukstis, 1982).

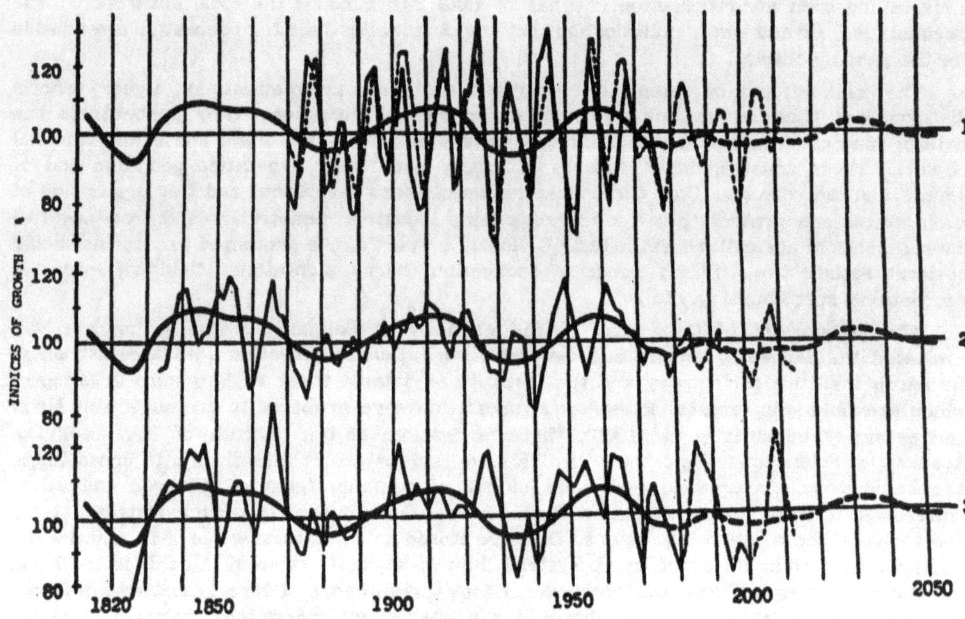

Figure 3: Climatic background fluctuations in the Lithuanian SSR.

—— actual data; ---- model forecast.

1 = Average for moisture soils; 2 = Average for dry soils, South Lithuania;
3 = Average for dry soils on the Baltic Sea.

Economic models (Buracas, Rajackas, 1982) simulating the development of the main branches of the national economy influencing the environment (agriculture, forestry, water management) and also of the industrial complexes (chemical, energy, etc.) through the Models of Interbranch Reproduction of Public Products and Land Use (Rutkauskas, 1978) give an opportunity to forecast and optimize the (regional) resources utilization.

The ecological consequences caused by the industrial processes in the course of public product reproduction are evaluated in the typified geographical landscapes. Besides, the model of environment factor dynamics suggested by the All Union Scientific Research Institute of Systems Analysis (Krutko, Pegov et al., 1982) and the CREAMS (Knisel, 1980) are also used. The adapted models are based on the suggestion that separate branches of the national economy, depending upon the management systems and technologies dominating in them, bear a different influence on the separate landscapes (farm lands, waters, urbanized territories, etc.). This influence is revealed in the changes of hydrological, erosive and chemical processes and parameters in soil, water and atmosphere. By simulating possible changes of farm lands and technologies, the desirable and undesirable tendencies of the economic influence on the environment are revealed. Those tendencies are evaluated according to the dynamic concept of the environment.

Analogously, the model adopted evaluates the ecological influence of the atmosphere on flora, fauna and man. In accordance with the functional destination of landscapes according to separate pollution sources (chemical plants, power stations, transportation networks, etc.), the necessary technological changes are made.

The results of economic development integrally determine the welfare of the population. The ecological consequences of economic development are determined and evaluated for each landscape with functional destination. These consequences determine productivity and genetic stability conditions of vegetation and animals, including man.

The decisionmaking (planning) bodies of the Republic select the optimal scenarios of natural resources utilization and regional environment formation. They analyze the alternatives and pass their decisions on to leading bodies--ministries, departments and industrial-agricultural complexes. The departments and complexes, carrying out the orders of the leading bodies through their economic systems influence the natural resources and environment on the specific territory. The annual and five-year results of this influence in the areas monitored, together with the amount of Gross National Product through the information channels (statistical) are returned to the GMS of the State Planning Committee. The information, bearing in mind the dendroclimatic prognoses of natural change of ecological conditions with the help of the mentioned models, makes it possible to correct the previously established annual and five-year scenarios for the optimal use of natural resources and environment formation.

For example, a preliminary optimization of farm lands shows that in the Lithuanian SSR the agrarian territories should occupy no more than 55–57% of the territory including about 17% of meadows stable to erosion. Forest areas should be widened up to about 31–33% of the country's territory. In regions with a small percentage of forests, as well as near highways and industrial complexes, it is advisable to grow antierosion protective forests. To keep a stable moisture regime, boggy areas and inner water reservoirs must occupy about 5–6%. The remaining territory (about 7%) will be inevitably urbanized. Conservation areas of various kinds, national and natural parks, recreational territories and territories under sanitary protection according to specificity, should occupy the most wooded areas, places of sandy dunes and with the greater number of lakes. Because these places in some cases are adjacent to farming and industrial-urban territories, the necessity for interstitial or buffer territories are needed. A special management of these territories is then determined. Therefore, the farming or forestry activity on specific territories are determined not only by the soil and climatic peculiarities, but also by the functional landscape destination. For example, forestry, in accordance with the dominating landscape destination specializes in:

1. Industrial wood growth – exploitable forests subsector;
2. Soil and water protection – agroprotectional forests subsector;
3. Recreation and rehabilitation needs – recreational forests subsector;
4. Conservation of flora, fauna and genofund – conservation forests subsector;
5. For scientific and hunting fauna needs – recreational hunting forests subsector.

For all these forestry subsectors, modules and programs are prepared for specialized economy management (Kairiukstis et al., 1983).

Optimizing landscape farm land will also show its specificity. Highly specialized industrial agriculture is retained in the main agricultural regions. In recreational and conservational areas, agriculture must be specialized to grow the so-called "bioproduction", unpolluted by chemicals.

On the other hand, in suburbs or lands adjacent to industrial centers as well as in other areas strongly polluted by industrial waste materials and transportation networking, state farms should not grow food (vegetables, cereals, etc.) but "technical" cultures necessary for industry, as well as flowers. As these territories are presently, and will be even more polluted in the future, their production, especially of vegetables and fruits, will be less suitable for food.

The division of the territory into zones according to their functional destination and economic regime will affect industries and the development of some towns. As we know, previously the growth of towns was determined by demographic structure, the availability of fresh water sources and other natural resources. Data on the state of atmospheric pollution, soil and water, also on the anthropoclimatic resources, prove that it is not advisable to evenly develop all the regions or a small state. It would complicate the struggle against industrial pollution, and would make the use of purification equipment and industrial waste materials more costly. In general, with the uniform development of industrial urbanization, the natural medium of pollution of the region is greater. All the more so as the even distribution of large towns and industrial plants in the region or state make it impossible to preserve non-polluted conservational and recreational territories which are distinguished by the most suitable conditions for health recreation.

In conclusion, it should be stressed that in the future, when the nature protection scheme of the Republic has been completed, the territory distribution has been carried out according to its functional destination, the economic activity in nature has been specified, a reasonable combination of branch and territory management fulfilled, only then will it be possible to escape the detrimental consequences in nature caused by branch management. It will promote a more rational use of natural resources, a stable development of agriculture, specialized water and forest husbandry with less expense in funds and labour force, to achieve a higher production; it will also help to lessen food cultivation pollution by industrial waste materials, to widen the urbanized areas and at the same time to retain clean recreational zones to ensure genofund and conservation zones for many years. Mitigation of the negative impact of human activities on the environment at a regional level will help to achieve a more sustainable development of systems of a higher level, including the biosphere as a whole.

5. CONCLUSIONS

The cumulative impact of industrial, agricultural and social development on the environment has approached a level where it is dangerous for particular regions and can be dangerous to the biosphere as a whole. At the same time, the environmental transformation process is still an inevitable precondition and consequence of the development of the material, cultural and social welfare of mankind.

In the process of general penetration of man into nature and the spontaneous destruction of ecological systems an alternative is made--the rational use of natural resources and reproduction united with an optimal and purposeful environmental formation. To realize this task, a system approach was developed to integrate the biospheric subsystems. Using this approach, it is possible to investigate local, regional, zonal and global systems related to each other, according to the level of the impact they bear on each other. This approach permitted the use of both complete and incomplete models in combination with a geographical information system and the generation of *time series* (dendrochronology, history of landscape transformation, etc.). Using the multi-faceted approach to resources and environment management, the situation in Lithuania was analyzed (regional level). Enlisting the rich heritage of scientific research and in collaboration with IIASA, the General Models System (GMS) was prepared on an interbranch basis. GMS should be of great use for the management and regional development planning agencies of the Republic. It consists of three blocks of models: economic models, models for the ecological evaluation of a medium, and models for the evaluation of social consequences of economic development and ecological changes. The cycle is modeled as follows: economy changes the environment (air and water pollution, destruction of natural ecosystems, etc.); the altered environment influences the biota including man, and through this by inverse link it influences the economic and social development of society (micropopulation).

According to the GMS the territory distribution optimization of the Republic conforming with the main landscapes was carried out beforehand, the specificity of the main landscapes' functional destination was revealed, the buffer zones between the landscapes of contradictory destination was pointed out, and tasks for the specificity of economical management (agricultural, forestry, etc.) were defined.

The use of the above-mentioned systems approach and the multi-faceted analysis of the rational use of natural resources and environment formation, when completed, will help to avoid detrimental actions in nature, to achieve the sustainability of regional systems and mitigate biospheric collisions.

REFERENCES

Alcamo, J., P. Kauppi, M. Posch and E. Runca (1984) Acid Rain in Europe: A Framework to Assist Decision Making. Laxenburg, Austria: International Institute for Applied Systems Analysis.

Atlavinyte, O. (1978) Ecology of Earthworms and their Influence on Soil Fertility in the Lithuanian SSR. Tartu, Estonia, USSR (in Russian).

Brouwer F., J-P. Hettling and L. Hordijk (1983) An Integrated Regional Model for Economic-Ecological Demographic Facility Interaction. Papers of the Twenty-second European Congress of the Regional Science Association, Vol.52.

Buracas, A. and R. Rajackas. (Eds.) (1982) Regional Reproduction Modelling. Vilnius, Lithuanian SSR (in Russian).

Clark, W.C. and C.S. Holling (1985) Sustainable Development of the Biosphere: Human Activities and Global Change. In: Global Change (The Proceedings of a Symposium sponsored by the International Council of Scientific Unions (ICSU) during its 20th General Assembly in Ottawa, Canada on September 25, 1984). T.F. Malone and J.G. Roederer (Eds.). Cambridge University Press.

Deltuvas, R. (1982) Timber Production Planning in the Lithuanian SSR. Helsinki, Finland: University of Helsinki. Research Notes No.16.

Eitmanaviciute, I. (1982) Formation Regularities of Complex Soil Non-vertebrates under the Influence of Anthropogenic Factors in Sod-podzolic Soil Zone. Moscow (in Russian).

Grossmann, W-D. (1984) Methods for Analyzing Multi-faceted Problems Applied to Forest Die-off. WP-84-65, Laxenburg, Austria: International Institute for Applied Systems Analysis.

Grossmann, W-D. (1985) POLLAPSE--a Model on Pollution and Forest Damage. WP-85-xx. Laxenburg, Austria: International Institute for Applied Systems Analysis (forthcoming).

Gvishiani, D. (1977) Methodological Problems of World-wide Development Modelling. VNIISI. Moscow (in Russian).

Kairiukstis, L. (Ed.) (1977) The Lithuanian Republic's Program on Man and the Biosphere (MAB). Academy of Sciences of the Lithuanian SSR, Vilnius, Lithuanian SSR.

Kairiukstis, L. et al. (1980) Formation of Forest Plants System in the Agrolandscape of the Lithuanian SSR. Science for Agriculture, Vilnius, Lithuanian SSR, (in Lithuanian).

Kairiukstis, L. (1981) Optimization of Forest Growth. Academy of Sciences of the Lithuanian SSR. Kaunas, Lithuanian SSR.

Kairiukstis, L. (1982) Outline of Environment Optimization in the Lithuanian SSR by Modeling of Regional Development. Vilnius, Kaunas, Lithuanian SSR, (in Russian).

Kairiukstis, L. and G. Golubev (1982) Application of the CREAMS Model as Part of an Overall System for Optimizing Environmental Management in Lithuania, USSR: First Experiments. European and United States' Case Studies in Application of the CREAMS Model. V. Svetlosanov and W. Knisel (Eds.). CP-82-S11. Laxenburg, Austria: International Institute for Applied Systems Analysis, pp:99-119.

Kairiukstis, L., J. Grigaliunas and S. Mizaras (1983) Forest Sector Modelling. Forests, No.2 Tiesa Publishing House, Vilnius, Lithuanian SSR, (in Lithuanian).

Kairiukstis, L. (1985) Interaction between Scientific Experts and Managers on Decision Making in the Lithuanian SSR. Paper presented at the International Forum Science for Public Policy, (16-20 Jan. 1984), Laxenburg, Austria: International Institute for Applied Systems Analysis.

Kairiukstis, L. and A. Umnov (1985) Integrative Approach to the Modeling of Environmental Systems. Proceedings of the IASTED International Symposium "Modelling and Simulation", Lugano, Switzerland, June 24-26, 1985; Acte Press (Anheim, Calgary, Zurich).

Krelle, W. (1983) Waves of Entrepreneurial Activity Induced by Transfer of Information and Valuation. Paper presented at the Bonn- Harvard Conference on Entrepreneurship commemorating the 100th anniversary of Joseph A. Schumpeter, Bonn, September 16.

Knisel, W.G. (Ed.) (1980) CREAMS: A Field-scale Model for Chemicals, Runoff and Erosion from Agricultural Management Systems. U.S. Dept. of Agriculture, Conservation Report No.26.

Krutko, V.N., S.A. Pegov, D.M. Chomiakov and P.M. Chomiakov (1982) Model Dinamiki Sredoobrazujushchix Faktorov, (Dynamic Model of Factors Creating an Environmental Medium). Moscow (in Russian).

Lee, D. B. (1973) Requiem for Large-scale Models. Journal of the American Institute of Planners, No.34.

Lekevicius, R. (1980) Ecological Genetics. Science, Vilnius, Lithuanian SSR, (in Lithuanian).

Moiseev, N.N., V.V. Alexandrov, U.F. Krapivin, A.V. Lotov, Yu.M. Svirezhev and A.M. Tarko. (1984) Global Models: the Biospheric Approach. Paper presented at the workshop on "Sustainable Development of the Biosphere" (27-31 Aug. 1984), Laxenburg, Austria: International Institute for Applied Systems Analysis).

Munn, R.E. (1979) Environmental Impact Assessment in Developing Countries: Principles and Procedures. Toronto, Canada: Institute for Environmental Studies, University of Toronto.

Parry, M.L. and T.R. Carter (1984) Assessing the Impact of Climatic Change in Cold Regions. SR-84-1. Laxenburg, Austria: International Institute for Applied Systems Analysis.

Pauliukevicius G. (1982) Ecological Role of Forests. Science and Life. No.3, Vilnius, Lithuanian SSR (in Lithuanian).

W. Mende, K.-F. Albrecht

Application of the evolon model on evolution and energy growth processes

Introduction

There exist different kinds of irreversible processes. Such which either use information nor accumulate information locally do not leave behind long-living traces. On the other hand there exist irreversible processes which have long-living consequences in the form of ordering and structure building. Important subclasses of these processes are such which accumulate information during their evolution closely connected with learning processes and such which decode already accumulated information and use it for structure building. These processes are accompanied by growth processes. From this it comes out that growth processes not only accumulate quantities but are connected with inner organizing processes. In general it is however possible to find out an appropriate growth indicator which characterizes this complex process in a simplified but effective manner. For instance the baby-weight is a very aggregated parameter but in spite that it is the result of a huge amount of metabolic processes it gives a highly condensed valuable information about the wellbeing of the baby. Growth indicators for other growth processes are for instance: biomass, energy consumption, population numbers, concentrations and so on. The pure phenomenological description of the time dependence of such indicators has its own value. But very often it is also possible to conlude from the outer growth behaviour to the underlying inner mechanisms (to-down approach).

The exponential tower for generating hyperbolic and parabolic growth

In /Me1, Pe1/ we used the equation $\dot{x} = cx^k$ for describing unsaturated growth processes. With $k < 1$ we get parabolic growth and with $k > 1$ we get hyperbolic growth. The exponential tower concept shows how these two types of growth can be generated by concatenation of exponential growth processes. The exponential tower is defined as follows:

$$\dot{x}_i = K_i x_i x_{i+1} \quad \text{with } x_i(0) = 1 \text{ and } i = 0, 1, 2, \ldots, n$$

(The initial condition $x_i(0) = 1$ can be always fullfilled by normalisation on each level i). For $n \rightarrow \infty$ and the K_i-sequences $K_o, -K, -K, -K \ldots$; $K_o, 0, 0, 0 \ldots$ and $K_o, K, K, K \ldots$ one gets parabolic: exponential and hyperbolic growth for x_o respectively /Me1/. The corresponding saturation processes one gets by the substitution $x' = B - x$. That means x_o and x'_o can be described by $\dot{x}_o = cx_o^{K_o}$ and $\dot{x}'_o = c' x_o'^{K'_o}$ respectively. Multiplying the growth and the saturation process and renaming K_o by k and

K'_o by 1 one gets the evolon model: $\dot{x}_i = Cx^k(B-x)^l$. From the K_i-sequences follows that we have inhibiting couplings on all higher levels in the case of parabolic growth and stimulating couplings in the case of hyperbolic growth. In the case of exponential growth there are no couplings. The parameters k and l we call cooperativity exponents (in analogy to chemical kinetics). Their values do not depend on linear transformations in x and t. These parameters can be used for classification of growth processes. Our condensed experience in interpretation of these parameters in investigations of growth processes in reality is given in table 1.

<u>Table 1</u>: Classification of growth and saturation processes with cooperativity exponents j, (j = k or j = l)

Parabolic: $j < 1$	exponential: $j = 1$	hyperbolic: $j > 1$
K_i=j,-K,-K,-K,...	K_i=j,0,0,0,...	K_i=j,+K,+K,+K,...
target directed		open searching processes, no target
prototype known		prototype unknown
repeatedly occurring process		unique process
reconstruction and repair		pioneer innovation process
planned, instructed process		unplanned selforganizing process
attractor and trajectory determined		attractor and trajectory undetermined
cooperative inhibiting couplings		cooperative stimulating couplings
decoding of information		accumulation of information
examples: ontogenesis, succession		examples: evolution, phylogenesis

In the following we examine the hypotheses given in the table for real growth processes for which we could find appropriate time series in literature. These data contain unfortunately only few information about the saturation process. Therefore the saturation exponent l cannot be determined as well as the growth exponent k from these data. So we should concentrate our examination on the initial growth phase which is controlled by the cooperativity exponent k. The identification of the evolon parameters is done using the function minimizing program system MINUIT/Ja1/. For the goodness function we use the square sum of the model related relative distances between data and model values in the fit region. The goodness of the fits we characterize by the relative mean deviation R /A12/ related to the number of degrees of freedom.

<u>Cases of target directed growth</u>

A typical target directed growth process is the development of a biological individuum from a fertilized egg cell to an adult organism. The process which will lead to the target of the growth is in this case encoded in the DNA-genome. During the growth process this information will be decoded in interaction with the expected normal environment by the decoding device in form of living cells. That means that genetic control damps the exponential growth of the zygote near cells. Already Schmalhausen/Sc1/ claimed that for embryonic growth the parabolic growth law is a good description. Reevaluating data of chicken embryo weight /Te1/ we found in accordance to Terskova 4 periods of parabolic growth with different exponents k (fig. 1 and table 2).

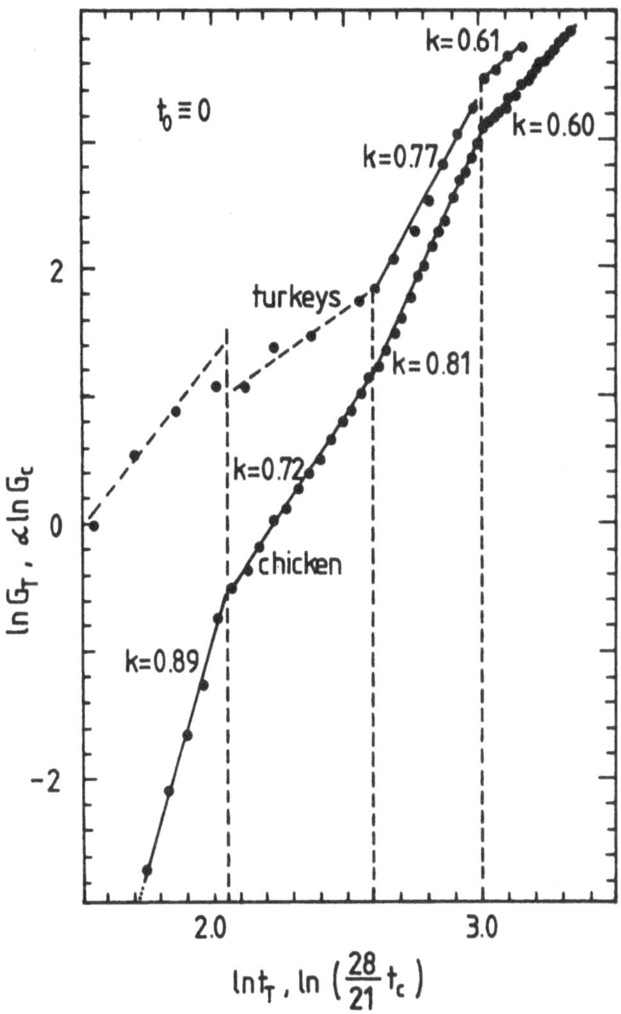

<u>Fig.1</u> The parabolic growth periods of birds in dependence on time normalized to the breeding time and fitted parabolic model curves corresponding to $\dot{x}=cx^k$; $x=dx^r$ with $r=1/(1-k)$. For chicken: $r_1=8.73$, $r_2=3.60$, $r_3=5.48$, $r_4=2.53$; turkeys: $r_3=4.26$, $r_4=2.53$. (see table 2)

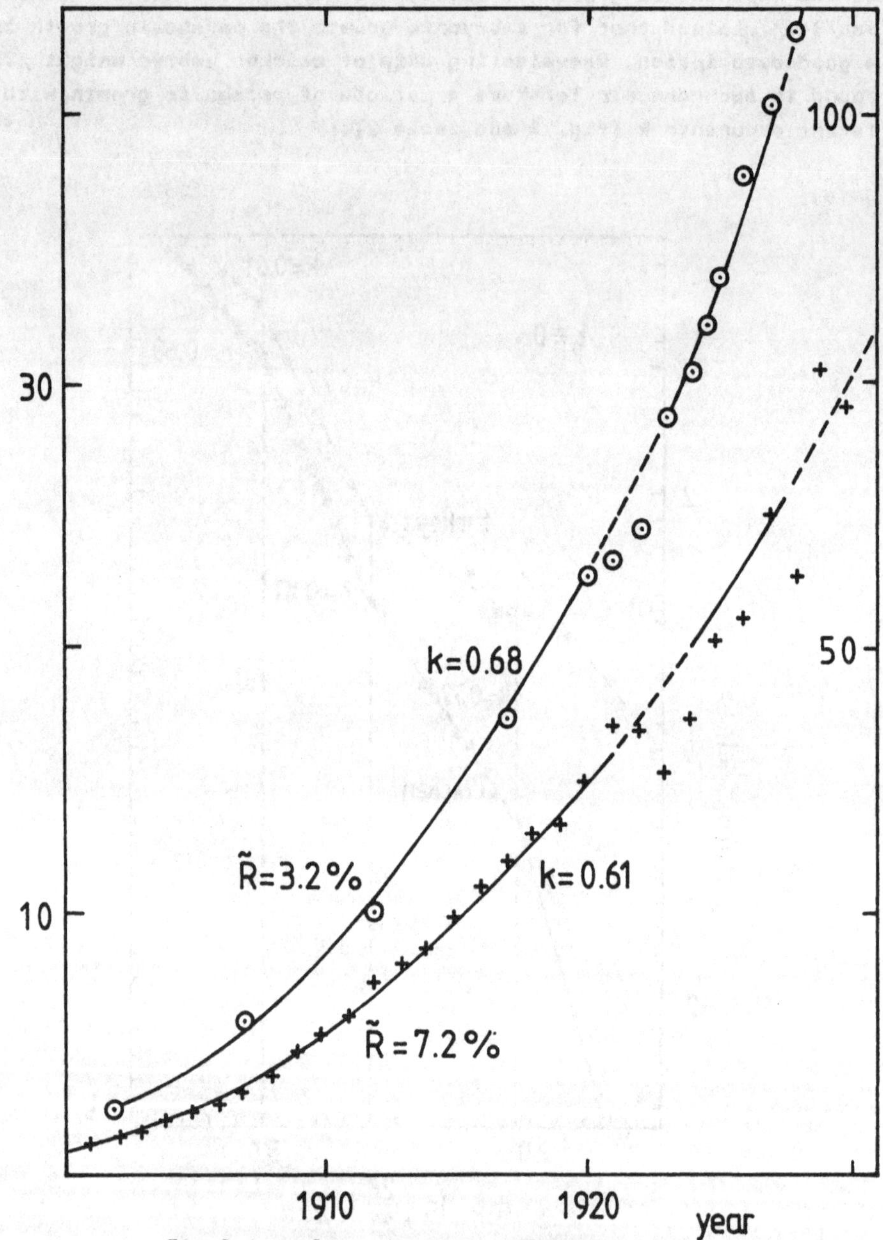

<u>Fig.2</u> Annual production of electric energy during the initial periods dominated by substitution processes for Germany (crosses) and the USA (circles). Data for Germany 1900-1927 with 1921-1924 excluded and for the USA 1902-1927with 1921-1922 excluded from fitting. (see table 2)

The errors of the found k-values show that exponential growth and nevertheless hyperbolic growth can be ruled out. In cooperation with Nitechmann /Ni1/ we found the same behavior in the case of turkey-hens. The r-values (r = 1/(1-K)) could be connected with the number of cellgenerations from certain proliferation cells /Me1/. Many growth processes in ontogenesis are succeśfully modeled by parabolic growth laws also after the embryonic period. The widely used Bertalanfy-function exibits parabolic initial growth /Sc1, Te1, Ni1/. A similar parabolic growth process we can expect in ecological successions because the involved species are already prepared for a special kind of interactions. The final climax state of the succession is predetermined by the involved species and the geophysical and geochemical environmental (climate, soil etc.) conditions. Infact there are strong functional analogies between succession and ontogenesis: "Ėcological succession involves the development of ecosystems it has many paralelles in the developmental biology of organisms...", Odum /Od1/. It is remarkable that this analogy is reflected in the growth dynamics also. From data for a deciduoue forest succession taken from a diagram of Lickens and Bormann /Bo1/ we got a cooperativity exponent $1 \approx 0.81$ for the biomass growth. Parabolic growth processes could be found in socio-economic processes also. Typical processes of this kind are for example substitution processes. Here a certain subsystem is already evolved and a certain function of that system well be fullfilled by a new kind of ressource or commodity. (This correeponds in some sense to the engagement of an ecofunction or an ecological niche by an other species). When the use of electricity for lighting begun other methods for comercial lighting were already evolved.
The early stimulus for the increase of electric energy production was the substitution of older light sources by electric lamps. Even at the beginning of the use of electricity for the supply of mechanical energy the steam engines were only replacee by central electric motors without structural changes (to many motors in one device) in the factories. Therefore we can expect parabolic growth in the production of electric energy at the beginning of the century as it is demonstrated in fig. 2. Attempts to describe these data by hyperbolas failed giving considerably higher chi-square.
A similiar process as substitution is the process of reconstruction ('wound healing'). In this case the target and the way to the target of the growth process is even more preciseely predetermined. Therefore the reconstruction of electricity production after the war shows parabolic growth (see table 2). This is the reason why the data in this period lie clearly below the hyperbolic evolon model curve, which is extracted

from the data of a period after the reconstruction phase is finished (see fig. 5a).

Table 2: Target directed processes for which parabolic growth (k > 1) is expected

Nr	process and	period of fitting	k	l	$\tilde{R}$ %	Reference
	Embryos of	0%- 32%	0.89			
1	chicken	32%- 52% of breeding	0.72			/Te1/
	and	52%- 77% time	0.81 (0,77)			
	turkeys	77%-100%	0.60 (0.61)			/Ni1/
2	Succession		0.81			/Bo1/
3	Electro	Germany 1900-1927	0.61	(=1.0)	7.2	/Mi1/
4	energy	USA 1902-1927	0.68	(= k)	3.2	/Hi1/
5	production	GDR 1946-1956	0.82		2.5	/Sg1/

Cases of true evolution processes

While parabolic target directed growth processes are going on repeatedly the true evolution processes are unique and the final attractor is not 'known' in this case. Structur building in true evolution processes is always connected with innovations.

One of the most complex systems is the world population. A high degree of cooperativity and interdependence is already established among the various groups of people in all structural levels such as nations, towns, organisations etc. During the whole history of mankind the long term growth-rates are slowly increasing in such a way that the world population increases nearly hyperbolically. The cooperativity of the world population system is reflected by the relative high cooperativity exponent k = 1.8 (tabl. 3). A similar argumentation holds for the population growth in towns. In our investigation of the growth of the averaged population number from 10 big German towns /A12/ we got clear hyperbolic initial growth behavior (tabl. 3). Parallel to the human population evolves the energy consumption. Energy is used in all subsystems of human population and in general these subsystems are in cooperative interaction. Energy consumption is a very suitable indicator for the evolutionary state of a complex system. Because it is a physical quantity, it es well defined, non substitutable by other physical quantities and in many cases easily measurable. Energy production and consumption reflects a complex network of human activities with many feedbacks between generation and application.

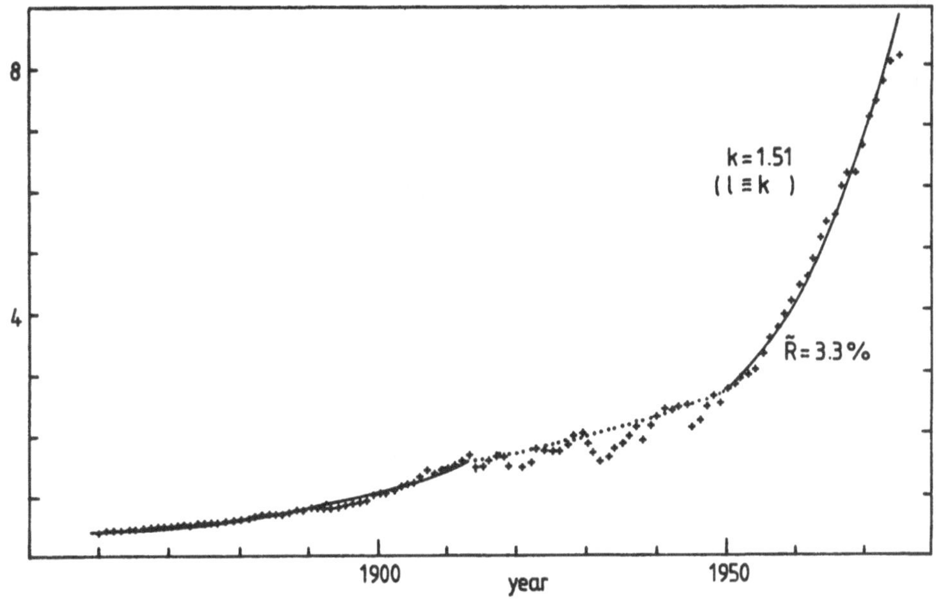

Fig.3 Annual world primary energy consumption and fitted Evolon model curve under the assumption l=k and maximum of relative growth rate at t_m=1974. The curve shows a hyperbolic initial growth phase (tabl. 3). Used data: 1860-1973 with 1914-1949 excluded from fitting.

So we should expect a high cooperativity exponent for the growth of the human energy system indicated by the world primary energy consumption. The obtained value K=1.51 for the growth cooperativity exponent confirms this expection of hyperbolic growth /Kr1/ (fig.3, tabl.3).
In the nonfitted region 1914-1949 strongly influenced by the world wars and economic crisises the whole evolution process goes on with diminished velocity and additional some already developed structure is switched off and is reactivated during recovery from war destructions. This explains the pits below the evolon model curve /Kr1/. It es remarkable that the same evolon model parameters describe the growth process before and after the disturbed unfitted region.
The most valuable energy form used by mankind is electricity. As Lenin recognized already in an early stage of electrification the use of electricity is an essential for all technical and social-economical progress. The evolution of the electrical energy system is strongly coupled with the evolution of other branches in economy. These couplings act symbietically (cooperative) and therefore we expect hyperbolic

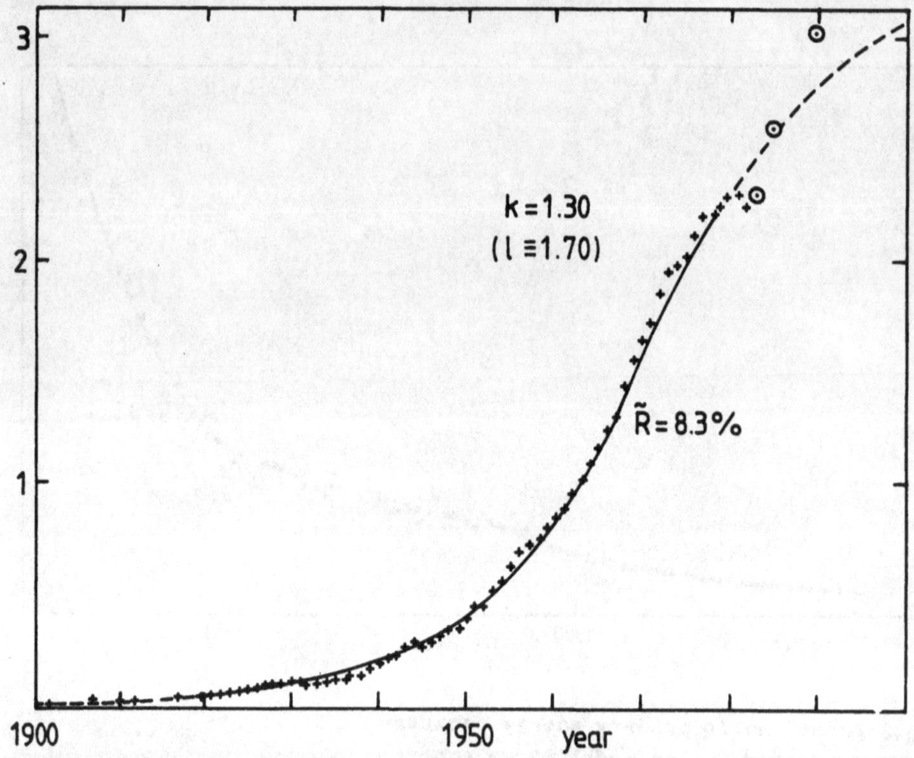

Fig.4 Annual production of electric energy of the USA. Used data from 1920 to 1978 in which period the process is dominated by a true evolution process yielding hyperbolic growth (tabl.3). l is fixed to l=1.7 (see text). The 'prognostic data (indicated by circles) indicate that l may be even greater.

growth in the period after the substitution phase when a true evolution process with a specific Eigendynamics becomes dominant. The indicators of the electrical energy system have the advantage of good measurability. The pioneer process in using electric energy took place in Germany and the USA starting 100 years ago.

The electric energy production data for the USA show a rather disturbed behavior but behind the disturbances the regular evolution process seems to be extractable. For the period of true evolution the obtained evolon modell parameters indicate clearly hyperbolic growth (fig.4 and tabl.3). The relative mean deviation $\tilde{R}$=8.3% of the data from the model values reflects the irregularities of this time series. Nevertheless the obtained growth cooperativity exponent is in the hyperbolic region. The l-value is fixed (preliminary to l=1.7) because of lack of data in the saturation phase. The 'prognostic' estimations for 1983, 1985 and 1990 /Sa1/ indicate that l may be even greater.

Table 3: True evolution processes for which hyperbolic growth ($k > 1$, $l > 1$) is expected

Nr	Process and period of fitting			k	l	$\tilde{R}$ %	References
1	world population		-10^6–1970	1.8			/Me1, Pe1/
2	population in big German towns		1871-1910 1955-1970	$1.60 \pm .07$	=k	1.7	/Al1/
3	world energy consumtion		1860-1913 1950-1973	$1.51^{+.01}_{-.04}$	(:=k)+)	1.3	/Kr1, Pe1/
4	electro	USA	1920-1978 1956-1980	1.30x)	-1.7x)	8.3	/Hi1, Sa1/
5	energy	UdSSR	1983-1985	$1.03^{+.04}_{-.05}$	$2,2^{+}_{-1,1}$	1.2	/St1, Na1/
6	production	GDR	1956-1984	1.11x)	1.30x)	1.5	/Sg1/

x) preliminary; +) max $\frac{\dot{x}}{x}$ at $t_m = 1974$

Because the growth of electric energy production in the USSR and the GDR is neither a typical pioneer process nor a typical repeated process but unique under special conditions we expect hyperbolic growth but near to exponential growth. Parabolic growth should be excluded, because there is no 'known' final target. In contrast to the USA data the Soviet data for electric energy production show a very regular smooth shape. This may be caused by the great attention Lenin already directed in 1920 to the planful and proportional development of the whole socio-economic system. These data are especially well documented and source-consistent. The only visible disturbance of the data behavior was during the world war 2-nd.

The resulting parameters for the fitted period after reconstruction from 1956-1985 correspond as expected to faint hyperbolic growth and strong hyperbolic saturation (fig.5a, tabl.3). The mean deviation of the data from the evolon model curve is only $\tilde{R} = 1.2\%$. Even the data from 7 years ago up to the fit period are surprisingly good described. (The corresponding R-value for these 7 years would be R = 2 %). The 9 % downwards deviation of the data in the period 1942-1949 is plausible because we expect a parabolic reconstruction phase during this time. The upward movement of the data in the thirtees is connected with the enormeous growth acceleration during the GOELRO x) period. The success of this enthusiastic period was unfortunately destroyed by the war. If we extend the fitregion only to 1980 the data after 1980 are described in a similar good manner. The data show regular oscillations around the model curve. Spectral analysis gives maxima at 11.0;20.7 and 4.0 years. Adding

x) Gouvernmental plan for electrification of Russia

an oscillation term to the model an even more precise model description of the data can be expected.
This opens the possibility to shrink the confidence intervals even for the saturation cooperativity exponent l. Without this additional analysis l can only be confined to $1,1 < l < \infty$. This large interval we get considering the data deviations as white noise, but this is obvious not

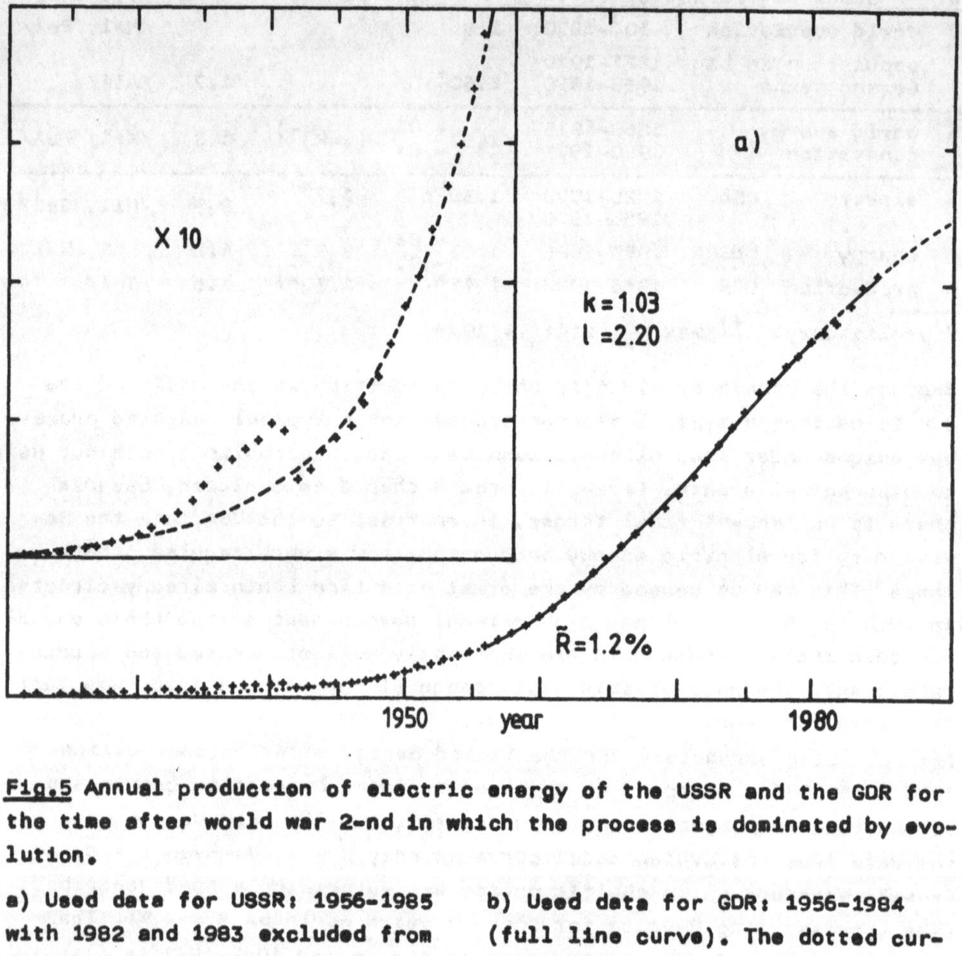

Fig.5 Annual production of electric energy of the USSR and the GDR for the time after world war 2-nd in which the process is dominated by evolution.

a) Used data for USSR: 1956-1985 with 1982 and 1983 excluded from fitting. In this case the description of the data is especially good. Even the data from 14 years ago up to the fit period are surprisingly good described by the backward projected model curve (see insert).

(Fig. 5b) see next side)

b) Used data for GDR: 1956-1984 (full line curve). The dotted curve shows the evolon fit with data 1946-1981. Here the reconstruction phase from 1946-1955 is locally parabolic (tabl.2). The influence of that reconstruction phase declines the k-value for the dotted curve and yields a complete unrealistic saturation behavior for the dotted curve.

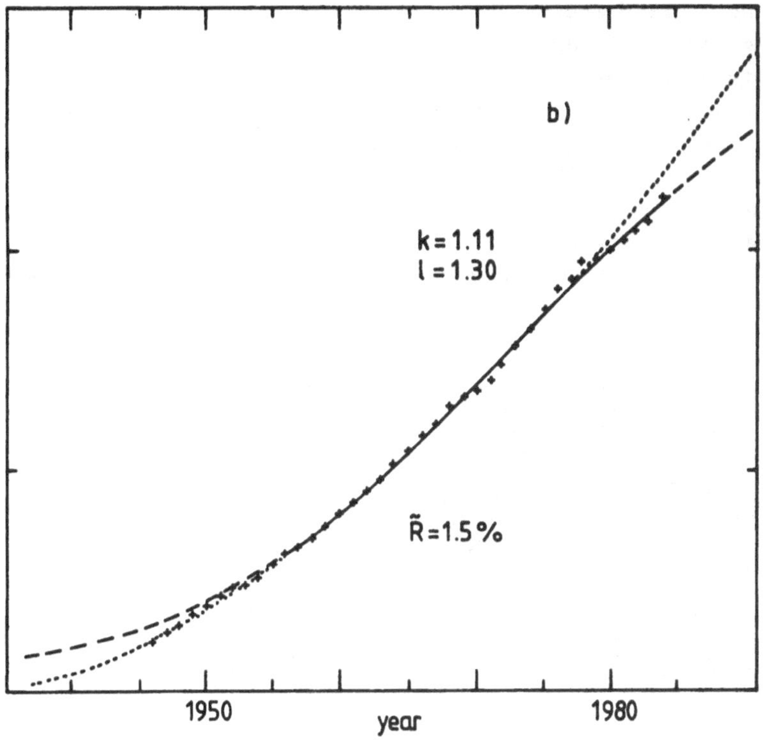

the case. We expect that the additional analysis yields upper limits for the confidence intervall of l.
The data for the GDR show a relative regular shape also. The evolon model fits these data within R = 1,5 % giving faint hyperbolic growth. Preliminary found parameters (corresponding to the solid curve in fig. 5b) are given in table 3. To get final results (including confidence intervals) would be time-consuming but valuable. As already mentioned above the data of the reconstruction phase from 1946–1955 show clearly parabolic growth (see tabl.2). Including these data in fitting we get the dotted curve in fig. 5b with a decreased k-value which yields simultaneously to complete unrealistic saturation behavior. This stresses the importance of selecting the data for fitting as free as possible from disturbances according to the criterion Karl Marx /Ma1/ mentioned in characterizing the investigation method of physicists: "Der Physiker beobachtet Naturprozesse entweder dort, wo sie in der prägnantesten Form und von störenden Einflüssen mindestens getrübt erscheinen, oder, wo möglich, macht er Experimente unter Bedingungen, welche den reinen Vorgang des Prozesses sichern". We strove for proceeding according to this maxime.

Conclusions:

The macrodynamics of a number of growth processes is investigated using highly aggregated indicators:

1. **The exponential tower concept offers the possibility to classify growth processes by cooperativity exponents.**
2. **The results of the investigation of the cooperativity exponents (for a number of real growth processes) are in accordance with our expectations given in table 1.**
3. The parabolic and hyperbolic behaviour of growth and saturation seems to be fundamentally different and closely connected with the dichotomy of target oriented processes on one hand and pioneer-evolution processes on the other hand.

References:

/Al1/ Albrecht,K.-F.; Mende,.: A New Possible Description and Interpretation of Population Growth Data in German Towns, Syst. Anal. Model. Simul.1 (1984), 3,p 237-249.

/Al2/ Albrecht,K.-F.; Mende, W.; Peschel, M.: Lotka Volterra-Approach to Cooperation and Competition in Dynamic Systems, ed. by W. Ebeling and M. Peschel, Akademieverlag, Berlin 1985,236-240

/Bo1/ Bormann,F.H. and Likens, G.E.: Pattern and Process in a Forested Ecosystem, Springer-Verlag, New York, 1979, 253 pp.

/Hi1/ Historical Statistics of the United States, p. 820, 828

/Ja1/ James, F.; Roos,M.: MINUIT, A system for function minimisation and analysis of the parameter errors and correlations. CERN COMPUTER CENTRE, PROGRAM LIBRARY, LONG WRITTE-UP, D 506, Geneve 1977 or Computer Physics Comunications 10 (1975), 343-367

/Kr1/ Kriegel,U.; Mende,W.; Grauer,M.: Discussion of world primary energy consumption and world population data in terms of a Hyperlogistic Evolution Model, Syst. Anal. Model. Simul 1 (1984) 3, p.225-236.

/Ma1/ Marx,K.: Das Kapital,in MEGA II/r,S.12,Dietz-Verlag,Berlin 1983.

/Me1/ Mende,W.: Lotka-Volterra-Approach to Cooperation and Competition in Dynamic Systems, ed. W. Ebeling and M. Peschel, Akademieverlag, Berlin 1985, p. 261-271.

/Mi1/ Mitchell: European Historical Statistics 1750 - 1970.

/Na1/ National Economy of the UdSSR (in russian), Moscow

/Ne1/ Neefe, M.: Statistisches Jahrbuch deutscher Städte 1890-1980

/Ni1/ Nitschmann, K.-F. Albrecht; W. Mende: Die Allantoisentwicklung beim Hausgeflügel, to be published.

/Od1/ Odum,E.P.: The strategy of ecosystem development, science 164 (1969), p. 262-270.

/Pe1/ Peschel,M.; Mende,W.: The Predator-Prey Model, Akademieverlag, Berlin, 1986.

/Sa1/ Statistical Abstract of the United States (1985

/Sc1/ Schmalhausen,I.: Beiträge zur quantitativen Analyse der Formbildung. Über die Gesetzmäßigkeiten des embryonalen Wachstums, Roux' Arch.Entw.-mech., 1927, 109, N4, S. 455-512.

/St1/ Statistisches Jahrbuch der DDR for different years.

/St1/ Steklov, W.Yu.: Development of the Electroenergetic Industry in the UdSSR (in russian), Energija, Moscow.

/Te1/ Terskova,N.I.: Analysis of the growth dynamics of biological objects (in russian) p. 5-15, Publishing house 'Science', Moscow, 1978.

NON-LINEAR DYNAMICS OF THE EPIDEMICS OF SOME CHILDHOOD DISEASES IN COPENHAGEN, DENMARK.

Lars F. Olsen.

Introduction

The time evolution of an infectious disease in a population has three possible outcomes [1]: (i) following a burst the incidence rate drops to zero and the disease becomes extinct; (ii) the incidence rate approaches a constant value, sometimes through a damped oscillation and the disease becomes endemic; (iii) the incidence rate does not converge on a fixed value but continues to fluctuate, sometimes in an apparently random way. In this case the disease has become epidemic. In the two former cases the dynamics are well understood. In the latter case there has been some controversy as to the mechanisms responsible for the perpetuation of the disease and the apparent randomness in incidence rates. In the earlier epidemiological literature the models proposed to explain such behaviour were built on primarily stochastic elements [3,4]. However, the suggestion by London and Yorke [10] to take into account seasonally dependent contact rates has led to models where stochastic factors are no longer necessary for the perpetuation of epidemic diseases [2,5]. Recently Aron and Schwartz [2] demonstrated that the socalled susceptible-exposed-infected-recovered (SEIR) model with seasonal variations in contact rates may exhibit period doubling bifurcations leading to chaos as one parameter is varied. If fluctuations in incidence rates could be shown to be chaotic this would further eliminate the requirements for stochastic factors to explain the irregularities in the disease data [11,14]. Such a demonstration was first given by Schaffer and Kot [17] who analysed the monthly reported cases of measles in New York and Baltimore from 1927-1963 by embedding the data in three dimensions using the reconstruction scheme of Takens [18]. They showed that the resulting flows were almost two-dimensional. Poincaré sections made by slicing the flows yielded almost one-dimensional maps exhibiting chaotic dynamics. Very recently Schaffer [15] has demonstrated a close relationship between the New York and Baltimore measles data and the dynamics of the SEIR model in the chaotic parameter region. In the present work I have analysed the monthly reported cases of four childhood diseases - measles, mumps, German measles and chicken pox - in Copenhagen, Denmark using essentially the same approach as Schaffer and Kot [17]. The results strongly favour an interpretation of measles, mumps and German measles epidemics as chaotic whilst the dynamics of chicken pox are those of a simple periodic oscillation perturbed by noise.

Time series and correlation exponents

The monthly reported cases of measles, mumps, German measles and chicken pox in Copenhagen for the periods 1927-1967 and 1938-1968 are shown in Fig. 1 [12]. Chicken pox and German measles were not recorded monthly until 1938. We note that for the first three diseases the outbreaks of epidemics occurred with dramatic variations in incidence rates. The fourth disease, chicken pox, exhibits considerably less variation in the number of cases than the other diseases. According to current beliefs the time evolutions of epidemics as those shown in Fig. 1 may either reflect periodic oscillations perturbed by noise or chaotic dynamics. In order to discriminate between these two possibilities the

Institute of Biochemistry, Odense University, Campusvej 55, DK-5230 Odense M, Denmark.

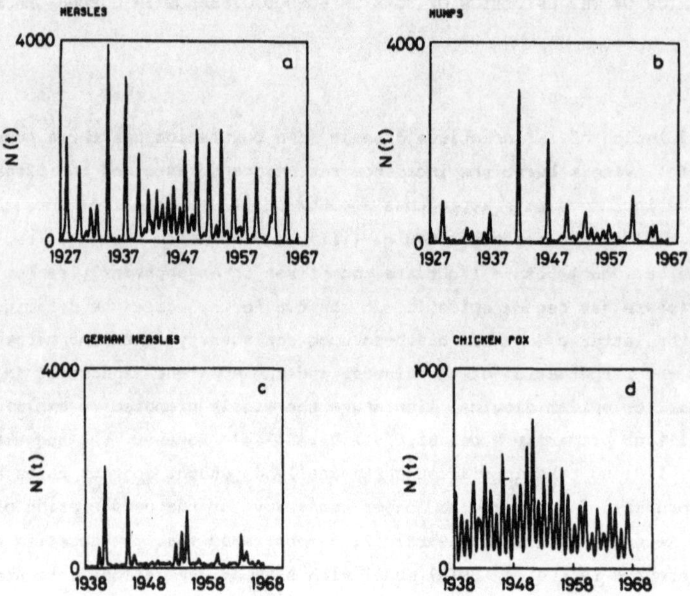

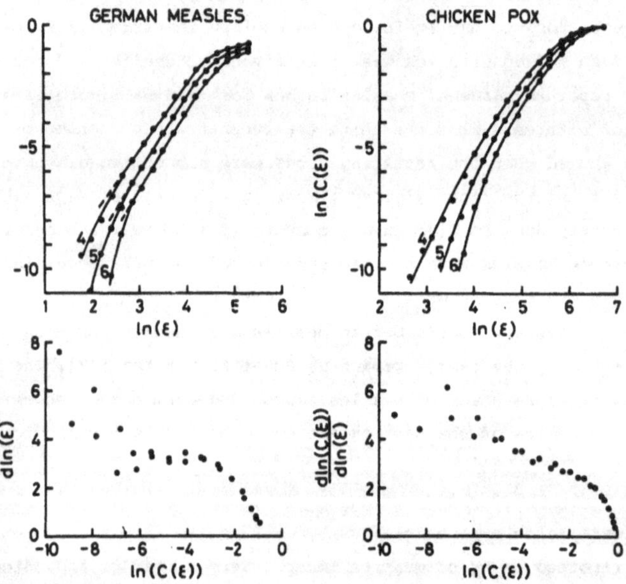

<u>Fig. 1</u>. The monthly reported cases of measles, mumps, German measles and chicken pox in Copenhagen, Denmark. The data obtained from [12] have been subjected to a three point smoothing.

<u>Fig. 2</u>. Estimation of correlation exponents for German measles and chicken pox. Top curves: Plot of ln C(ε) against ln ε for embedding dimensions d = 4,5 and 6. Bottom curves: Plot of the slopes dln C(ε)/dln ε against ln C(ε) for d = 5 and d = 6.

Table I. Correlation exponents for the disease data

Disease	no. of points	ν
Measles	480	1.9 ± 0.2
Mumps	480	1.75 ± 0.3
German measles	360	3.2 ± 0.2
Chicken pox	360	-

dimensions of the attractors of the flows of the four diseases were estimated. The dimension is an extremely useful measure in the characterisation of attractors, in particular chaotic attractors [6]. Although an infinite set of dimensions exist only two of these can be computed for a time series of a single observable. One is the information dimension which is believed to be identical to the Lyapunov dimension [6]; the other is the correlation exponent [8]. Calculation of the information dimension requires very large numbers of measured data which unfortunately are not available here. Instead it was decided to probe the disease data by calculating the correlation exponent, ν [8]. The exponent is calculated using a time series of points on the attractor $X_1, X_2, \ldots, X_N$ from the correlation integral which is defined as

$$C(\varepsilon) = \frac{1}{N^2} \sum_{i \neq j} \Theta\,(\varepsilon - |X_i - X_j|) \qquad (1)$$

where Θ is the Heaviside function. $C(\varepsilon)$ counts the number of point pairs X_i, X_j whose distance is smaller than ε. It was shown that $C(\varepsilon)$ scales as

$$C(\varepsilon) \sim \varepsilon^{\nu}$$

where ν serves as a measure of the attractor in question. ν is related to the information dimension, D_I, by the inequality $\nu \leq D_I$ [8]. Furthermore for systems perturbed by noise a plot of log $C(\varepsilon)$ against log ε has two regions. For length scales, ε, above the noise level $C(\varepsilon)$ should continue to scale as ε^{ν} whereas for length scales comparable to or smaller than the noise level $C(\varepsilon)$ should scale as ε^{d} where d is the dimension of the state space in which the flow was embedded [8]. To calculate the correlation exponents of the four diseases the monthly incidence numbers were embedded in successively higher dimensions d = 2, 3, 4 ... using the reconstruction scheme of Takens [18]. The vectors X_i were chosen as $(N(t), N(t+h), \ldots, N(t + (d-1)h))$ where $N(t)$ is the number of cases in month t and h is a time delay (usually three months). Figs. 2 a and c show plots of log $C(\varepsilon)$ vs. log ε for German measles and chicken pox and Figs. 2 b and d show the slope vs. log $C(\varepsilon)$ plots. We note for German measles that the slope vs. log $C(\varepsilon)$ plot has a plateau corresponding to $\nu = 3.2 \pm 0.2$ for $-6 < \log C(\varepsilon) < -3$ whereas the corresponding plot for chicken pox continues to increase for decreasing log $C(\varepsilon)$. We may thus conclude that the dynamics of German measles contain a chaotic attractor whilst the dynamics of chicken pox do not. The correlation integrals of measles and mumps also revealed chaotic attractors with $\nu = 1.9 \pm 0.2$ and $\nu = 1.75 \pm 0.3$ respectively. The value for measles should be compared with the recent value obtained for measles in New York of

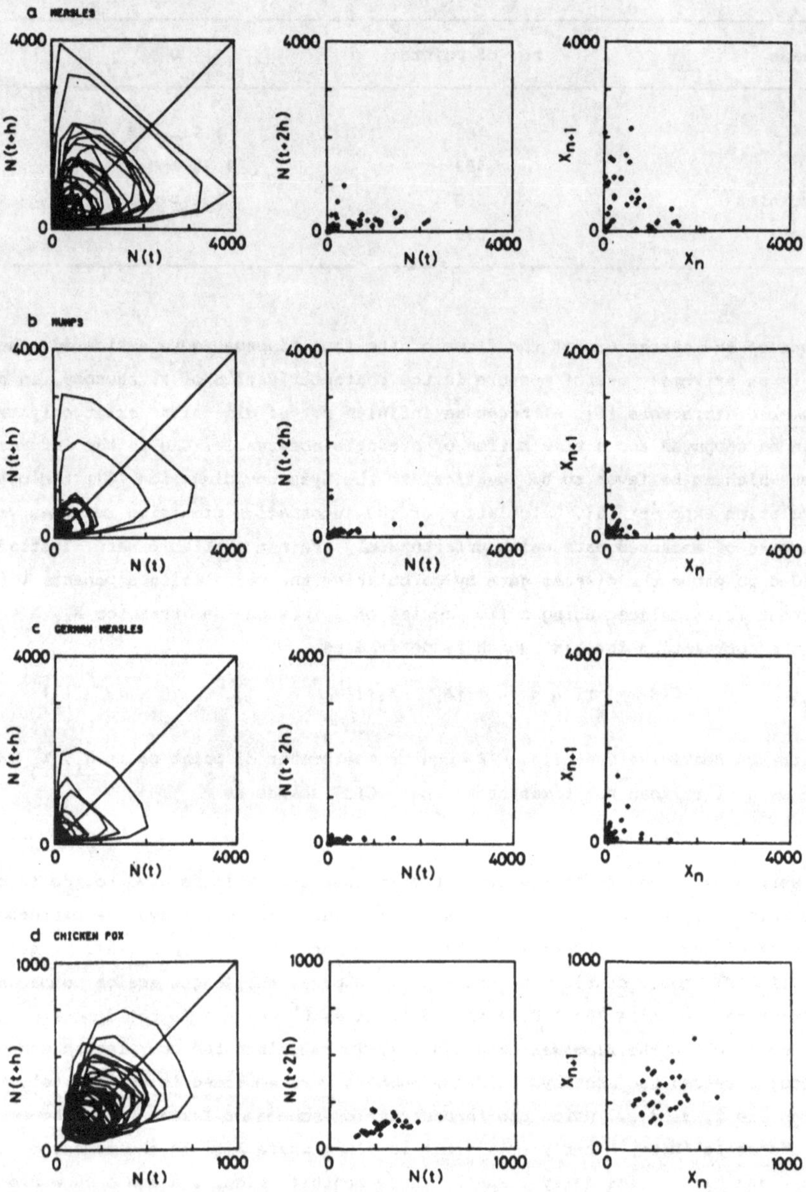

Fig. 3. Flows, Poincaré sections and return maps of the four diseases following reconstruction of the data in three dimensions. Left column: Projection of the flow of N(t), N(t+h), N(t+2h) (h = 3 months) in the N(t), N(t+h) plane. The solid line indicates the plane (normal to the page) slicing the flow. Middle column: one half of the Poincaré sections. Right column: Return maps obtained by plotting each point from the Poincaré section against the preceding point.

$\nu = 2.55$ [15]. However, it is worth emphasizing that these values of ν should be taken with some precaution due to the relatively small size of the samples. Moreover the value of ν estimated for German measles is almost certainly too high because the linear region of the log $C(\varepsilon)$ vs. log ε plot with slope ν is small compared to the corresponding intervals for measles and mumps. In a recent work by Franaszek [7] it was shown that when the linear region of the log $C(\varepsilon)$ vs. log ε plot with slope ν decreases due to addition of increasing noise levels, the slope will increase and consequently the value for ν will be overestimated. With these precautions in mind, the plots of log $C(\varepsilon)$ vs. log ε suggest that the dynamics of measles, mumps and German measles are chaotic.

3-dimensional flows, Poincaré sections and return maps

The data for the four diseases were reconstructed in 3 dimensions using the reconstruction scheme of Takens [18] and time delays of 3 months. The resulting plots indicated that the flows were all confined to almost 2-dimensional surfaces. For measles, mumps and German measles these surfaces defined cones with their vertex near the origin. The projections of the flows in the $N(t)$, $N(t+h)$ plane are shown in the left column of Fig. 3. Poincaré sections were made by slicing the flows with a plane normal to their direction as indicated by the solid lines. The resulting Poincaré sections consisted of two half lines, in the cases of measles, mumps and German measles joined to form a "V" whereas for chicken pox the two half lines were unconnected. The Poincaré sections are shown as one half line in the middle column of Fig. 3. The points on the Poincaré sections were used to construct return maps by plotting each point against the preceding point as shown in the right column of Fig. 3. For measles, mumps and German measles the maps have a superficial resemblance to one-dimensional maps of the form

$$X_{n+1} = aX_n \exp(-bX_n) \tag{2}$$

(a and b are constants) perturbed by noise [17]. For the mumps data, however, the compression of the map is high. The return map constructed from chicken pox data appears to consist of randomly scattered points. If we interpret the return maps of measles, mumps and German measles as maps defined by eq (2) with additional noise, we may conclude that these epidemics evolve with chaotic dynamics. In the following section we shall compare the disease data with numerical solutions of the SEIR model with seasonal variations in contact rates.

The SEIR model with seasonal variations in contact rates

The SEIR model consists of a set of three non-linear differential equations [2,5]

$$\frac{d\,S(t)}{dt} = \mu\,(1-S(t)) - \beta(t)\,S(t)\,I(t)$$

$$\frac{d\,E(t)}{dt} = \beta(t)\,S(t)\,I(t) - (\mu+\alpha)\,E(t) \tag{3}$$

$$\frac{d\,I(t)}{dt} = \alpha\,E(t) - (\mu+\gamma)\,I(t)$$

where $S(t)$, $E(t)$ and $I(t)$ denote the fractions of the population which are susceptible to the disease (S), infected but not yet infectious (E), and infectious and capable of

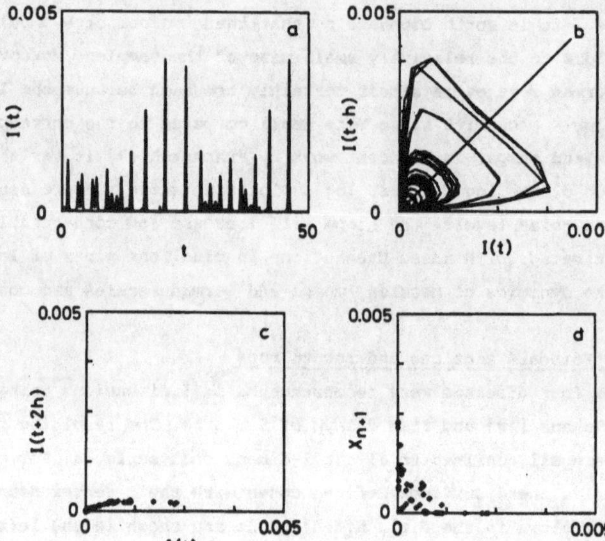

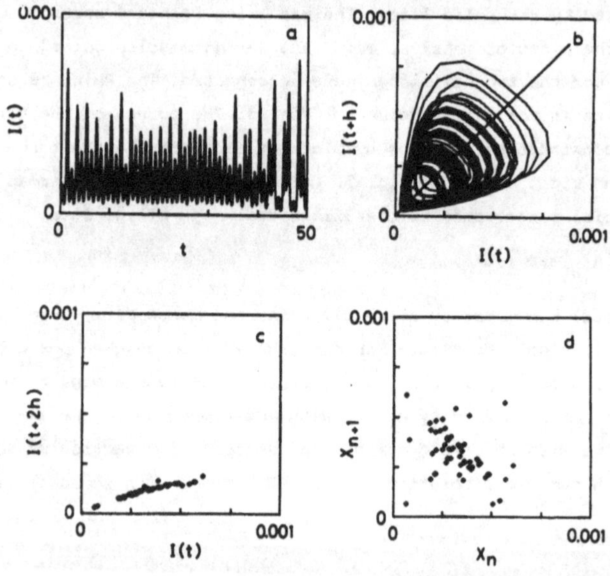

Fig. 4. Numerical simulation of the SEIR model, α = 36 years^{-1}. Time series, flow, Poincaré section and return map using I(t) and reconstructing the data in 3 dimensions (h = 0.15 years). Parameters are α = 36 years^{-1}, β_0 = 1800 years^{-1}, β_1 = 0.28, μ = 0.02 years^{-1} and γ = 100 years^{-1}. During the numerical integration S(t) was perturbed by a low level of noise by adding a random number to S(t) at every 0.05 year. The random number is selected uniformly in an interval of ± 1% of the value of S(t).

Fig. 5. Numerical simulation of the SEIR model, α = 22 years^{-1}. Time series, flow, Poincaré section and return map using I(t) and reconstructing the data in 3 dimensions (h = 0.2 years). Parameters as in Fig. 4 except for α = 22 years^{-1}. S(t) was perturbed by noise as in Fig. 4, the noise level corresponding to ± 2% of the value of S(t).

transmitting the disease (I), respectively. μ represents the birth and death rates which are assumed to be equal, α^{-1} is the mean latent period after initial contact, $\beta(t)$ is the contact rate and γ^{-1} is the mean infectious period. The model furthermore assumes that the fractions of susceptibles (S), exposed (E), infected (I) and recovered (R) are smooth functions of time such that $S(t) + E(t) + I(t) + R(t) = 1$ for all time [5]. The contact rate $\beta(t)$ is assumed to be seasonally dependent through the relation

$$\beta(t) = \beta_0 (1 + \beta_1 \cos 2\pi t) \tag{4}$$

where β_0 and β_1 are constants.

Figs. 4 and 5 show the computed time series of I(t), the reconstructed trajectories of I(t) in 3 dimensions projected in the I(t), I(t+h) plane, the Poincaré sections and the return maps for two different values of α. The solutions have been perturbed by low levels of noise. The two solutions differ by the unperturbed solution corresponding to Fig. 4 being chaotic and the unperturbed solution corresponding to Fig. 5 being a simple periodic oscillation. We note that the return map in Fig. 4 closely resembles that computed for mumps, whereas the return map in Fig. 5 resembles that computed for chicken pox. Other solutions with chaotic dynamics produced return maps similar to those computed for measles and German measles. The numerical solutions thus support the previous interpretations of the dynamics of measles, mumps and German measles as being chaotic and in addition they offer an explanation of chicken pox epidemics in terms of a periodic oscillation perturbed by low levels of noise. Schaffer [15] has made a more thorough numerical analysis of the SEIR model and shown that the return map constructed from the embedded flow of I(t) in 3 dimensions is in fact not one-dimensional. It has a fractal substructure reminiscent of the structure found in the Hénon map [9]. Furthermore, Schaffer [15] has calculated the Lyapunov dimension for parameters very close to those used in Fig. 4 obtaining values of 2.3-2.5. These values should be compared with the correlation exponents listed in Table I. The interpretation of chicken pox in terms of a periodic oscillation perturbed by noise also explains why no finite correlation exponent could be obtained for this disease. The Lyapunov dimension, D_L, of a periodic oscillation is zero [6] and from the inequality $\nu < D_I = D_L$ we infer that the correlation exponent for chicken pox in the absence of perturbations should be zero.

Discussion

From the results presented in the previous sections we may conclude that the epidemics of measles, mumps and German measles in Copenhagen, Denmark before 1968 were chaotic whilst the epidemics of chicken pox were those of a simple periodic oscillation perturbed by noise. The results for measles and chicken pox are similar to those previously obtained for these diseases in New York and Baltimore [17]. However, the epidemics of mumps, diagnosed as chaotic here, were shown to correspond to a simple periodic oscillation perturbed by noise in New York [17]. No explanation for this difference in the dynamics of mumps can be offered at present. Although the SEIR model seems to provide a good qualitative explanation of the epidemics data there is still a good deal of numerical work left before this model can be used quantitatively in the description of the dynamics of epidemic diseases. The present analysis of the Copenhagen childhood diseases like the previous analysis by Schaffer and Kot [17] of these diseases in New York and Baltimore suggest that stochastic factors are not necessary for the perpetuation of epidemics. In fact stochastic effects only account for a few percent of the fluctuations in incidence rates as inferred from the logarithmic plots of correlation integrals (Fig. 2). Nevertheless, these few percent are sufficient to completely obscure an otherwise periodic oscillation of chicken pox epidemics. By contrast the chaotic dynamics of measles, mumps and German measles are relatively robust to such stochastic perturbations. These findings are consistent with previous numerical studies of the effects of noise on periodic and chaotic attractors [13,16] which conclude that periodic oscillations are highly sensitive to low levels of random perturbations whereas chaotic behaviour is relatively unaffected by such perturbations.

Acknowledgements

This research was supported by the Danish Natural Science Research Council (J.No. 11-5378). The author wishes to thank Dr. W.M. Schaffer, University of Arizona, for communicating his results prior to publication.

Literature

1. Anderson, R.M. and R.M. May: Population biology of infectious diseases: Part I. Nature 280 (1979), 361-367.
2. Aron, J.L. and I.B. Schwartz: Seasonality and period-doubling bifurcations in an epidemic model. J. Theor.Biol. 110 (1984), 665-679.
3. Bailey, N.T.J.: The Mathematical Theory of Infectious Diseases and its Applications. Charles Griffin & Company, London, 1975.
4. Bartlett, M.S.: Stochastic Population Models in Ecology and Epidemiology. Methuen & Co., London, 1960.
5. Dietz, K.: The incidence of infectious diseases under the influence of seasonal fluctuations. Lect. Notes Biomath. 11 (1976), 1-15.
6. Farmer, J.D., E. Ott and J.A. Yorke: The dimension of chaotic attractors. Physica 7D (1983), 153-180.
7. Franaszek, M.: Effect of random noise on the deterministic chaos in a dissipative system. Phys. Lett. 105A (1984) 8, 383-386.
8. Grassberger, P. and I. Procaccia: Measuring the strangeness of strange attractors. Physica 9D (1983), 189-208.
9. Hénon, M.: A two-dimensional mapping with a strange attractor. Commun. math. Phys. 50 (1976), 69-77.
10. London, W.P. and J.A. Yorke: Recurrent outbreaks of measles, chickenpox and mumps I: seasonal variations in contact rates. Am. J. Epidemiol., 98, 453-468.
11. May, R.M.: Simple mathematical models with very complicated dynamics. Nature 261 (1976), 459-467.
12. Medical report for the Kingdom of Denmark. The National Health Service of Denmark, Copenhagen 1927-1968.
13. Olsen, L.F.: Effects of periodic and stochastic perturbations on oscillations and chaos in a model of the peroxidase-oxidase reaction Z. Naturforsch., 40a (1985), 1283-1288.
14. Olsen, L.F. and H. Degn: Chaos in biological systems. Q. Rev. Bicphys. 18 (1985) 165-225.
15. Schaffer, W.M.: Can nonlinear dynamics help us infer mechanisms in ecology and epidemiology. IMA J. Math. Appl. Med. & Biol. (1986) (in press).
16. Schaffer, W.M., S. Ellner and M. Kot: Effects of noise on dynamics in ecology and epidemiology. J. Math. Biol. (1986) (in press).
17. Schaffer, W.M. and M. Kot: Nearly one-dimensional dynamics in an epidemic. J. Theor. Biol. 112 (1985) 403-427.
18. Takens, F.: Detecting strange attractors in turbulence. Lect. Notes in Math. 898 (1981), 366-381.

MODELLING AND ESTIMATING TRANSPORT AND FATE OF HEAVY METALS IN WATER COLUMN AND SEDIMENT LAYER IN SOME ENCLOSED BRANCHES OF THE SEA IN THE S.W.NETHERLANDS 1).

E. M. van de Vrie 2).

ABSTRACT.

In the Delta area in the S.W.Netherlands 3 of the main European rivers enter the sea. For protection against storm floods a complex network of dikes, dams and sluices is constructed, changing estuaries in stagnant or semi-stagnant, saline, brackish or freshwater lakes. Some of the new created lakes are sedimentation areas of the rivers Rijn, Maas and Schelde. Beside suspended particles these rivers carry also a considerable amount of heavy metals partly remaining in the former estuaries.

To calculate the concentrations of heavy metals in the water column and the sediment layer a compartment model 'SIMMV' is developed. Heavy metals are present in the dissolved phase and bound to particles. Transport between water and sediments takes place bounded on settling particles and in the dissolved phase. Using measurements over 9 years of zinc, cadmium and lead from the Hollandsch Diep/Haringvliet the model has been calibrated. The values, uncertainties and mutual correlation of the parameters are estimated, using Monte Carlo methods. Estimated distribution coefficients were about 115, 80 and 310 l/g for zinc, cadmium and lead in the water column and about 20, 20 and 40 in the sediment layer. Thickness of the active top sediment layer was estimated between 25 and 40 cm.

With the results predictions are made for the loading process in the future Volkerakmeer and Zoommeer. When input concentrations remain unchanged, it will be seriously contaminated within several decades. In the sediment layer the 50% adaptation times range from 24 years for zinc to 70 years for lead.

1. INTRODUCTION

Transport and fate of heavy metals in aquatic systems are influenced by complex fysical, chemical and biological processes (3). A lot of research is done on details of these processes (6). Synthesizing these parts of knowledge to an overall theory is complicated by differences in laboratory and environmental conditions, unpredictability and interdependence of boundary conditions, input variables and system parameters. Such a bottom-top approach never resulted in a usefull model for description and prediction of concentrations of heavy metals in aquatic systems.

---

1) Communication no. 342 of the Delta Institute for Hydrobiological Research.

2) Delta Insitute for Hydrobiological Research, Vierstraat 28, 4401 EA Yerseke, The Netherlands.

A different approach is looking at the system as an entity, analysing the most important processes determining its behaviour. Unknown effects of processes of minor importance are accepted, introducing an uncertainty in model results. This approach is usefull when there are long time-series of data available and application of results is needed in the same or similar environment. Such a situation exists in the S.W.Netherlands. This part is the Delta area of 3 of the main European rivers, Rijn, Maas and Schelde. For protection against storm floods a complex network of dikes, dams and sluices is constructed, changing estuaries in stagnant or semi-stagnant, saline, brackish or freshwater lakes.

In 1986 and 1987 two secondary dams will be closed, creating the Volkerakmeer and Zoommeer, parts of a tidal-free shipping-lane between Antwerp and Rotterdam (Fig. 1). The inflow of water from the rivers Rijn and Maas can be controlled and varies between 0 and 100 m3/s. Beside suspended particles these rivers carry also a considerable amount of heavy metals partly remaining in the former estuaries. Water column and sediment layer of the new created freshwater lake will slowly be contaminated, causing serious risks for the water and bottom ecosystems.

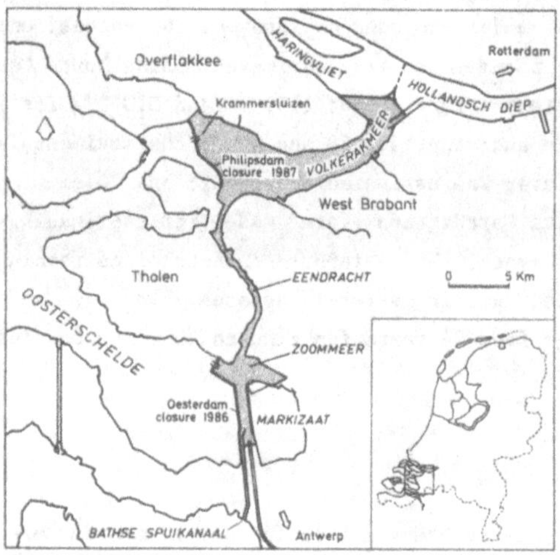

Fig 1. Volkerakmeer and Zoommeer.

In order of a responsible management concerning development and control it is necessary to be able to describe and predict the transport and fate of heavy metals and other micropollutants. Effects of different management strategies must be analysed and compared to find an optimal water control policy. Other aspects of the optimal water control policy are the nutrient (eutrophication) and chlorinity (agriculture irrigation) concentrations. Furthermore, the slow contamination offers a unique scientific opportunity to test models for the behaviour of heavy metals by comparing the predictions with the measurements after closure in 1987.

This article describes the construction and calibration of a model 'SIMMV' to calculate the concentrations of heavy metals in the water column and sediment layer.

The model is used to predict the future concentrations in the Volkerakmeer and Zoommeer.

2. STRUCTURE OF THE MODEL.

Similar to the concepts of Di Toro et al. (2) and Halfon (5) a compartment model 'SIMMV' is constructed. The basin under consideration is divided in several interconnected compartments. Each compartment contains a water column and a top sediment layer of fixed limited thickness (Fig. 2). We assume the distributions of the heavy metals to be homogeneous in the entire compartment in the water column as well as in the top sediment layer. Inflow containing contaminants can be from the proceeding compartment or from external origin.

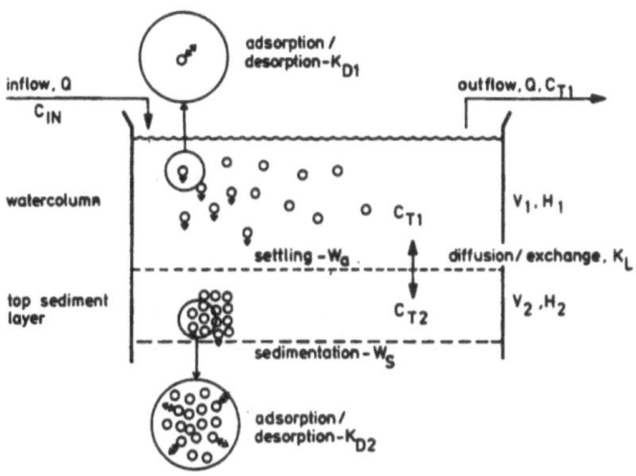

Fig 2. Structure and processes in one compartment.

The total concentration in the water column, C_{T1}, as well as the total concentration in the top sediment layer, C_{T2}, are partly dissolved and partly adsorbed or precipitated on the suspended or settled particles. The dissolved fractions are respectively $f_{d1}C_{T1}$ and $f_{d2}C_{T2}$ and the fractions bounded on the particles are $f_{p1}C_{T1}$ and $f_{p2}C_{T2}$. The fractions f_d and f_p are defined by the concentrations of the particles, m, and the distribution coefficient K_D.

$$f_d = \frac{1}{1 + m\,K_D} \tag{1}$$

and

$$f_p = \frac{m\,K_D}{1 + m\,K_D} \tag{2}$$

Part of the suspended particles settles with a velocity W_a and enters the top sediment layer, transporting the particle bounded contaminant from the water column to the homogeneous mixed top sediment layer. Here a new equilibrium between the dissolved and the particulate phase establishes, determined by different values of m and K_D. Due to differences in concentrations of the dissolved fractions direct transport of the contaminant in the dissolved phase is possible between water column and top sediment layer defined by a diffusion/exchange coefficient, K_L.

From the active top sediment layer with constant thickness permanent sedimentation takes place to the deeper layers with a velocity W_s. This part of the contaminant leaves the system and is outside the boundaries of the model. Volatilation or decay processes are not modelled.

When V is the volume of the compartment, H_1 and H_2 the thicknesses of water column and top sediment layer, Q the in- and outflow and C_{IN} the contaminant input concentration, the model can be mathematically formulated as a set of inhomogeneous first order differential equations.

$$\frac{d\,C_{T1}}{dt} = -\frac{Q}{V_1} C_{T1} - \frac{K_L}{H_1}(f_{d1}C_{T1} - f_{d2}C_{T2}) - \frac{W_a}{H_1} f_{p1}C_{T1} + \frac{Q}{V_1} C_{IN} \tag{3}$$

and

$$\frac{d\,C_{T2}}{dt} = \frac{K_L}{H_2}(f_{d1}C_{T1} - f_{d2}C_{T2}) + \frac{W_a}{H_2} f_{p1}C_{T1} - \frac{W_s}{H_2} f_{p2}C_{T2} \tag{4}$$

A compartment model with n compartments results in a set of 2n differential equations. Knowing parameters, initial and boundary conditions, this problem can be solved analytical.

3. ESTIMATION OF PARAMETERS.

For the estimation of parameters in the model 'SIMMV' results of measurements are used from the Hollandsch Diep and Haringvliet, closed of from the sea in 1970. This basin also receives water from the rivers Rijn and Maas and is of similar dimensions. It differs from the future Volkerakmeer and Zoommeer by its hydrological features. Through the Rotterdam harbour there is a tidal influence and discharge into the North Sea depends on the discharges of the rivers Rijn and Maas.

In the period 1974-1982 there are taken monthly samples in the water column. From these samples total and dissolved concentrations are determined for zinc, cadmium and lead, as well as the concentrations of suspended sediments. In the same period 5 times the concentrations of heavy metals in the sediments are analysed. Based on the concentrations of particulate organic carbon the total contamination of the top sediment layer is estimated.

Concentrations in the water column fluctuate rapid, while changes in the top sediment layer are only significant at a much longer time scale. As a compromise one-year periods are taken as the time step in the model. This means that for all input concentrations, flow quantities and settling velocities yearly means are taken. Based on the knowledge of the concentration gradients, locations of the sampling points and the hydrological boundaries the Hollandsch Diep and Haringvliet basin is divided in three compartments. From these compartments dimensions, stream velocities, suspended sediment concentrations and settling velocities are known.

For the calibration of the model several process parameters must been estimated: distribution coefficients in water column and top sediment layer, diffusion/exchange coefficients for transport in the dissolved phase and the thickness of the active top sediment layer. These parameters can have different values in the various compartments.

For estimation of these parameters a Monte Carlo method is used (4). Within boundaries, based on a-priori knowledge, values for the parameters are chosen at random. With these parameter values the system response is calculated. The response is compared with the measurements and thereafter accepted or rejected. Based on the sets of accepted parameters estimates can be calculated for their values, standard deviations and correlation.

For acceptance or rejection several criteria are possible. In this application a set of parameters is accepted when for all measurements the corresponding calculated values differ less then a fixed number times the standard deviations from these measurements.

Table 1. Estimated values, standard deviations and significant correlations for the process parameters in the model 'SIMMV' applied to the data of the Hollandsch Diep and Haringvliet.

Parameters: 1 - thickness of top sediment layer (m)
2 - distribution coefficient water column (1/g)
3 - distribution coefficient top sediment layer (1/g)
4 - diffusion exchange coefficient (m/d)

parameter	estimated value	standard deviation	correlation with 1	2	3
zinc					
1	.39	.10			
2	114.	8.3			
3	18.8	1.8			
4	1.	.7			
cadmium					
1	.37	.10			
2	81.	27.			
3	22.	4.3		.96	
4	8.1	4.3			
lead					
1	.26	.09			
2	309.	63.			
3	38.	8.5		.86	
4	.88	.25			

4. RESULTS.

For zinc, cadmium and lead the parameters of the model were estimated based on results of measurements over nine years in the Hollandsch Diep and Haringvliet. When different parameter values in the various compartments were accepted some results diverged to not realistic values. Reason for this can be an insufficient amount of data.

When all values for the same parameters in the various compartments were equal, reasonable estimates were found. Results are presented in table 1 and are in accordance with literature values. Values for the distribution coefficients in the top sediment layer are about 10-20% of those in the water column. Thickness of the active top sediment layer is found to be several decimeters. Only the estimate for the diffusion/exchange coefficient for cadmium was not realistic.

In table 1 only significant correlations are presented. There is a positive correlation found between the distribution coefficients in the water column and top sediment layer for cadmium and lead. Other correlations were not significant.

After this parameter estimation proceaure slight differences in parameter values for different compartments were allowed. Thus a real optimum is not found, but an image emerged from the differences between the various compartments. Only for zinc the distribution coefficient in the water column raises. This coincides with higher values of pH, what is also found elsewhere (1) and (7). Using these parameter values the system response for lead is calculated and presented in Fig. 3.

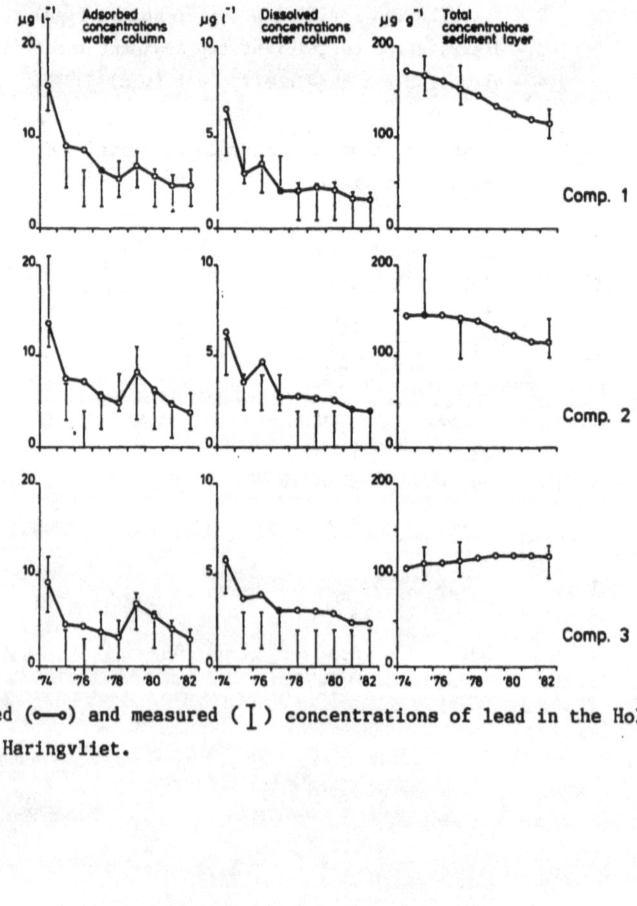

Fig 3. Calculated (o—o) and measured (I) concentrations of lead in the Hollandsch Diep and Haringvliet.

The results obtained in the calibration process for the Hollandsch Diep and Haringvliet are used to predict the concentrations of heavy metals in the future Volkerakmeer and Zoommeer. Until its closure in 1987 it is in open connection with the sea and is relatively uncontaminated. After closure of the Oesterdam and the Philipsdam water column and top sediment layer will slowly be contaminated. For the predictions of the concentrations of heavy metals it is divided in two compartments, the Volkerakmeer in the north and the Zoommeer in the south. Similar parameter values are chosen as in the Hollandsch Diep and Haringvliet.

With constant contaminant input the concentrations will raise until an equilibrium value. In reality the loading process will be more dynamic. To get an expression reflecting the sensitivity of the system to changes in inputs the 50% adaptation time is introduced. This is the time, a system needs to bridge for 50% the difference between the steady-state situations, when a change in input has occured.

Table 2. 50% Adaptation times and final concentrations of heavy metals in the future Volkerakmeer and Zoommeer.

	comp.	t-50% (years)	standard deviation	final concentrations (ug/l)	(ug/g)
zinc					
water column	1	18.	8.	48.	
	2	37.	17.	48.	
sediment layer	1	24.	11.		400.
	2	45.	21.		404.
cadmium					
water column	1	34.	17.	.38	
	2	68.	34.	.37	
sediment layer	1	35.	18.		5.4
	2	69.	35.		5.6
lead					
water column	1	28.	13.	4.8	
	2	59.	28.	4.7	
sediment layer	1	36.	17.		66.
	2	70.	33.		67.

Results for the final concentrations and the 50% adaptation times are given in table 2. It turns out that the concentration in water column and top sediment layer are highly interdependent. When loading remains unchanged the future Volkerakmeer and Zoommeer will be seriously contaminated within several decades.

5. DISCUSSIÓN.

For the calibration of the model data were used from the Hollandsch Diep and Haringvliet. These data were not gathered for this purpose but resulted from a routine monitoring program. This is the reason that locations of the sampling sites and sampling frequencies were not optimal. Only at the end of the monitoring program it was decided to do this model exercise.

The chosen model structure for transport and fate of heavy metals was limited. More complex models introduce higher uncertainties in parameter estimations with the same data available. Most important processes not included are biological and chemical processes.

For the model construction the hypothesis of a homogeneous mixed top sediment layer was needed. In reality the sediments are nor homogeneous mixed nor limited in thickness. Even so estimates for the thickness of the active top sediment layer gave the same results for the indepent datasets of zinc, cadmium and lead. The hypothesis of the homogeneous mixed top sediment layer showed to be usefull.

The model is calibrated on data from the Hollandsch Diep and Haringvliet. There was no other independent dataset available for the verification of the model. After creating the Volkerakmeer and Zoommeer an extended research program is planned to gather data for testing models of heavy metals, nutrients etc.

Application of the results in the future Volkerakmeer and Zoommeer must be handled with great care. The different hydrological situation is responsible for essential different processes. The lower stream velocity causes higher settling and higher production of suspended material. The effect of these changes on the process parameters is unknown. Water pollution control activities upstream will have its effect on the future concentrations of heavy metals in the inflowing water. Therefore the final concentrations of heavy metals can be lower than in our calculated hypothetical situation.

6. CONCLUSION.

The aim of the research was to develop a model for description of fate and transport of heavy metals in sediment and water column. Although there were limitations in model structure and available data there could been made reasonable estimates for the process parameters. Satisfactory simulations are made of concentration in sediments and water column. Application of the model indicates that the future Volkerakmeer and Zoommeer will be seriously contaminated in several decades.

A real examination of the model was not possible. Uncertainties in the estimates for the model parameters can be caused by insufficient and uncertain data or by an incorrect model concept. Future research will give a better basis for evaluation of the model.

ACKNOWLEDGEMENT.

This research was supported by the Delta Department, Environmental Division Rijkswaterstaat, Min. of Transport and Public Works.

REFERENCES.

(1) Bourg,A.C.M. and C.Mouvet: A heterogeneous complexation model of the adsorption of trace metals on natural particulate matter. In: Complexation of trace metals in natural waters, Ed. C.J.M.Kramer and J.C.Duinker, Nijhoff/Junk Publ., the Hague, 1984, pp. 267-278.

(2) Di Toro,D.M., D.J.O'Conner, R.V.Thomann and J.P.St.John: Simplified model of the fate of partitioning chemicals in lakes and streams. In: Modelling the fate of chemicals in the aquatic environment, Ed. K.L.Dickson. Ann Arbor, Science Publications, 1982, pp. 165-190.

(3) Duursma,E.K. and M.Smies: Sediments and transfer at and in the bottom interfacial layer. In: Pollutant transfer and transport in the sea, Ed. G.Kullenberg, CRC Press, Florida, Vol.2, 1982, pp. 101-139.

(4) Fedra,K.: A Monte Carlo approach to estimation and prediction. In: Uncertainty and forecasting of water quality, Ed. M.B.Beck and G.van Straten, Springer Verlag, Berlin, 1983, pp. 259-291.

(5) Halfon,E.: Is there a best model structure? I. Modelling the fate of a toxic substance in a lake. Ecological Modelling, 20(1983)2,3 135-152.

(6) Salomons,W. and U.Forstner: Metals in the Hydrocycle. Springer Verlag, Berlin, 1984.

(7) Salomons,W. and W.G.Mook: Biogeochemical processes affecting metal concentrations in lake sediments (IJsselmeer, The Netherlands). The science of the total environment 16(1980)217-229.

Mathematical Software developed by the Karl-Weierstraß-Institute of Mathematics of the Academy of Sciences of the GDR

T. Cierzynski [1)]

The International Workshop on Dynamical Systems and Environmental Models 1986 exhibited a poster demonstration on basic software in numerical mathematics and mathematical statistics developed by the Karl Weierstraß Institute of Mathematics. A poster sesson was also devoted to these programs.

PP-NUMATH-2 is a comprehensive program package for methods of numerical mathematics developed under the direction of Dr. G. Hebermehl. Modern methods of numerical mathematics are implemented in 500 routines with a total of 50.000 FORTRAN statements for the following problem categories: matrix manipulation, linear algebra, integrals, optimization, ordinary differential equations, integral equations, partial differential equations, nonlinear equations, special functions and approximation, interpolation and differentiation of functions.
The program package is available for the following equipments: ESER- and IBM equipments, SM4 and successor models, K 1630, BC A5120/30, PCA7100, PDP 11, IBM- and OLIVETTI-PC.

The statistical software includes individual programs which were developed under the direction of Dr. J. Läuter.

DIST is a universal and programmable interface for data organization with the aim of coupling different programs and computer systems. Data can be varied in code, format and structure. Data recoding and transformations are easy to do.

SMOOTHER is a program package for smoothing by polynomials and splines as well as median calculation. The program processes equidistant as well as non-equidistant measurements and accepts multiple measuring values in the observation points.

DATORG prepares data for further processing with analysis of variance techniques. It stores the necessary statistics in dependence of a statistical model and in an efficient way for computation.

---

1) Karl-Weierstraß-Institute of Mathematics of the Academy of Sciences of the GDR, Mohrenstrasse 39, DDR-1086 Berlin

AMP performs by means of modern statistical methods the multiway analysis of variance in nested and cross-classified models with fixed and random effects for balanced and unbalanced designs. The program can be implemented in every computer provided it has a FORTRAN compiler.

MVD/MDA accomplishes the multivariate analysis of variance and discriminant analysis including selection of variables. It processes continuous and categorical data. MDA permits many interactive interventions while the program is running.

CLUSTER implements 4 partitioning and 12 hierarchical algorithms for cluster analysis in a comprehensive program package. It processes continuous as well as categorical data.

CORRAN implements the correspondence factor analysis and principal components analysis. It makes possible the simple representation of multidimensional relationships.

Zeitfracht Medien GmbH
Ferdinand-Jühlke-Straße 7
99095 Erfurt, Deutschland
produktsicherheit@kolibri360.de